T0093845

Undergraduate Lecture Notes in Physics

Undergraduate Lecture Notes in Physics (ULNP) publishes authoritative texts covering topics throughout pure and applied physics. Each title in the series is suitable as a basis for undergraduate instruction, typically containing practice problems, worked examples, chapter summaries, and suggestions for further reading.

ULNP titles must provide at least one of the following:

- An exceptionally clear and concise treatment of a standard undergraduate subject.
- A solid undergraduate-level introduction to a graduate, advanced, or non-standard subject.
- A novel perspective or an unusual approach to teaching a subject.

ULNP especially encourages new, original, and idiosyncratic approaches to physics teaching at the undergraduate level.

The purpose of ULNP is to provide intriguing, absorbing books that will continue to be the reader's preferred reference throughout their academic career.

More information about this series at http://www.springer.com/series/8917

Volker Ziemann

Physics and Finance

 Springer

Volker Ziemann 🆔
Department of Physics and Astronomy
Uppsala University
Uppsala, Sweden

ISSN 2192-4791 ISSN 2192-4805 (electronic)
Undergraduate Lecture Notes in Physics
ISBN 978-3-030-63645-6 ISBN 978-3-030-63643-2 (eBook)
https://doi.org/10.1007/978-3-030-63643-2

This Springer imprint is published by the registered company Springer Nature Switzerland AG
The registered company address is: Gewerbestrasse 11, 6330 Cham, Switzerland

Preface

One fine day, my son, who was studying economics at the time, brought along books on the use of stochastic differential equations in finance. Following a dialog along the lines of "Oh, I know that, its a Fokker-Planck equation"—"No dad, that's Black-Scholes," I got curious. After all, there might be some fun in economics and finance, besides the money. So, I borrowed Hull's book about the basics of financial economics, because I wanted to understand the basic concepts and the lingo. Just looking through the book, I recognized those differential equations that look so similar to a diffusion equation with a drift term. So, I set out to understand what finance has to do with diffusion.

I later presented some lectures about my explorations to a few interested students and colleagues, which was very stimulating and caused me to explore the subject further. That was how the later chapters came about. They all deal with some aspect of random processes and have some overlap between physics, finance, and other neighboring disciplines.

At that point, I prepared a 5 ECTS (European transfer credits) lecture series for masters students at Uppsala University and expanded the manuscript to serve as lecture notes for this course, which ran for the first time in the spring of 2019, with about 15 interested students. The feedback after the course was rather positive such that I gave the course again in the spring of 2020. This time with 24 students, who provided much more feedback and criticism, which caused me to revise parts of the manuscript to bring it to its present form.

Obviously, many people helped to improve the manuscript. First, I have to thank my son Ingvar. He stimulated my interest in finance and also critically read parts of the manuscript. Likewise, I am indebted to my colleagues and the students who participated in the early lectures and in the course later. Many of them gave valuable criticism and feedback on the growing manuscript. I want to single out a few students, who were particularly diligent: Joe and Martin from the course in 2019; Friedrich, Sebastian, and Elias from 2020. They helped me weed out many ambiguities and errors. They are, however, not to blame for any remaining bugs, those are my responsibility alone. I also need to thank our director of studies, Lisa Freyhult, for her support at the faculty to include this course in the curriculum. Finally, I want to thank my family for their patience with me when I was "a bit" overfocused on the manuscript.

Uppsala, Sweden   Volker Ziemann

Contents

Chapter 1
Introduction

Abstract This chapter sets the stage for the book when it establishes a common theme in many physical and financial systems; both deal with dynamical systems subject to external random forces. A brief discussion of the book's target audience follows, before an overview over its contents is given.

What do physics and finance have in common? The short answer is: they both deal with dynamical systems that are subject to external random forces.

In physics, an example is the random walk of pollen floating on a liquid, the Brownian motion first interpreted and theoretically analyzed by Einstein [1]. A modern example is the startup of conventional lasers and free-electron laser from noise. In general, most of the sub-domain of statistical physics treats systems that are subject to random forces and are described by distributions of the state variables. Many diffusion processes fall into this group.

In finance, the dynamics of the stocks or other financial quantities can often be described by an average drift towards higher values that is superimposed by large fluctuations. Bachelier's analysis [2] was the first application of such random processes to financial systems.

This correspondence was, of course, noted earlier and led to a large body of literature, some under the name of *Econophysics* with notable books being [3–6]. The resemblance of the general features makes it possible to use similar mathematical methods in order to describe and understand these systems. The target audience of these volumes are mature physicists, eager to explore a new field.

In this book, however, we address students earlier in their career, typically in their third or fourth year and use examples from finance to highlight concepts known from physics with the intent to deepen their understanding of the methodology. Frequently, we juxtapose systems from physics and systems from finance that use very similar methods, albeit in different contexts. This approach should aid the students to see methods from their physics education from a different angle, which should lead to their increased appreciation. Seeing the same method applied to problems in rather different contexts should deepen their understanding.

Moreover, some of the mathematical methods, such as stochastic differential equations or path integrals, are not part of the core curriculum. Presenting simple applications of these methods in two fields—physics and finance—side by side will fill

© The Author(s), under exclusive license to Springer Nature Switzerland AG 2021
V. Ziemann, *Physics and Finance*, Undergraduate Lecture Notes in Physics,
https://doi.org/10.1007/978-3-030-63643-2_1

the students' "toolbox" with additional ways to solve problems. Basically, it should open their eyes to all the exciting methods out there that just wait to be applied to problems that come their way.

Before delving into the subject matter, however, we have to familiarize ourselves with the language and the concepts used in finance, to which the second chapter is devoted. In the third chapter, we discuss how to pick a collection of stocks—a portfolio—in such a way that the risk of losses is minimized. This is a wonderful application of variational methods with constraints, the same methods used to analyze mechanical systems. This analysis is quasi-static, a restriction we relax in the fourth chapter, where we consider stochastic processes and their description by Langevin and Fokker-Planck equations. This chapter sets the stage to derive and solve the Black-Scholes equation and to some of its applications in the fifth and sixth chapter. Since stock values are recorded over time, they constitute time series of data, from which we try to extract information by fitting regression models. This fitting of models to data is very common in physics and is the topic of the seventh and eights chapter, where we explore and illustrate reliability and robustness of the fitting as well as forecasting into the future. In Chap. 9 we have a look at financial bubbles and crashes by first discussing historical crashes and their possible reasons. This leads to questioning the basis of the previous analysis and we encounter fat-tailed distributions, fractals, and the central-limit theorem in attempts to understand these often dramatic events. In Chap. 10 we return to less disruptive topics and apply Feynman's path integrals to quantum and to financial systems and solve them using Monte-Carlo methods. All systems covered up to this point followed some intrinsic dynamics. In the eleventh chapter we develop methods to influence and control these system ourselves by adjusting external parameters in order to optimize some performance measure. This gives us a chance to review the relation between Lagrangian and Hamiltonian mechanics and leads to the subject of optimal control theory. Here we pick examples from mechanics and simple macroeconomic models. In the final chapter we focus on cryptocurrencies and explore the underlying concepts from information theory and cryptography before discussing two currencies: Bitcoin and Ethereum. We close the chapter with a discussion of the threat that quantum computers might pose to the cryptographic methods. MATLAB® is used throughout the book to illustrate various numerical methods. Useful functions are collected in an appendix and are available as electronic supplementary material (ESM) from the book's web page at https://www.springer.com/9783030636425. At the end of each chapter, the reader will find a number of exercises whose solutions are discussed in the final chapter. The associated MATLAB examples are also available from the book's web page.

With the plan laid out, let's jump right into the main part of the book and review basic concepts and the language used in finance.

For product information on MATLAB®, please contact:

The Mathworks, Inc.
3 Apple Hill Drive
Natick, MA 01760-2098 USA
Tel: 508-647-7000

Fax: 508-607-7001
Email: info@mathworks.com
Web: www.mathworks.com

References

1. A. Einstein, Über die von der molekularkinetischen Theorie der Wärme geforderte Bewegung von in ruhenden Flüssigkeiten suspendierten Teilchen. Annalen der Physik **322**, 549 (1905)
2. L. Bachelier, Theorie de la speculation. Annales Scientifiquies de l'Ecole Normale Superieure **17**, 21 (1900)
3. R. Mantegna, H. Stanley, *Econophysics* (Cambridge University Press, Cambridge, 2004)
4. J. Voit, *The Statistical Mechanics of Financial Markets* (Springer, Berlin, 2005)
5. A. Schmidt, *Quantitative Finance for Physicists* (Elsevier, Amsterdam, 2005)
6. W. Paul, J. Baschnagel, *Stochastic Processes—from Physics to Finance*, 2nd edn. (Springer, Berlin, 2013)

Chapter 2
Concepts of Finance

Abstract After introducing the volatility of stocks as a central feature, the concepts of hedging, short selling, and discounting are introduced, as well as future contracts and options as functions—derivatives—of the underlying stocks. Before discussing the various participants in the ensuing markets, the efficient market hypothesis and some simplifying assumptions, in order to make the theoretical treatment feasible, are covered.

Let us start by familiarizing ourselves with some of the concepts and the lingo used in the financial world. One of the central items is to no one's surprise—stocks.

2.1 Stocks and Other Tradeable Goods

Stocks or *shares* S in companies are examples of underlying assets that are at the heart of financial economics. Other assets are large sums of money or even basic commodities such as halves of hogs or grain. Think of Kellogs buying corn for their flakes! These assets are typically traded in (stock or other) exchanges. Big ones are in Frankfurt, London, New York, or in Chicago at the Mercantile Exchange (CME). These exchanges often publish composite indices, such as DAX, FTSE, or the Dow Jones to track the overall behavior of trading.

A crucial feature is the fluctuating value of these assets. It depends on many and varying influences, such as a competitor introducing a better product, the company losing a lawsuit, or a natural catastrophe such as an earth quake that damages production facilities. Even the expectations of market analysts affects their value. They try to figure out whether shares are worth buying or not and if their expectations are not satisfied, the share prices fall. All of these, and many more, factors contribute to fluctuations of the stock values. These fluctuations are normally quantified by their relative variation or variance $V = \sigma^2 = \langle (\Delta S/S)^2 \rangle$. Here the angle brackets denote the average over a suitable period of time, such as a month or a year. The variability or *volatility* σ is commonly associated with *risk* and plays a crucial role when dealing with stocks. Thus random fluctuations of tradeable quantities are perceived as a risk. And managing risk is the predominant occupation of financial economists. They build a collection of tradeable quantities—*a portfolio*—with the aim to minimize the risk, or equivalently, the variance of the portfolio. This balancing of risks is called *hedging*.

2.2 Hedging and Shorting

At first sight, adding several fluctuating variables results in a variable that fluctuates even more than its constituents. This is true, unless two or more constituents are anti-correlated. We therefore need a quantity that increases in value, when a stock decreases in value, and vice-versa. This could be an asset that is in high demand, while another asset is in low demand. As a first example consider investing in the stock of a consumer good, such as Apple® Inc., and in Gold. In times of global unrest, the value of the Gold will increase and in times of tranquility people will buy more consumer goods, which increases the value of the Apple stock. A second, more down-to-earth, example is based on the stock of a maker of raincoats and umbrellas and the stock of a soft-drink maker, such as Pepsi. In a wet summer the maker of rain-gear will thrive and in a sunny summer, the maker of soft-drinks. The stocks in these examples are only moderately anti-correlated. The third example is perfectly anti-correlated and is based on the possibility of betting on a falling stock value. This mechanism is called *shorting* and is based on me first borrowing (at no or little cost) the stock from a broker with the promise to give it back at a later time. Upon receiving the borrowed stock, I immediately sell it at market value, the spot-price. At the later time I need to purchased back the stock at market value before giving it back to the original owner. But if the stock value has decreased in the intervening time, I pay less than I initially received when I sold the borrowed stock. Thus I make a profit, given by the difference in price. This mechanism constitutes a derived commodity, which is anti-correlated with the underlying stock price.

Note that this procedure—shorting—of using an underlying asset, the stock constitutes a *financial tool,* that allows us to achieve a particular financial goal, here hedging a risk. Such tools or mechanisms, which are functions of the basic assets, the stocks or other commodities, are called *derivatives*. They are the tools of financial engineering, the mechanics of managing risk. We will discuss a number of them in the following sections.

2.3 Derivatives

The most straightforward financial tool is a *forward contract,* which is an agreement between two parties and as such it is called an *over-the-counter* (OTC) product. In a forward contract two parties agree to trade an asset at a later time for an agreed-upon price that may be different from today's price. It protects the producer against the volatility of the market and guarantees a known, albeit potentially lower, price. This makes the producer's planning safer. And the purchaser of the forward contract has the chance to make a profit in case the market swings upwards and the later market- or spot-price is higher. The agreed-upon price of the forward contract remains fixed for the duration of the contract.

A relative to the forward contract is a *future contract*, which also stipulates a trade at a later time, but the trading is done at an exchange, such as the CME, rather than in a two-party contract. Whereas the price of a forward contract is fixed, the value of a future contract varies with time and approaches the spot price as the delivery time approaches.

Arguably the most prominent tools of financial engineering are *options,* which are contracts that give the owner of the option the right to sell or to buy an asset at a later time for a fixed, agree-upon price. The owner has the choice to exercise the option, but can forfeit as well. There are many types of options, where the two simplest ones are: *call options*, which give the owner the right to buy an asset at time T at price K. The second is a *put option*, which gives the owner the right to sell at time T at price K. At this point in time the option reaches *maturity* and K is called the *strike price* of the option.

Options come in two main flavors; *European* options have a fixed expiration date and *American* options have flexible expiration dates. The names European and American have nothing to do with geographic prevalence, they are just used as labels, so European means fixed and American means flexible expiration date.

Let us consider an example: I sell a *European call option* for a price c that will permit the purchaser of the option at to buy some asset at the strike price K from me at maturity T. The purchaser of the option therefore limits the maximum price he will later have to pay for the asset to K. At time T the price of the asset S_T may be below the agreed strike price K. Then it is more advantageous for the purchaser to directly buy at the market price S_T and forfeit the call option. In that case I, the seller, have made a profit c. If the market price S_T is above K, the purchaser will exercise the option and I, the seller have to provide the asset at price K and might incur a loss, unless I already own a share. Note that the seller of the call option receives some money up front, the purchasing price c of the option, and hopes that the market price drops so he can keep the money. If, on the other hand, the market price increases, the seller has to invest c, and potentially a lot more in order to satisfy the option, when the purchaser exercises it at maturity. Note that at maturity the value of the call option is zero, if the price of the asset is below the strike price K and it is $S-K$ if the price of the asset is above the strike price.

So, why do I want to issue an option? One reason is to hedge the acquisition of the underlying asset, say some share. Assume that I buy a share and, at the same time, sell a call option. If the share price increases, but stays below the strike price K, I make a profit by keeping the money I received for the call option. If the share price increases above the strike price K, I have to give the share I own to the purchaser of the option. I will make a loss, but a least not more than what I own anyway—the share. Balancing the number of options with the acquired asset is part of hedging and depends on correct valuation of the option, i.e. setting the sale price c of the option. We will return to this topic in Chaps. 5 and 6.

Note how owning an option constitutes a value, because it can be used at a later time to do something useful, for example, to hedge a risk. Therefore options have a price. In later chapters we will discuss how this price can be determined. Derivatives, such as forwards, futures and options all depend on underlying assets, but they constitute independent values themselves, so they can be traded. In this way derivatives are full members of the market pool, having the useful property of being highly correlated with the underlying stocks. And that makes them particularly suitable for hedging and for balancing risks.

Apart from the options, forwards, and futures there are further derivatives, such as swaps and other credit derivatives, forward options, and accumulators, but their discussion is beyond the scope of this book. On the other hand, an asset that is central in many trades and that we therefore need to discuss, is *money*.

2.4 Money

Historically, trades were based on *bartering*, which involves the direct exchange of goods or services between trading partners. Since this exchange is necessarily based on a *double coincident of wants*, where one trading partner desires what another has to offer, it is often difficult to find such a partner. Moreover, it is often inconvenient, because only entire animals can change hands. Therefore, our predecessors introduced a medium of exchange—money. Initially silver, gold, or other rare commodities were used and cast into standardized units—coins—which served as a reference for trades. Note that coins have an intrinsic value, which corresponds to a number of cows or eggs. As trade expanded, however, the limited supply of the rare commodities began to limit trade.

This problem can be circumvented, if some authority—historically a king, today a central bank—guarantees to exchange a piece of paper, having no value in itself, for something else, for example, gold coins or other currencies. This paper money, based on a promise from an authority, is called *fiat-money*, because it is introduced by decree.[1] Note that the value of the paper money is closely linked to our *trust in the authority* to ensure that the money is very difficult to counterfeit and anyone, who forges the money, is prosecuted and punished. This trust in a central authority that manages the flow of money is the linchpin on which fiat money hinges.

On a national level, central banks are the authorities that regulate the availability of money by setting the interest rate at which commercial banks can borrow money from the central bank, such as the European Central Bank in the Eurozone or the Federal Reserve Bank in the US. Until August 1971 the US, and many other countries, linked to the US through the Bretton-Woods agreement, used gold reserves to back their currencies. Notably, a large fraction of the US reserves was stored in Fort Knox.

---

[1] Such as "fiat lux"—let there be light (Genesis 1:3).

Since 1971 the dollar, and the currencies of the other members of the Bretton-Woods agreement, are pure fiat-monies and the Federal Reserve does not have to limit the amount of money they loan to commercial banks, which stimulates growth. This increased flow of money, however, has a down-side; it reduces the value of each dollar in circulation and thereby increases inflation.

Cryptocurrencies, such as Bitcoin, Ethereum, or Litecoin are currencies that do not depend on central authorities to guarantee their integrity. Instead, they maintain a distributed ledger, a database where all transactions are stored. These transactions are digitally signed using strong cryptography and miners compete to verify the transactions. The process is organized to make it practically impossible to reverse a transaction and this renders the database tamper-proof and immutable. How this is implemented in practice is the topic of Chap. 12. For cryptocurrencies the trust in a central authority is replaced by the trust in the integrity of the ledger. As long as trading partners trust this mechanism, a cryptocurrency has a value, but this is a decision that all trading partners collectively reach. As long as all think that bitcoins are a great idea and that they serve a purpose, bitcoins carry a value. This communal sentiment, however, changes over time and that causes the exchange rate of cryptocurrencies to fluctuate more than most other currencies.

2.5 Discounting and Liquidity

When calculating the price of options and other derivatives, we compare values at different times, for example, now and at maturity, which requires some additional thought. Let us therefore work out how much money we can actually earn from shorting an asset. At first sight, it is just the difference of the selling price immediately after borrowing the stocks and the re-purchasing price at a later date. At this point we compare money at different times, which makes little sense, because money at the earlier time can be invested in a *risk free* asset, such as savings bonds or the *London Interbank Offered Rate* (LIBOR). The LIBOR is the rate at which banks borrow and loan money to each other and is considered risk-free. If invested at the LIBOR rate, the initial money would grow at the *risk-free interest rate* r_f by a factor $e^{r_f t}$, where we assume continuously compounding interest. Conversely, this implies that money at a later time is worth less compared to the earlier time, and the factor by which it needs to be *discounted* is $e^{-r_f t}$, again assuming continuously compounding interest. Essentially discounting means to back-propagate a value for the time t with the discount factor $e^{-r_f t}$.

One of the underlying assumptions in financial theory is that there always exists an institution, which provides *liquidity*. This implies that we can always borrow unlimited amounts of money at the risk-free interest rate r_f and we can always deposit surplus funds at the same rate. In this way there is always the chance to increase the value of ones assets by depositing at the bank with rate r_f, but the hope is, of course, to invest in some other assets that will provide a higher rate of return ρ on the invested money, albeit at a higher risk. The gamble is getting the highest return at the lowest tolerable risk.

2.6   Efficient Market Hypothesis

Another basic assumption is that in a large market with many tradeable goods and many trading agents there always will be someone willing to participate in a trade. I may try to short large quantities of a stock because I believe its value will decrease, but there is still someone believing that the value increases and buys all the stock I want to sell. Likewise, we assume that a farmer, who is interested in a smaller, but guaranteed, later price to sell his wheat, always finds a trading partner who accepts the other side of the trade.

A consequence of the large number of eager trading agents and the existence of an investment opportunity at the risk-free rate r_f, the market as a whole will grow at the risk-free rate r_f on average, though large fluctuations can occur. Why is that so? Well, if the average rate of the entire market were lower then all traders would invest money in risk-free bonds and nobody would invest in companies. The companies would need to react by promising a higher rate of return, albeit at some risk for investors. But some investors will take the risk and some will make money and some will lose, but on average the overall gain will be at the risk-free rate. Note, that the same mechanism will work between other trading partners. As soon as there is an imbalance of rate of return and risk, someone will exploit it and the average growth is restored by supply and demand. This is called the *efficient market hypothesis,* namely that supply and demand will always exploit any imbalance until it is removed. The exploitation of a small trading imbalance without incurring a risk is called *arbitration,* an example is the difference in gold price in the US and UK. If it gets larger than the handling costs, it makes sense to buy Gold where it is cheaper, transport it across the Atlantic and sell it at a profit. This only works until the demand on gold is so large that the Gold price rises to a level that makes the deal unprofitable.

After this discussion it should become a bit clearer that stocks fluctuate randomly, but the market as a whole maintains an average growth rate equal to the risk-free interest rate r_f. This already hints at the diffusion equation with a drift term, which will occupy us in the coming chapters.

Before progressing, let us briefly summarize the "rules of the trade," a number of basic assumptions the characterize markets in a somewhat idealized form, mainly to make the theoretical analysis in later chapters feasible.

2.7   Theoretical Markets

The markets and mechanisms are based on assumptions that we list in the following. Some of the rules are rather generic to enable fair trading with equal information available to all participants and others are somewhat idealized in order to make analytic methods feasible. These rules are nevertheless largely valid even in real markets.

- **Liquidity**: Borrowing and lending money in unlimited quantities at the risk-free rate r_f is always possible. It is always possible to find a trading partner for buying and for short selling.
- **Equilibrium and no arbitrage**: The market is in equilibrium and any possibilities for arbitrage are removed immediately by a response of the market as a whole.
- **Rational investors**: all traders are rational and attempt to avoid risks or at least maximize their gain for a given amount of risk.
- **No transaction costs**: normally every financial transaction incurs some fee to be paid to the broker or bank. This effect can be minimized by trading large quantities and we will ignore it.
- **No taxation**: taxation can affect decisions regarding stock markets, because incurring a loss on the stocks can be used to offset gains in other markets and thus minimizing taxes. We ignore this effect.
- **Transparency**: all information is available to all traders. No inside information is available.

In reality these idealized rules are at least approximately valid even in the real world, whose occupants we consider in the next section.

2.8 Market Participants

Before discussing the participants we need to introduce a bit of lingo, because it is frequently used in the context of finance. A trader, who sells a good and therefore no longer owns it, is short of the good and assumes the *short position* in the trade. Conversely, the buyer, who now owns the good, assumes the *long position*.

First we consider the *hedgers*. They try to balance risk of their portfolio and are mostly interested in avoiding large losses while, at the same time, being forced to accept risks. An example is a treasurer in a manufacturing company that converts raw material into a more valuable product. The risks the company is exposed to are the varying price of raw material and possibly fluctuating currency exchange rates, if the raw material must be imported from abroad. The risk due to the price of raw materials can be alleviated by buying options or futures in parallel to the raw material and in this way reduce their price fluctuations. The currency risk can be hedged by off-setting sales of the final product in the raw material's country of origin. In fact, most traders use some form of hedging to limit their exposure to risks outside their realm of control. Note, however, that hedging has a price. Besides canceling risks it also limits profits. Even if one side of hedging gives a large payoff, the other incurs a loss.

A second large group of traders are *speculators*. They actually seek to increase their exposure to risk in order to increase their chance of making a large profit, albeit at the expense of losing big-time, in case the market develops in unwanted and unexpected ways.

The third large group are the *arbitrageurs,* who try to exploit small imbalances in the market fluctuations to make a risk-free profit. One example was already mentioned above, namely different prices of the same asset, for example Gold, in different countries. Arbitrageurs buy at the inexpensive place an sell at the expensive place, thereby driving demand up at the inexpensive place and leveling the price difference. A further example are traders that utilize small differences in the value of two assets that normally track each other. Let us consider the share prices of soft-drink makers Pepsi® and Coca Cola®. Both follow each other rather closely, because the determining influences are very similar. If, by chance, a small discrepancy appears, the arbitrageurs can be rather sure that at a later time the stock prices will converge again. By exploiting this information, they can buy the temporarily too cheap asset and short the expensive one. Arbitrageurs play an important role in the market. They ensure that any imbalancing fluctuations are reacted upon and thereby reduced. They keep the market in balance at its equilibrium.

Of course *banks* are participants in the market, and especially the central banks. They set the discount rates, by which commercial banks can borrow money themselves and this determines all other interest rates. Often affiliated to banks are *brokerage firms.* They are financial advisors or stock-brokers, who assist other persons in facilitating their trades.

Options are traded by *market makers*, which are often banks, but can be independent brokers. They provide the liquidity, mentioned above, to the market, because they immediately react to external offers and purchase requests and thereby guarantee that stocks and options actually *can* be traded. They ensure that buying and selling options, often for the purpose of continuous adaption of the hedging (see below), always works.

The rules for trading are defined and enforced by national *trade commissions* that supervise the exchanges. In the US this is the *U.S. Securities and Exchange Commission.* In the Euro-zone as recent as 2014, the supervisory authority for large banks was transferred to the *European Central Bank.*

Clearing houses are intermediaries between traders. They guarantee that the contracts between other market participants are honored. They normally require the traders to maintain a *margin account*, which contains sufficient funds to cover any expected and foreseeable losses.

After having discussed the prerequisites we can now turn to managing risk and determine a portfolio that balances our appetite for profit and our distaste to the exposure to risk. It turns out that this discussion is closely linked to the valuation of stocks. Namely, the question of what determines the value of a given asset.

Exercises

1. Write a short essay about what G. Soros did in 1992 when he "shorted the Bank of England."
2. Research what a hedge fond is and write a short essay about it. Give a few examples.
3. Find out where you can trade in hog meat, where in stocks of the manufacturer of your smart phone?
4. Where and how can you trade in derivatives based on the weather? Why would someone want to do so?
5. Who owns and operates the main stock exchanges in Frankfurt, New York, and Stockholm?
6. Find out why selling is called "going short," and why is purchasing called "going long?"
7. Write a short essay about the Bretton-Woods agreement.
8. Which authorities oversee payments by credit cards?

Chapter 3
Portfolio Theory and CAPM

Abstract This chapter deals with the compilation of stocks into a portfolio that suits an investor's hunger for profit, while minimizing the risk of losing the initial investment. Since the theory is based on variational methods and on Lagrange multipliers, applications of these concepts in physics are juxtaposed to their use in finance. As an extension of the portfolio theory, the basic equations of the capital asset pricing model are derived and used to value companies.

Let us assume that we want to invest a sum of money in shares. So, how do we split the available money among shares in a way that maximizes our profit, while minimizing the risk to actually lose money? This question was addressed by Markowitz in the 1950s, which led to his winning a share of the 1990 memorial prize in A. Nobel's memory. We start by considering his analysis of choosing a portfolio [1], which relies heavily on methods we use in physics: Lagrange multipliers and variational methods.

Let us look at different traders having different degrees of eagerness to make a profit and varying degrees of acceptance to risk. A very risk-averse trader, for example, would simply place all his available cash in a bank account that provides the risk-free rate r_f, which does not have any volatility σ and poses no risk of losing any money. A more ambitious investor, a speculator, will request a higher return rate than the risk-free rate r_f, but at the expense of having to tolerate some level of risk or volatility, which may potentially lead to a reduction of the value of the original investment. One usually only has a finite amount of money available to start trading. This poses a constraint on one's ability to select stocks. Such a constraint is efficiently handled by Lagrange multipliers. Moreover, optimizing a portfolio is based on methods from variational calculus. We therefore take the opportunity to briefly touch upon the use of both concepts in classical mechanics [2, 3].

Electronic supplementary material The online version of this chapter (https://doi.org/10.1007/978-3-030-63643-2_3) contains supplementary material, which is available to authorized users.

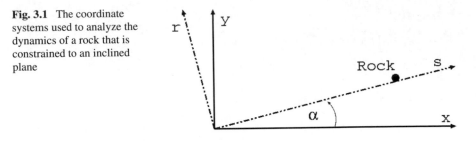

Fig. 3.1 The coordinate systems used to analyze the dynamics of a rock that is constrained to an inclined plane

3.1 Variational Calculus and Lagrange Multipliers

The equations of motion for mechanical systems follow from Hamilton's principle of maximizing the action functional S, which implies $\delta S = 0$. The action S is defined as the time-integral of the Lagrangian $L(q, \dot{q})$, which depends on the position q and velocity $\dot{q}$ of, for example, a point mass. Here we ignore any explicit dependence of the Lagrangian on the time t. Minimizing the action S over a fixed time interval from t_1 to t_2 is thus given by

$$0 = \delta S = \delta \int_{t_1}^{t_2} L(q, \dot{q})dt = \int_{t_1}^{t_2} \left(\frac{\partial L}{\partial q}\delta q + \frac{\partial L}{\partial \dot{q}}\delta \dot{q} \right) dt \,, \tag{3.1}$$

where we varied $L(q, \dot{q})$ with respect to its two arguments q and $\dot{q}$. These two quantities, however, are not independent, but related through $\delta\dot{q} = \frac{d}{dt}\delta q$, which allows us to rewrite the preceeding equation as

$$0 = \frac{\partial L}{\partial \dot{q}}\delta q \Big|_{t_1}^{t_2} + \int_{t_1}^{t_2} \left(\frac{\partial L}{\partial q} - \frac{d}{dt}\frac{\partial L}{\partial \dot{q}} \right) \delta q dt \,, \tag{3.2}$$

where we use partial integration to write $\frac{\partial L}{\partial \dot{q}}\delta\dot{q} = \frac{d}{dt}\left(\frac{\partial L}{\partial \dot{q}}\delta q \right) - \frac{d}{dt}\frac{\partial L}{\partial \dot{q}}$. We then integrate the total derivative to obtain the first term, which vanishes, because the trajectory $q(t_1)$ and $q(t_2)$ at the end point is fixed. This requires the second term to vanish for all variations δq, which can only satisfied if the expression in the brackets is zero. This leads to the *Euler-Lagrange equations*

$$0 = \frac{d}{dt}\frac{\partial L}{\partial \dot{q}} - \frac{\partial L}{\partial q} \,, \tag{3.3}$$

from which the equations of motion follow in which all (generalized) coordinates are assumed to be independent. If there are additional constraints among them, we can use *Lagrange multipliers* to take these constraints into account.

We discuss the Lagrange multipliers by considering the simple mechanical model of a falling rock and then constraining the rock's motion to an inclined plane. Figure 3.1 illustrates the geometry. We describe the unconstrained motion of the rock in two dimensions x and y by the Lagrangian $L = T - U$, where $T = (m/2)(\dot{x}^2 + \dot{y}^2)$ is the kinetic energy and $U = mgy$ is the potential energy. Constraining the "falling" motion to the inclined plane, given by the equation $y = x \tan \alpha$, results in the adapted Lagrange function

$$L = T - U + \lambda(y - x \tan \alpha)$$
$$= \frac{1}{2}m(\dot{x}^2 + \dot{y}^2) - mgy + \lambda(y - x \tan \alpha) \tag{3.4}$$

with the Lagrange multiplier λ, which we treat as an additional dynamic variable. There is no term with $\dot{\lambda}$, such that the Euler-Lagrange equations of motion are given by

$$0 = \frac{d}{dt}\frac{\partial L}{\partial \dot{x}} - \frac{\partial L}{\partial x} = \frac{d}{dt}m\dot{x} - [-\lambda \tan \alpha] = m\ddot{x} + \lambda \tan \alpha$$
$$0 = \frac{d}{dt}\frac{\partial L}{\partial \dot{y}} - \frac{\partial L}{\partial y} = \frac{d}{dt}m\dot{y} - [-mg + \lambda] = m\ddot{y} + mg - \lambda \tag{3.5}$$
$$0 = \frac{\partial L}{\partial \lambda} = y - x \tan \alpha \ .$$

The derivative with respect to λ recovers the constraint and the two first equations have the constraint built-in with the aid of the Lagrange multiplier λ. In the next step we eliminate λ from the equations by multiplying the first equation by $\cos \alpha$ and the second by $\sin \alpha$ and then adding the two equations. This leads us to $0 = \ddot{x} \cos \alpha + \ddot{y} \sin \alpha + g \sin \alpha$. Now we use the constraint to eliminate $\ddot{y} = \ddot{x} \tan \alpha$ and arrive at

$$0 = \frac{\ddot{x}}{\cos \alpha} + g \sin \alpha \qquad \text{or} \qquad 0 = \ddot{s} + g \sin \alpha \tag{3.6}$$

where $s = x/\cos \alpha$ is the distance measured along the inclined plane, which is the result one can expect from more elementary considerations.

Note that we simply patch the constraint onto the Lagrange function with the help of a Lagrange multiplier λ and then treat the latter as an additional dynamical parameter. This "recipe" then incorporates the constraint into the equation of motion and, at the same time, the Euler-Lagrange equation with respect to λ reproduces the constraint.

After this refresher to illustrate the use of Lagrange multipliers for treating constraints in variational problems we address the optimization of portfolios.

3.2 Portfolio with Risky Assets Only

We consider a portfolio of of shares that are weighted by w_i, with $\sum_i w_i = 1$, which implies that the available funds are split among shares and $w_i = 0.1$ means that 10% of our funds are allocated to share number i. The shares are assumed to produce some return r_i on the initial investment. The question is how to choose the w_i in order to maximize the averaged portfolio return $\rho = \sum_i w_i \langle r_i \rangle$ but do this with the minimum uncertainty. The latter requirement translates into minimizing the variance V of the return, which is given by

$$V(w) = \left\langle \left[\sum_i w_i (r_i - \langle r_i \rangle) \right] \left[\sum_j w_j (r_j - \langle r_j \rangle) \right] \right\rangle = \sum_i \sum_j C_{ij} w_i w_j \,, \quad (3.7)$$

where C_{ij} is given by

$$C_{ij} = \langle [r_i - \langle r_i \rangle][r_j - \langle r_j \rangle] \rangle \qquad (3.8)$$

and where the angle brackets $\langle \cdot \rangle$ denote averages of the return ρ over a suitably chosen time span in the past. Note that the covariance matrix C is symmetric by construction. The requirement to minimize V and and simultaneously achieve a return ρ, while maintaining $\sum w_i = 1$, can be cast into the form to minimize a cost-function, often denoted by χ^2, given by

$$\chi^2 = \frac{1}{2} \sum_i \sum_j C_{ij} w_i w_j + \lambda_1 \left[\rho - \sum_i w_i \langle r_i \rangle \right] + \lambda_2 \left[1 - \sum_i w_i \right] . \quad (3.9)$$

Here λ_1 and λ_2 are Lagrange multipliers to accommodate the constraints to make the weighted average return equal to ρ and keep the sum of the weights w_i equal to unity. This is analogous to incorporating the constraint to stay on the inclined plane into the Lagrangian for the mechanical system in (3.4).

We can find the optimum portfolio distribution of shares w_k for a given profit eagerness ρ by determining the minimum of χ^2 with respect to the w_k as a function of ρ and write

$$0 = \frac{\partial \chi^2}{\partial w_k} = \sum_j C_{kj} w_j - \lambda_1 \langle r_k \rangle - \lambda_2 \qquad (3.10)$$

or, more conveniently in matrix form

$$0 = C\mathbf{w} - \lambda_1 \langle \mathbf{r} \rangle - \lambda_2 \mathbf{e} \,, \qquad (3.11)$$

where we introduced the vector $\mathbf{e}$ with all components equal to unity. This equation is easily solved by

$$\mathbf{w} = C^{-1} [\lambda_1 \langle \mathbf{r} \rangle + \lambda_2 \mathbf{e}] \qquad (3.12)$$

and its transpose

$$\mathbf{w}^T = \langle \mathbf{r}^T \rangle C^{-1} \lambda_1 + \mathbf{e}^T C^{-1} \lambda_2 \,, \tag{3.13}$$

provided that the covariance matrix is invertible. This is the case if there are no risk-free assets in the portfolio, a point we relax in the next section. But for now we insert the expression for the weights of the portfolio vector $\mathbf{w}$ into the condition for the desired return ρ and obtain

$$\rho = \mathbf{w}^T \langle \mathbf{r} \rangle = \langle \mathbf{r}^T \rangle C^{-1} \langle \mathbf{r} \rangle \lambda_1 + \mathbf{e}^T C^{-1} \langle \mathbf{r} \rangle \lambda_2 \,. \tag{3.14}$$

The normalizing condition can be written as $1 = \mathbf{w}^T \cdot \mathbf{e}$, which yields

$$1 = \mathbf{w}^T \mathbf{e} = \langle \mathbf{r}^T \rangle C^{-1} \mathbf{e} \lambda_1 + \mathbf{e}^T C^{-1} \mathbf{e} \lambda_2 \,. \tag{3.15}$$

These two equations form a set of linear equations that can be cast into a matrix-valued equation

$$\begin{pmatrix} \rho \\ 1 \end{pmatrix} = \begin{pmatrix} \langle \mathbf{r}^T \rangle C^{-1} \langle \mathbf{r} \rangle & \mathbf{e}^T C^{-1} \langle \mathbf{r} \rangle \\ \langle \mathbf{r}^T \rangle C^{-1} \mathbf{e} & \mathbf{e}^T C^{-1} \mathbf{e} \end{pmatrix} \begin{pmatrix} \lambda_1 \\ \lambda_2 \end{pmatrix} \,, \tag{3.16}$$

which is easily solved for the Lagrange multipliers λ_1 and λ_2 by inverting the 2×2 matrix appearing in (3.16)

$$\begin{pmatrix} \lambda_1 \\ \lambda_2 \end{pmatrix} = \frac{1}{\Delta} \begin{pmatrix} \mathbf{e}^T C^{-1} \mathbf{e} & -\mathbf{e}^T C^{-1} \langle \mathbf{r} \rangle \\ -\langle \mathbf{r}^T \rangle C^{-1} \mathbf{e} & \langle \mathbf{r}^T \rangle C^{-1} \langle \mathbf{r} \rangle \end{pmatrix} \begin{pmatrix} \rho \\ 1 \end{pmatrix} \,, \tag{3.17}$$

where Δ is the determinant of the 2×2 matrix. Inserting λ_1 and λ_2 into (3.12), we obtain the optimal weight vector $\mathbf{w}$ for the portfolio as a function of the desired portfolio return ρ

$$\mathbf{w} = \lambda_1 C^{-1} \langle \mathbf{r} \rangle + \lambda_2 C^{-1} \mathbf{e} \,, \tag{3.18}$$

which contains only known quantities such as the average return of shares $\langle \mathbf{r} \rangle$ and their variance as well as the desired portfolio return ρ specifying the investor's eagerness to make a profit. The resulting volatility σ of the portfolio, given by $\mathbf{w}$, is $\sigma^2 = \mathbf{w}^T C \mathbf{w}$.

We can now determine the relation between desired return ρ and the resulting volatility by plotting ρ versus σ, which is done in Fig. 3.2 for a sample portfolio with three assets, approximately having returns $\langle r_1 \rangle \approx 1\%$, $\langle r_2 \rangle \approx 2\%$, and $\langle r_3 \rangle \approx 3\%$ with approximate volatilities of 1, 2, and 3%, respectively. The plot was generated with the MATLAB script from Appendix B.1 and uses sequences of correlated random numbers with returns and volatilities shown as the three black asterisks in Fig. 3.2. We also show the relation between the desired portfolio return ρ and the portfolio volatility σ as the solid blue line for our three-stock portfolio. The portfolio with a minimum volatility given by the left-most point in the knee of the blue hyperbola. The coordinates of this point can be calculated from the requirement that

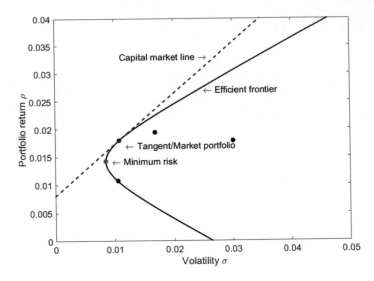

Fig. 3.2 The locus of pairs (ρ, σ) in the portfolio plane. The blue hyperbola relates the desired return ρ to the volatility σ for a purely risky portfolio. The red line relates the same quantities if a risk-free investment option is available

$\sigma^2 = \mathbf{w}^T C \mathbf{w}$, where $\mathbf{w}$ is given by (3.18) and (3.17) in terms of the desired return ρ. Explicitly inserting (3.18), we find

$$
\begin{aligned}
\sigma^2 &= \lambda_1^2 \mathbf{r}^T C^{-1} \mathbf{r} + \lambda_1 \lambda_2 \mathbf{e}^T C^{-1} \mathbf{r} + \lambda_1 \lambda_2 \mathbf{r}^T C^{-1} \mathbf{e} + \lambda_2^2 \mathbf{e}^T C^{-1} \mathbf{e} \\
&= \begin{pmatrix} \lambda_1 & \lambda_2 \end{pmatrix} \begin{pmatrix} \mathbf{r}^T C^{-1} \mathbf{r} & \mathbf{e}^T C^{-1} \mathbf{r} \\ \mathbf{r}^T C^{-1} \mathbf{e} & \mathbf{e}^T C^{-1} \mathbf{e} \end{pmatrix} \begin{pmatrix} \lambda_1 \\ \lambda_2 \end{pmatrix} \\
&= \begin{pmatrix} \lambda_1 & \lambda_2 \end{pmatrix} \begin{pmatrix} \rho \\ 1 \end{pmatrix} \\
&= a_{11} \rho^2 + 2 a_{12} \rho + a_{22} \,,
\end{aligned}
\tag{3.19}
$$

where we used (3.16). Moreover, we denote the matrix elements of the symmetric matrix in (3.17) by a_{ij}. Differentiating σ^2 with respect to ρ and equating the result to zero yields the return ρ^* at the point of minimum risk, given by

$$
\rho^* = -\frac{a_{12}}{a_{11}} = \frac{\mathbf{e}^T C^{-1} \mathbf{r}}{\mathbf{e}^T C^{-1} \mathbf{e}} \,.
\tag{3.20}
$$

The corresponding minimum volatility σ^* is given by

$$
\sigma^{*2} = a_{11} \rho^{*2} + 2 a_{12} \rho^* + a_{22} = \frac{a_{11} a_{22} - a_{12}^2}{a_{11}} = \frac{1}{\mathbf{e}^T C^{-1} \mathbf{e}} \,.
\tag{3.21}
$$

The point (σ^*, ρ^*) is indicated in Fig. 3.2 by an asterisk in the knee of the blue curve. Note that the minimum is actually smaller than the smallest volatility of the underlying assets, which was 1%. We find that mixing different shares can reduce the volatility of the combined portfolio, though not eliminate it completely. Note also that in some cases some of the weights w_k can become negative, which corresponds to shorting the asset with the negative weight.

In the derivation of the optimum portfolio distribution of shares $\mathbf{w}$, we assumed that the covariance matrix C is invertible, which is not the case if we include risk-free assets such as government bonds in our portfolio. In the following section we relax this constraint.

3.3 Portfolio with a Risk-Free Asset

If we have a risk-free asset available in our portfolio, the discussion in the previous section up to (3.11) is unaffected, but inverting the covariance matrix C, which led to (3.12) is not possible. In order to inspect the problem closer, we write out the equivalent equation to (3.11)

$$
\begin{pmatrix} 0 \\ 0 \\ \vdots \\ 0 \end{pmatrix} = \begin{pmatrix} & & & 0 \\ & \hat{C} & & 0 \\ & & & \vdots \\ 0 & 0 & 0 & 0 \end{pmatrix} \begin{pmatrix} w_1 \\ w_2 \\ \vdots \\ w_f \end{pmatrix} - \lambda_1 \begin{pmatrix} \langle r_1 \rangle \\ \langle r_2 \rangle \\ \vdots \\ r_f \end{pmatrix} - \lambda_2 \begin{pmatrix} 1 \\ 1 \\ \vdots \\ 1 \end{pmatrix} ,
\tag{3.22}
$$

where we assume that we have N risky assets and the one with number $N+1$ is risk-free, which accounts for the zeros in the last column and row. We denote the covariance matrix, restricted to the top left $N \times N$ non-zero part, by $\hat{C}$.

From the last row of (3.22) we immediately find

$$
\lambda_2 = -\lambda_1 r_f ,
\tag{3.23}
$$

which, after inserting it in the upper N rows of (3.22), leads to

$$
\begin{pmatrix} 0 \\ \vdots \\ 0 \end{pmatrix} = \hat{C} \begin{pmatrix} w_1 \\ \vdots \\ w_N \end{pmatrix} - \lambda_1 \begin{pmatrix} \langle r_1 \rangle \\ \vdots \\ \langle r_N \rangle \end{pmatrix} + \lambda_1 r_f \begin{pmatrix} 1 \\ \vdots \\ 1 \end{pmatrix} .
\tag{3.24}
$$

Solving for the w_i yields

$$
\begin{pmatrix} w_1 \\ \vdots \\ w_N \end{pmatrix} = \lambda_1 \hat{C}^{-1} \begin{pmatrix} \langle r_1 \rangle - r_f \\ \vdots \\ \langle r_N \rangle - r_f \end{pmatrix} = \lambda_1 \hat{C}^{-1} \Delta \mathbf{r} ,
\tag{3.25}
$$

where we introduced the abbreviation $\Delta \mathbf{r} = \langle \mathbf{r} \rangle - r_f \mathbf{e}$. The Lagrange multiplier λ_1 can be determined from the requirement that the return shall be ρ. This brings us to

$$
\begin{aligned}
\rho &= \mathbf{w}^T \langle \mathbf{r} \rangle + w_f r_f \\
&= \mathbf{w}^T \langle \mathbf{r} \rangle + \left(1 - \mathbf{w}^T \mathbf{e}\right) r_f \\
&= r_f + \mathbf{w}^T \left(\langle \mathbf{r} \rangle - r_f \mathbf{e}\right) \\
&= r_f + \lambda_1 \Delta \mathbf{r}^T \hat{C}^{-1} \Delta \mathbf{r} \ .
\end{aligned}
\tag{3.26}
$$

Solving for λ_1, we obtain

$$
\lambda_1 = \frac{\rho - r_f}{\Delta \mathbf{r}^T \hat{C}^{-1} \Delta \mathbf{r}}
\tag{3.27}
$$

and for the first N components of the portfolio vector $\mathbf{w}$ we find the following expression upon inserting λ_1 in (3.25)

$$
\mathbf{w} = \frac{\rho - r_f}{\Delta \mathbf{r}^T \hat{C}^{-1} \Delta \mathbf{r}} \, \hat{C}^{-1} \Delta \mathbf{r} \ .
\tag{3.28}
$$

The fraction invested in the risk-free asset w_f can be determined from the normalization requirement $w_f + \sum_i^N w_i = 1$, here written by splitting up the sum in the contribution to the risky assets $1, \ldots, N$ and the risk-free asset.

The variance (or volatility) of the portfolio consisting of a mixture of the risk free asset and the risky assets is given by

$$
\sigma^2 = \mathbf{w}^T \hat{C} \mathbf{w}
\tag{3.29}
$$

where $\mathbf{w}$ is given by (3.28). Inserting $\mathbf{w}$ and simplifying the resulting expression, we arrive at

$$
\sigma^2 = \frac{(\rho - r_f)^2}{\Delta \mathbf{r}^T \hat{C}^{-1} \Delta \mathbf{r}} \quad \text{or} \quad \frac{(\rho - r_f)}{\sigma} = \sqrt{\Delta \mathbf{r}^T \hat{C}^{-1} \Delta \mathbf{r}}
\tag{3.30}
$$

from which we derive that the excess return above the risk free rate $\rho - r_f$ is proportional to the volatility σ. The constant of proportionality $\sqrt{\Delta \mathbf{r}^T \hat{C}^{-1} \Delta \mathbf{r}}$ encapsulates all knowledge about our stock portfolio. It defines the dashed red line, called the *capital allocation line,* in Fig. 3.2, whose slope is referred to as the *Sharpe ratio*.

3.4 Capital Market Line and Sharpe Ratio

Let us now generalize the discussion from the previous section by assuming that the underlying assets of the portfolio comprise a representative mixture of all available assets in the market, such as the S&P 500. In that case we express the right-hand-side of the right equation of (3.30) in terms of the return r_M and volatility σ_M of *the market* by noting that $C^{-1} = 1/\sigma_M^2$ and $\Delta r = r_M - r_f$. Inserting these simplifications we obtain

$$\frac{\rho - r_f}{\sigma} = \frac{r_M - r_f}{\sigma_M} \tag{3.31}$$

which shows that the Sharpe ratio for our portfolio on the left-hand-side of the equation equals that of the entire market, but we can adjust our desired return ρ to our preference, albeit linked to a specific risk or volatility σ. Therefore, the Sharpe ratio is sometimes called the *price of risk;* if we want a higher pay-off or return on our investment, we have to accept a higher exposure to risk.

The line, that was the capital allocation line for a specific portfolio is called the *capital market line* when we consider the whole market. It derives from a portfolio that is representative of the market. Note that the (red) market allocation line in Fig. 3.2 always lies above the (blue) efficient frontier, as given by investing in risky assets only. This is no surprise, because we have an extra option; either investing in the risk-free asset or borrowing at the risk-free rate and investing the borrowed money in stocks. Optionality is good!

Note also that the intercept of the capital market with the vertical axis describes a situation where all our available funds are invested in the risk-free asset. Conversely, the point where the capital market line touches the blue efficiency frontier line describes the situation where all our funds are invested in risky assets of the market portfolio. The coordinates of this point at which the capital market line is tangent to the efficient frontier can be easily calculated from the requirement that all available funds are invested in risky assets, namely $\mathbf{e}^T \mathbf{w} = 1$. Using (3.28) and solving for the return of the tangent portfolio ρ_t, we find

$$\rho_t = r_f + \frac{\Delta \mathbf{r}^T \hat{C}^{-1} \Delta \mathbf{r}}{\mathbf{e}^T \hat{C}^{-1} \Delta \mathbf{r}} \tag{3.32}$$

and inserting in (3.30), we find for the risk σ_t of the tangent portfolio

$$\sigma_t^2 = \frac{\Delta \mathbf{r}^T \hat{C}^{-1} \Delta \mathbf{r}}{\left(\mathbf{e}^T \hat{C}^{-1} \Delta \mathbf{r} \right)^2} . \tag{3.33}$$

In Fig. 3.2 the tangent point is indicated by a red asterisk. The section of the capital market line between the intercept and tangent point describes a mixture of risk-free and optimal market portfolio that matches our taste for risk. If we are risk-averse, we use a mix that corresponds to a point on the line located towards the left, near the intercept where the return ρ is close to the risk-free rate r_f. If we are more risk-tolerant, we choose a mix on the line located closer to the tangent point. If we are even more adventurous, we can borrow money at the risk-free rate (negative weight w_f) and invest the borrowed money into the market portfolio. This is called leveraged investment, which however, carries a high risk.

Note that all points on the capital market line constitute the most efficient portfolios, and are convex combinations of investing a fraction of ones wealth in the risk-free asset and the rest in the tangent portfolio, which incidentally is the same as the market portfolio discussed in the beginning of this section. That all efficient portfolios on the capital market line are linear combinations of just two investment assets was first found by Tobin [4] and was called *separation theorem*.

In order to compare a previously unknown asset with the market, we can calculate its Sharpe ratio and compare it to that of the market. If it is higher, the new asset is worth buying. This is equivalent of placing the new asset in the diagram with the portfolio space Fig. 3.2, where the properly valued assets should lie close to the (red) capital market line. If stocks lie above the line, they are undervalued and constitute a potential investment option. Conversely, if they lie below the capital market line, they are over-valued and provide too little return for the risk that investing in them would entail.

In the following section we discuss a slightly more refined method to figure out whether we want to include a new asset in a portfolio.

3.5 Capital Asset Pricing Model

The *Capital asset pricing model* or CAPM, as it is known, describes the expected return r_1 of an asset, that was not previously included in a portfolio, here the market portfolio. The model was initially introduced by Sharpe [5] and Lintner [6]. Since any efficient portfolio can be expressed as a sum of the risk-free asset and the market portfolio, we consider a market with these three assets. All information about this mini-market is encoded in the covariance matrix

$$C = \begin{pmatrix} \sigma_1^2 & \sigma_{1M} & 0 \\ \sigma_{1M} & \sigma_M^2 & 0 \\ 0 & 0 & 0 \end{pmatrix} \tag{3.34}$$

and the returns r_1, r_M, r_f for the new asset, the market, and the risk-free asset, respectively. We now perform the same analysis for the mini-market that we did in Sect. 3.3 and use (3.28) to determine the weights for the allocation of funds to the assets. They turn out to be

$$\begin{pmatrix} w_1 \\ w_M \end{pmatrix} = \frac{\rho - r_f}{\Delta \mathbf{r}^T \hat{C}^{-1} \Delta \mathbf{r}} \, \hat{C}^{-1} \Delta \mathbf{r} , \tag{3.35}$$

where $\hat{C}$ is the 2×2 matrix at the top left corner of the covariance matrix C and $\Delta \mathbf{r}$ is given by the following relations

$$\hat{C} = \begin{pmatrix} \sigma_1^2 & \sigma_{1M} \\ \sigma_{1M} & \sigma_M^2 \end{pmatrix} \quad \text{and} \quad \Delta \mathbf{r} = \begin{pmatrix} r_1 - r_f \\ r_M - r_f \end{pmatrix} . \tag{3.36}$$

Fig. 3.3 The security market line shown as the expected return as a function of the systematic risk β.

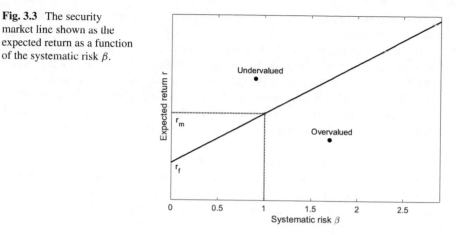

In order to be considered as an investment, we require the new asset to be at least as profitable as the market as a whole. Otherwise, we would not pick it and invest in the market portfolio instead. This implies that the new asset will have a weight w_1 larger than zero. Considering that the fore-factor $(\rho - r_f)/(\Delta \mathbf{r}^T \hat{C}^{-1} \Delta \mathbf{r})$ is positive this implies that the first component of $\hat{C}^{-1} \Delta \mathbf{r}$ must be larger than zero. Evaluating this expression, we find

$$\hat{C}^{-1} \Delta \mathbf{r} = \frac{1}{\sigma_1^2 \sigma_M^2 - \sigma_{1M}^2} \begin{pmatrix} \sigma_M^2 (r_1 - r_f) - \sigma_{1M} (r_M - r_f) \\ -\sigma_{1M} (r_1 - r_f) + \sigma_1^2 (r_M - r_f) \end{pmatrix} . \tag{3.37}$$

Upon reordering the first component, this leads to

$$r_1 \geq r_f + \frac{\sigma_{1M}}{\sigma_M^2} (r_M - r_f) = r_f + \beta_1 (r_M - r_f) \tag{3.38}$$

where we introduced the beta β_1 of the new asset

$$\beta_1 = \frac{\sigma_{1M}}{\sigma_M^2} . \tag{3.39}$$

Equation 3.38 is the Sharpe-Lintner version of the key feature of the CAPM. Interpreting (3.38) as a linear equation between the expected return r_1 and the risk, as given by β_1, is often called the *security market line,* shown in Fig. 3.3. Plotting a new asset's r_1 vs β_1 one can determine if the new asset is over- or under-valued, depending on r_1 lying above or below $r_f + \beta_1(r_M - r_f)$.

Whereas r_1 describes the expected return of a new asset, the beta describes how much the asset is correlated to the movement of the market as whole. If β_1 is unity it implies that the new asset will rise or fall by the same percentage as the market as a whole. For example, a beta of 1.5 implies that the new asset moves up by 15%

should the market move up by 10%. If beta is negative it can be used for hedging and offsetting other risks.

Another interpretation is that a new asset with beta equal to unity behaves just like the market and its behavior is already covered and optimally hedged by the market portfolio. The expected return r_1 in that case is equal to the return r_M of the market portfolio. On the other hand, a positive beta, but different from unity, implies a risk, inherent or systematic to the new asset. This extra risk, which can not be hedged, therefore requires a higher return r_1. This extra cost of risk, as characterized by beta, is thus given by the higher demand on return. The return r_1 that is consistent with the assumption of efficient markets in equilibrium is the one given in (3.38).

We need to point out that the above results are part of the standard canon of economic theory, though not beyond reproach. The CAPM model is based on a number of simplifying assumptions, necessary for the analytic treatment. These assumptions can be and are questioned. See for example [7] for a comprehensive review of the criticism.

Closely linked to the question whether to buy an asset, such as a share in a company, is the question about the value of a company. The simple answer is "the sum of all shares at current market value." But that implies that the market "knows" the real value, which in turn is determined by supply and demand. This is a bit of a hen-and-egg problem and a way to escape the circular logic is to use other methods to put a value onto a company. We cover two methods in the next section.

3.6 Valuation

One way to determine the value of a company, for example, before acquiring a company or merging with it, is based on calculating the *market cap*, or market capitalization of the company. It is given by the value of the (outstanding) shares, traded at stock exchanges, multiplied by the value of the share. Note, however, that the market cap fluctuates with the share price from one day to the next and one might wonder how the "market" as a whole—or better, its participants—determines what the correct price actually is.

The second method is based on evaluating the expected performance of the company keeping in mind that an investor expects to recover his investment over a finite time, say $Y = 5$ years. One often used performance measure is the *discounted cash flow D*, given by

$$D = \sum_{k=1}^{Y} \frac{C_k}{(1+r)^k} , \qquad (3.40)$$

where r is the annual discount rate, which takes the time-value of money, discussed in Sect. 2.5, into account. C_k is the expected annual *cash flow* k years in the future.

The cash flow C_k in year k is essentially given by the difference $C_k = I_k - P_k$ of the income I_k the company receives for its services and the purchase price P_k of the raw materials. At the present time these values are unknown and need to be estimated. Big changes can be expected, if new production facilities are opened within the time horizon Y.

The discount rate r that is used to calculate the discounted cash flow describes the annual cost of sustaining the capital V in the company. Often the $WACC$, the *weighted average cost of capital*, is substituted for r. It is given by

$$WACC = \frac{V - B}{V} r_e + \frac{B}{V} r_d (1 - t) , \qquad (3.41)$$

where B/V is the fraction of the capital financed by borrowing money at interest rate r_d from a bank. Since taxes do not have to be paid on the interest payments, the latter is reduced by $1 - t$ where t is the corporate tax rate, presently 21% in the US and between 20 and 30% in most European countries. $(V - B)/V$ is the remaining fraction of the capital, which is financed through shares and r_e is the expected rate of return from investors, which can be calculated from CAPM with (3.38)

$$r_e = r_f + \beta (r_M - r_f) \qquad (3.42)$$

where β is the correlation of the company's shares with the a market portfolio, as discussed in Sect. 3.5, where r_f and r_M are already defined as the risk-free rate and the average growth rate of the market, respectively.

Now we have all ingredients available in order to calculate the discounted cash flow D and compare it with the price P_a that the present owner of the company is asking. For us, or any other investor, the difference between D and P_a, called *net present value* $N_0 = D - P_a$ of the company, is the crucial quantity to assess. If it is positive, we have a decent chance to recover our investment over the next Y years. If N_0 is negative, the investment is questionable. Note, however, that there are many ad-hoc assumptions entering the calculation of the discounted cash flow D, such that this analysis should be complemented by additional methods, such as evaluating the price/earnings ratio of the company over the past few years in order to assess its performance and whether it actually used a positive past cash flow to pay dividends to the investors.

So far we considered static properties of stocks, but in the following sections we will investigate how stocks evolve in time.

Exercises

1. A particle with mass m moves in two dimensions x and y in the potential $U = (k/2)(x^2 + xy + y^2)$, but is constrained to a line given by $x + y = 1$. Calculate the resonance frequency and the equilibrium point around which the particle oscillates.

2. Unexpectedly, you have inherited some money from an unknown aunt and you want to invest it in the stock market. To make the problem manageable, you decide to invest in two items, Apple® Inc. and an index fund, based on a mix of SP500 companies. A table of closing values at the end of every trading day for one year (27.3.2018–27.3.2019) is available in the file stocks.dat from this book's web site. This text file contains three columns: the first one indicates the trading day t_d, the second contains the Apple data S_1, and the third the SP500 data S_2. There is a short MATLAB script get_stocks.m available to load this file.

   a. Plot the stock values S versus the trading day to obtain an impression of how the stocks evolve.
   b. Calculate the day-to-day returns $r_k = (S_{k+1} - S_k)/S_k$, plot the values and calculate their average. Reminder: scale the average daily returns with the trading days to obtain the annual returns.
   c. Then you decide to use Markowitz's theory to calculate how to partition your money between the two stocks in order to obtain an annual return of 5, 10, and 15%. Interpret the results and how to implement them.

3. For the highest annual return of $\rho = 0.15\%$, repeat the analysis from Exercise 2 with a risk-free asset a) with $r_f = 0\%$, b) with $r_f = 3\%$ included in the portfolio.
4. Based on the data in stocks.dat, use the SP500 data in the third column as a proxy for the market as a whole and assume an annual risk-free rate of $r_f = 5\%$. Use CAPM to determine whether the stock in the second column is over or undervalued.
5. Before buying a company, you analyze their financial reports, which state that the capital base is 16×10^6 € of which 7×10^6 € is financed by debt at an interest rate of 5%. The tax rate is $t = 30\%$. The annual rate of return to the investors historically was about 12% and the annual cash flow was typically 1.7×10^6 €. The asking price that the present owner demands is 8.1×10^6 €. Can you expect to recover your investment over 6 years?

References

1. H. Markowitz, Portfolio Selection. J. Finance **7**, 77 (1952)
2. L. Landau, E. Lifschitz, *Lehrbuch der theoretischen Physik, Band I: Mechanik* (Akademie Verlag, Berlin, 1979)
3. H. Goldstein, J. Safko, C. Poole, *Classical Mechanics* (Pearson, Harlow, 2014)
4. J. Tobin, Liquidity preference as behavior towards risk. Rev. Econ. Stud. **XXVI**, 65 (1958)
5. W. Sharpe, Capital asset prices: a theory of market equilibrium under conditions of risk. J. Finance **19**, 425 (1964)
6. J. Lintner, The valuation of risk assets and the selection of risky investments in stock portfolios and capital budgets. Rev. Econ. Stat. **47**, 13 (1965)
7. E. Fama, K. French, The capital asset pricing model: theory and evidence. J. Econ. Perspect. **18**, 25 (2004)

Chapter 4
Stochastic Processes

Abstract After introducing binomial trees as a time-discrete model to value options, this chapter moves on to discuss how a Wiener process gives rise to the diffusion equation, which is subsequently solved in terms of Green's functions. A short digression of the role of Green's functions in physics follows, before Ito's lemma is discussed and used to derive the Fokker-Planck equation for the continuous-time model that describes the evolution of stocks. Solving the Fokker-Planck equation then leads to the well-known log-normal distribution. The latter is then used to derive the expectation value of an option's value, should it be exercised at maturity. Examples from other uses of expectation values in physics concludes this chapter.

In the previous chapter we ignored the time evolution of the stocks, apart from assuming that the underlying dynamics is governed by an average growth with superimposed fluctuations. In the present and subsequent chapters we relax this constraint and will deal with the temporal evolution of financial assets; an elaborate system of guesstimating the unforeseeable. We will not only consider the evolution of stocks, but also that of other financial derivatives, such as options. Before dealing with quasi-continuous systems below, we first discuss a simple system that helps us understand the mechanisms. It is based on discretizing time and the method is commonly referred to as *binomial trees*.

4.1 Binomial Trees

In the theory of binomial trees, we consider different ways in which the value of a financial product, for example, an option can move, either up or down and assign probabilities to the different cases. The suitable initial price is then determined by the expectation value of the option price at the final time. Note that this will depend on the strike price K of the option. An option that carries a higher risk, but also the promise of making a higher profit, will be worth more.

For definiteness, let us consider the pricing of a call option with initial price c for an underlying asset, which we assume to be a share with initial price S. The tree is

© The Author(s), under exclusive license to Springer Nature Switzerland AG 2021
V. Ziemann, *Physics and Finance*, Undergraduate Lecture Notes in Physics,
https://doi.org/10.1007/978-3-030-63643-2_4

Fig. 4.1 A binomial tree.
The initial share price is S
and the value of the option is
c. After some time that is
characteristic for one step the
share price can either go up
by a factor f_1 or down by f_2.
The value of the option in the
respective cases is c_1 and c_2.

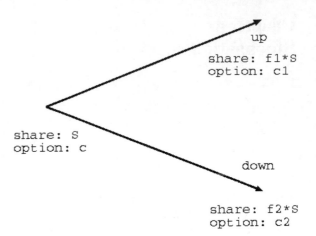

```
                                        up
                              share:  f1*S
                              option: c1


share:  S
option: c
                                        down


                              share:  f2*S
                              option: c2
```

shown in Fig. 4.1. Between the initial time, represented by the node on the left and
the later time, the price of the share can go either up or down, which is represented
by the up and downward arrows. If the price goes up, the value of the share is $f_1 S$
and the value of the option is c_1. If it goes down, the value of the share will be $f_2 S$
and that of the option c_2.

Let us start by estimating f_1 and f_2 if the share has a volatility σ, which means
that is has a relative spread in percent of increments from (typically) day to day by σ
as determined over (typically) a year. We then assume that the share price performs
a random walk similar to Brownian motion. But Brownian motion is described by
the diffusion equation for which we know that an initial point source spreads out
like a Gaussian whose width increases with the root of the time. We will validate
this statement in the following sections. For now, we rather heuristically consider
the width of the spreading Gaussian to be representative of the development of the
share price, we can estimate that we have

$$f_1 = e^{\sigma\sqrt{T}} \quad \text{and} \quad f_2 = e^{-\sigma\sqrt{T}}, \tag{4.1}$$

where T is measured in units of years if σ was determined to be the characteristic
spread over one year. Note that the assumptions in this model are far from being
beyond criticism, but they are reasonable.

Furthermore we can assume that the share price on average grows with rate ρ.
We can therefore ask ourselves what the probability p for an upward move is. For
a downward move it then is $1 - p$. Both p and $1 - p$ will be rather close to $1/2$
with a small advantage to go up, in order to produce the average increase at rate
ρ. To formalize this reasoning we equate the estimated average increase with the
expectation value for an up and downward move of the stock market

$$Se^{\rho T} = pf_1 S + (1 - p)f_2 S = pe^{\sigma\sqrt{T}}S + (1 - p)e^{-\sigma\sqrt{T}}S. \tag{4.2}$$

Note that the absolute value of the share price S drops out of the equation and we can solve for the probability p

$$p = \frac{e^{\rho T} - e^{-\sigma\sqrt{T}}}{e^{\sigma\sqrt{T}} - e^{-\sigma\sqrt{T}}} , \qquad (4.3)$$

which then determines the probabilities for up (p) and for downward $(1 - p)$ moves as a consequence of the share volatility σ and the average expected growth rate ρ.

Since the call option participates in the same market, it is on average subjected to the same probabilities for up and down movement and we can estimate the expectation value $\bar{c}_T$ of the option at the later time T to be

$$\bar{c}_T = pc_1 + (1 - p)c_2 , \qquad (4.4)$$

but we wanted to calculate the value of the option at the initial time and therefore need to back-propagate the value using the discount factor $e^{-\rho T}$ and determine the value of the option c at the initial time

$$c = e^{-\rho T}[pc_1 + (1 - p)c_2] . \qquad (4.5)$$

Keep in mind that the values for c_1 and c_2 depend on the strike price of the option K. For example, the value of the call option at node i is c_i and is given by

$$c_i = \max(S_T^i - K, 0) , \qquad (4.6)$$

where S_T^i is the value of the underlying share S at node i at time T.

Obviously the development of the price of the option can be analyzed further by subdividing any interval of time in many, say n, sub-sections. This results in a tree where each section has 2^n nodes. Progressing from one section to the next one branch describes the probability for the share value to increase and the other to decrease. In the first pass through the tree, we fill in the expected values of the share price S at all later times by multiplying the previous share value with the appropriate of move-up factors $e^{\sigma\sqrt{T}}$ and move-down factors $e^{-\sigma\sqrt{T}}$. In this way we fill the entire tree with share value prices, including the values at the final nodes. These values we compare with the strike price K using (1.6) and calculate the value of the option on each of the final nodes. In a second pass through the tree, now backwards in time, we calculate the value of the option on each of the preceeding nodes using the single-branch (4.5). In this way we can trickle down backwards through the tree to arrive at the option price c at the initial time—now.

Up to this point the reasoning was based on discretized time steps, but in the limit of reducing the step length to zero, we approach a time-continuous model, which can be shown [1] to be equivalent to the model we discuss in subsequent sections.

4.2   Wiener Process

From the discussion in the previous section it should be obvious that the price of shares S and options have an average drift upwards rate ρ and a superimposed randomly fluctuating part, characterized by the rms spread or volatility σ. This directly implies to write the temporal evolution of S as a stochastic process and express it as

$$dS = \rho S dt + \sigma S dW(t) \quad \text{or} \quad ds = \rho s dt + \sigma s dW(t) , \qquad (4.7)$$

where we introduce $s = S/S_0$ and use the initial share value S_0 to normalize. dW is a white-noise (Wiener) process. An equation of this type, in physics often an equation of motion subject to additional random forces, is called a *Langevin equation*. Remarkably, one of the first examples appears in Bachelier's thesis [2] on finance, which precedes Einstein's famous analysis of Brownian motion [3] by a few years. Note that in the absence of noise ($\sigma = 0$) we recover a purely exponential growth with rate ρ. Equation 4.7 is an example of a stochastic differential equation. Since working with them is not everyday fare, we will discuss their ingredients and the rules of manipulating them in some detail. Let us start with the source of randomness in (4.7), the Wiener process dW itself.

   In order to discuss the Wiener process dW let us consider the simplified Langevin equation $dx = \sigma dW$, which basically means that the particle with coordinate x receives a sequence of random kicks $\sigma \xi$, where ξ has rms amplitude unity. The values of ξ are sampled from a Gaussian distribution

$$\phi(\xi) = \frac{1}{\sqrt{2\pi}} e^{-\xi^2/2} . \qquad (4.8)$$

In order to calculate the expected value $\langle g(\xi) \rangle$ of a function $g(\xi)$ that depends on the random variable ξ, we need to average over many realizations of random numbers, denoted by $\langle g(\xi) \rangle = \int g(\xi)\phi(\xi)d\xi$. By direct integration of $g(\xi) = \xi$ and $g(\xi) = \xi^2$ we find

$$\langle \xi \rangle = 0 \quad \text{and} \quad \langle \xi^2 \rangle = 1 , \qquad (4.9)$$

which shows us that the mean is zero and the rms amplitude is unity. If we introduce a discretized form of the equation of motion, we obtain

$$x(t_n) = x(t_{n-1}) + \sigma \xi(t_n) \quad \text{or} \quad x(t_n) = \sigma \sum_{i=1}^{n} \xi(t_i) , \qquad (4.10)$$

where we assume that time is discretized in small steps $\Delta t = t_n - t_{n-1}$. The second equation tells us that the value of x at time t_n is just the sum of n random numbers with the properties specified above. It is easy to see that for average position $\langle x(t_n) \rangle$ and second moment averaged over many realizations of the random numbers, are given by

$$\langle x(t_n) \rangle = 0 \tag{4.11}$$

$$\langle x(t_n)^2 \rangle = \sigma^2 \sum_{i=1}^{n} \sum_{j=1}^{n} \langle \xi(t_i)\xi(t_j) \rangle = \sigma^2 \sum_{i=1}^{n} \sum_{j=1}^{n} \delta_{ij} = n\sigma^2 = t_n \frac{\sigma^2}{\Delta t}.$$

The random numbers $\xi(t_i)$ for different time steps are independent, which causes $\langle \xi(t_i)\xi(t_j) \rangle$ to be zero for $i \neq j$ and to be unity for $i = j$, thus $\langle \xi(t_i)\xi(t_j) \rangle = \delta_{ij}$, where δ_{ij} is the Kronecker delta. From the second equation we see that the second moment, the squared width of the distribution of x, grows linearly with time t_n, or the width grows with $\sqrt{t_n}$. We see that the scale factor is the squared amplitude, divided by Δt. This indicates that the units of σ are the units of x divided by root of time, which also what we used in (4.1), where we argued that σ was the volatility calculated for one year and then scaled with $\sqrt{T}$ in order to estimate the expected volatility for a different period T. As we see, the reason for this argument lies in the second of the previous equations and the linear growth with time of the *square* of the width $\langle x^2 \rangle$.

This also explains the somewhat sloppy notation for dW and $dW = \xi\sqrt{dt}$ that is used in [1] but makes the equation $dW^2 = dt$ intuitively digestible. A formal proof of the latter can be found in [4].

4.3 Diffusion Processes and Green's Functions

The spreading of the rms expected position, implicit in (4.11), is characteristic for a diffusion process, which we will briefly explore in this section. Examples of diffusion processes are the spreading of a localized heat source through its surroundings, or the spreading of a dye injected in a liquid. After the discussion at the end of the previous section we can express the corresponding Langevin equation as $dx = \sigma\xi\sqrt{dt}$, where σ is the magnitude of the random jump and ξ can be visualized as a "random number generator" that produces random numbers with rms unity and mean zero.

We now ask ourselves, how does a distribution function $\psi(x, t)$ evolve in time if all the x evolve according to the Langevin equation in one dimension. At a time $d\tau$ later, $\psi(x, t + d\tau)$ can be approximated by its first-order Taylor expansion

$$\psi(x, t + d\tau) = \psi(x, t) + \frac{\partial \psi}{\partial t} d\tau. \tag{4.12}$$

A complementary view is to consider where the particles at x came from during the time interval $d\tau$. This is given by the so-called Master equation

$$\psi(x, t + d\tau) = \int_{-\infty}^{\infty} \psi(x - dx, t)\phi(\xi)d\xi \tag{4.13}$$

with $dx = \sigma\xi\sqrt{d\tau}$, where we assume that the distributions of jumps ξ is described by the Gaussian from (4.8). In the next step we expand $\psi(x, t)$ up to second order in the spatial variable x and find

$$\psi(x, t + d\tau) = \int_{-\infty}^{\infty} \left[\psi(x, t) - \frac{\partial\psi}{\partial x}\sigma\xi\sqrt{d\tau} + \frac{1}{2}\frac{\partial^2\psi}{\partial x^2}\sigma^2\xi^2 d\tau \right] \phi(\xi)d\xi . \quad (4.14)$$

Since the Gaussian $\phi(\xi)$ is normalized, has mean zero and unit rms, only the first and third term in the square bracket remain after performing the integral over $d\xi$. Equating the resulting equation with the left-hand side of (4.12), we finally obtain

$$\frac{\partial\psi}{\partial t} = \frac{\sigma^2}{2}\frac{\partial^2\psi}{\partial x^2} , \quad (4.15)$$

which is referred to as the the *diffusion equation*.

A special solution $G(x, t)$ of this linear partial differential equation can be found by introducing its Fourier-transform $\tilde{G}(k, t)$ with respect to the spatial variable x

$$G(x, t) = \frac{1}{2\pi} \int_{-\infty}^{\infty} e^{-ikx}\tilde{G}(k, t)dk \quad (4.16)$$

and inserting it in (4.15), which leads to

$$\frac{\partial\tilde{G}}{\partial t} = -\frac{\sigma^2}{2}k^2\tilde{G} \quad \text{or} \quad \frac{d\tilde{G}}{\tilde{G}} = -\frac{\sigma^2 k^2}{2}dt , \quad (4.17)$$

where we separated the variables. This equation is easily integrated and we obtain $\tilde{G}(k, t) = \tilde{G}_0 e^{-\sigma^2 k^2 t/2}$ with an integration constant $\tilde{G}_0$. Using (4.16) to calculate the solution $G(x, t)$ we find

$$G(x, t) = \frac{\tilde{G}_0}{2\pi} \int_{-\infty}^{\infty} e^{-ikx - \sigma^2 k^2 t/2}dk = \frac{\tilde{G}_0}{\sqrt{2\pi\sigma^2 t}}e^{-x^2/2\sigma^2 t} , \quad (4.18)$$

where we solved the integral by completing the square in the exponent and then using $\int_{-\infty}^{\infty} e^{-ay^2}dy = \sqrt{\pi/a}$.

This solution $G(x, t)$ has the well-known behavior of a spreading-out Gaussian already used in Sect. 4.1 for (4.1). Note also that $G(x, t)$ is a special solution that vanishes at $x \to \pm\infty$. Moreover, considering the solution in the limit $t \to 0$, we find that the Gaussian becomes increasingly more pointed and approaches Dirac's delta-function in the limit. It therefore mimics the spreading of, for example, heat injected at a single point. If we were to inject equal amounts of heat at two separate points x_1 and x_2 at the same time, we could model it as the spreading of two fundamental solutions, proportional to $G(x - x_1, t) + G(x - x_2, t)$, because the diffusion equation is linear. If we had a smooth distribution $\rho(x')$ of heat injected into the system at $t = 0$ we could describe the solution $H(x, t)$ as the sum over all individual heat sources, weighted by $\rho(x')$. But this is the convolution of the initial heat distribution with the spreading function $G(x, t)$

$$H(x, t) = \int_{-\infty}^{\infty} G(x - x')\rho(x')dx' , \tag{4.19}$$

where $G(x, t)$ is given by (4.18) and is called a *Green's function* for the diffusion equation.

Let us briefly digress on the concept of Green's functions by mentioning two examples, one from physics and one from engineering. In general, Green's functions are "helper functions"—integral kernels—that appear when constructing the inverses of differential operators. One prominent example is the Green's function of the Poisson equation for the electro-static potential U, given by

$$\Delta U(\mathbf{x}) = \frac{1}{\varepsilon_0}\rho(\mathbf{x}) \tag{4.20}$$

with the charge density $\rho(\mathbf{x})$ and the dielectric constant ε_0. Here, the Green's function is the potential of a point charge, located at $\mathbf{x}_0$. Its charge distribution is $\rho(\mathbf{x}) = e\delta(\mathbf{x} - \mathbf{x}_0)$ with Dirac's delta function $\delta(\mathbf{x})$ and the elementary charge e. It is well-known from elementary electro-dynamics lectures that the potential $G(x)$ of a point charge is given by

$$G(\mathbf{x} - \mathbf{x}_0) = \frac{e}{4\pi \varepsilon_0 |\mathbf{x} - \mathbf{x}_0|} . \tag{4.21}$$

For a point-like charge the symmetry of the system is spherical and we can introduce spherical coordinates. Then, to verify this statement, we set $\mathbf{x}_0$ to the origin, such that $|\mathbf{x}| = r$ and express the Laplace operator Δ and the solution $G(r) = 1/4\pi \varepsilon_0 r$ in spherical coordinates. As an exercise, calculate $\Delta G(r)$ and verify that it is zero everywhere, except at the origin, where it diverges.

The Laplace operator is linear in the charge density ρ and we find the potential from two point charges at r_1 and r_2 as the sum of the potentials $G_1 = \frac{e}{4\pi \varepsilon_0 |\mathbf{x} - \mathbf{x}_1|}$ and $G_2 = \frac{e}{4\pi \varepsilon_0 |\mathbf{x} - \mathbf{x}_2|}$ from the individual charges. As a generalization, we find the potential of a continuous charge distribution ρ as the sum over the, properly weighted, potentials of point charges. But this is the just convolution of the charge distribution $\rho(\mathbf{y})$ and the point-charge potential—the Green's function $G(\mathbf{x} - \mathbf{y})$. The can therefore write

$$U(\mathbf{x}) = \int_V G(\mathbf{x} - \mathbf{y})\rho(\mathbf{y})d^3y = \int_V \frac{\rho(\mathbf{y})d^3y}{4\pi \varepsilon_0 |\mathbf{x} - \mathbf{y}|} , \tag{4.22}$$

where the integration extends over a suitable volume V that covers all charges in $\rho(\mathbf{y})$.

Comparing (4.20) with (4.22), we observe that the first is a (partial) differential equation that specifies how to find the Laplace operator applied to the potential U. But we do not want to know $\triangle U$; we want to know $U(\mathbf{x})$, and (4.22) provides us with just that information. It determines U directly, albeit at the expense of calculating the integral over the charge distribution, weighted with an integral kernel, the Green's function. The latter has, however, the intuitive interpretation as the potential of a point charge.

Note that the Green's functions can be viewed as the response U to a stimulus ρ where we calculate U with a convolution integral over ρ. This is conceptionally similar to what happens in electrical filters. They are stimulated by external, possibly noisy, signals and then remove the noise by, for example, low-pass filtering. The filter function in these applications assumes the same role as the Green's function where the noisy input signal is convoluted with the filter function. Moreover the propagators in quantum field theory, which mediate the fundamental interactions, are Green's functions that are caused by one particle and mediate its stimulus to a second particle. We will encounter this concept repeatedly in the following chapters.

In the next sections, however, we will address the use of stochastic differential equations in a financial context.

4.4   Stochastic Integrals and Ito's Lemma

In (4.7) looms a second specialty of stochastic processes, namely the product of two random variables s and dW. Both are jumping and this requires some additional discussion about how to evaluate the product when it appears in an integral, which is what we need in order to step (4.7) forward in time. There are two commonly used ways how to handle the products [4], one according to Stratonovich and the other is due to Ito. The latter is commonly used in finance and the interpretation of the product is that the function s is taken just before a jump and integrating $s\,dW$ entails

$$\int s\,dW = \sum_i s(t_{i-1})\left[W(t_i) - W(t_{i-1})\right] . \qquad (4.23)$$

whereas in the Stratonovich interpretation the average of s at time t_i and t_{i-1} is used.

In (4.7) it is tempting to introduce $z = \ln(s)$ as a new variable and retrieve an equation with $d\ln(s)$ on the left hand side. But the fact that we are dealing with stochastic variables requires us to take special rules into account. This will become apparent in a minute. Let's start by calculating dz to second order

$$\begin{aligned}
dz &= z(s + ds) - z(s) \\
&= \frac{\partial z}{\partial s} ds + \frac{1}{2} \frac{\partial^2 z}{\partial s^2} ds^2 \\
&= \frac{\partial z}{\partial s} [\rho s dt + \sigma s dW] + \frac{1}{2} \frac{\partial^2 z}{\partial s^2} \sigma^2 s^2 dt ,
\end{aligned} \tag{4.24}$$

where we substituted ds from (4.7). We find that the term ds^2 is the product of two random variables and it turns out [4] that in the limit $dt \to 0$ the terms with dt^2 and $dW dt$ go to zero whereas $dW^2 = dt$. Evaluating the derivatives of $z = \ln s$ with respect to s, we finally find

$$dz = \left[\rho - \frac{1}{2}\sigma^2 \right] dt + \sigma dW . \tag{4.25}$$

This equation shows the additional factor $\sigma^2/2$ which is due to the fact that we are dealing with random variables. The appearance of this additional term proportional to dt is often referred to as *Ito's lemma*. Apart from transforming the variables in stochastic equations, Ito's lemma plays a central role in the description of derivatives like options and futures. They are functions of the underlying stocks S which are stochastic variables, and therefore are stochastic variables themselves.

The stochastic differential equations can, for example, be solved by Monte-Carlo simulations which entails to discretize a stochastic equation and use a random number generator to provide the random kicks ξ and average over many realizations of random numbers. We will address these methods in Chap. 10. A complementary way is to derive a Fokker-Planck equation that describes the time evolution of a distribution function of the random variable z, a path we follow in the following section.

4.5 Master and Fokker-Planck Equations

We consider the stochastic differential equation in (4.25) where the random variable z grows linearly with time and during the time interval dt receives a random kick $dW = \xi \sqrt{dt}$. Here we assume ξ to be sampled from a normalized distribution $\phi(\xi)$ that is centered at zero and symmetric around the origin $\phi(\xi) = \phi(-\xi)$ and has second moment equal to unity. As an example we may visualize the Gaussian given in (4.8).

The random variables z will be distributed according to a probability distribution function $\psi(z, t)$ in the sense that at time t the probability to find z in the interval between $z - dz/2$ and $z + dz/2$ is given by $\psi(z, t)dz$. We may now ask ourselves how $\psi(z, t)$ develops in time, provided that z obeys the Langevin equation (4.25). To do so, we follow the strategy from Sect. 4.3 that led us from the Wiener process to the diffusion equation. We therefore consider the probability distribution at time $t + d\tau$ and write its Taylor expansion to second order

$$\psi(z, t + d\tau) = \psi(z, t) + \frac{\partial \psi}{\partial t} d\tau . \tag{4.26}$$

At the same time we can consider how the distribution at time $t + d\tau$ was populated. This is given by the Master equation

$$\psi(z, t + d\tau) = \int_{-\infty}^{\infty} \psi(z - dz, t)\phi(\xi)d\xi \tag{4.27}$$

where dz implicitly depends on the random process dW or ξ through (4.25). Taylor-expanding up to second order in the spatial variable z we find

$$\psi(z, t + d\tau) = \int_{-\infty}^{\infty} \left[\psi(z, t) - \frac{\partial \psi}{\partial z} dz + \frac{1}{2} \frac{\partial^2 \psi}{\partial z^2} dz^2 \right] \phi(\xi)d\xi$$

$$= \int_{-\infty}^{\infty} \left[\psi(z, t) - \frac{\partial \psi}{\partial z} (\hat{\rho}d\tau + \sigma\xi\sqrt{d\tau}) \right. \tag{4.28}$$

$$\left. + \frac{1}{2} \frac{\partial^2 \psi}{\partial z^2} (\hat{\rho}^2 d\tau^2 + 2\xi \hat{\rho}d\tau^{3/2} + \sigma^2\xi^2 d\tau) \right] \phi(\xi)d\xi ,$$

where we used (4.25) to substitute dz and introduced the abbreviation $\hat{\rho} = \rho - \sigma^2/2$. Now we can evaluate the integral over ξ and use the fact that the distribution of $\phi(\xi)$ is symmetric

$$\psi(z, t + d\tau) = \psi(z, t) - \frac{\partial \psi}{\partial z} \hat{\rho}d\tau + \frac{1}{2} \frac{\partial^2 \psi}{\partial z^2} (\hat{\rho}^2 d\tau^2 + \sigma^2 d\tau) . \tag{4.29}$$

where the term $\hat{\rho}^2 d\tau^2$ vanishes in the limit $d\tau \to 0$. Furthermore, equating with (4.26) and rearranging terms, we obtain the Fokker-Planck equation for the distribution of $z = \ln s$

$$\frac{\partial \psi}{\partial t} = -\hat{\rho}\frac{\partial \psi}{\partial z} + \frac{\sigma^2}{2} \frac{\partial^2 \psi}{\partial z^2} . \tag{4.30}$$

Here it becomes obvious that (4.30) is a diffusion equation with an additional drift term where $\sigma^2/2$ plays the role of the diffusion constant and $\hat{\rho} = \rho - \sigma^2/2$ that of the drift velocity.

It is straightforward, but somewhat tedious to verify that the following distribution function

$$\psi(z, t)dz = \frac{1}{\sqrt{2\pi\sigma^2 t}} \exp\left[-\frac{(z - \hat{\rho}t)^2}{2\sigma^2 t} \right] dz \tag{4.31}$$

Fig. 4.2 The probability distribution function of share values after one month, six month, one year, and two years where the horizontal axis is normalized to the initial share value S_0. The parameters used in preparing the plot are $\sigma = 30\%/\sqrt{year}$ and $\rho = 10\%/year$

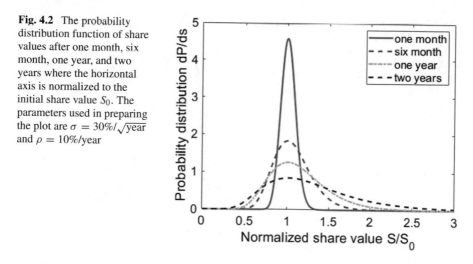

is normalized to unity and satisfies the Fokker-Planck equation in (4.30). This equation describes a diffusing Gaussian that moves towards positive values of z with "speed" $\hat{\rho}$.

We can now substitute back the original variable S through $z = \ln(S/S_0)$ and find for the distribution function Ψ of the share values after some time t

$$
\begin{aligned}
\Psi(S, t) &= \int \psi(z, t)\delta(S - S_0 e^z)dz \\
&= \int \psi(z, t)\frac{\delta(z - \ln(S/S_0))}{|S_0 e^z|}dz \\
&= \frac{1}{\sqrt{2\pi\sigma^2 t}}\frac{1}{S}\exp\left[-\frac{(\ln(S/S_0) - \hat{\rho}t)^2}{2\sigma^2 t}\right],
\end{aligned}
\tag{4.32}
$$

where $\delta(x)$ is Dirac's delta function. It is used to collect all values of z, such that $S = S_0 e^z$. In the second equality we use the property of the delta function $\delta(f(x)) = \sum_i (x - x_i)/|f'(x_i)|$. The last line in (4.32) describes the well-known log-normal distribution.

It is instructive to visualize the temporal evolution of the distribution. From an initial share value $s = 1$ or $S = S_0$ which is represented by a delta function $\delta(s - 1)$. The probability distribution function after one month, six month, one year and two years is shown in Fig. 4.2, where we assume a share volatility of $\sigma = 30\%/\sqrt{year}$ and an annual growth rate of $\rho = 10\%$. Note the distinct asymmetry due to the logarithm in the exponent and the $1/S$ dependence. The suppression of the tail on the left of the distribution can be explained by the fact that if the share value decreases all subsequent changes are based on the already smaller value.

We are now in a position to apply the reasoning from Sect. 4.1 on binomial trees to determine the value of options as expectation values of the share value to exceed a certain strike price, which is the topic of the next section.

Fig. 4.3 The probability distribution function of share values after one year with the areas of a call option with strike price of $K/S_0 = 1.5$ and a put options with $K/S_0 = 0.8$ indicated as shaded areas

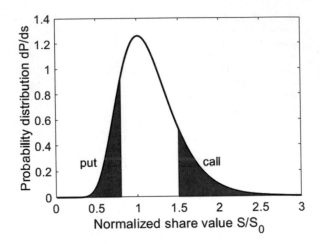

4.6 A First Look at Option Pricing

The value of an European option at the initial time is given by the expectation value of the payoff $S{-}K$ at maturity, back-propagated to the initial time with the discount factor $e^{-\rho t}$. This is the same process we discussed when back-propagating the option prices on the final nodes through the binomial tree, which is encoded in (4.5). Using distribution functions instead of binomial trees, the value of a call option c is given by the integral

$$c = e^{-\rho t} \int_K^\infty (S - K)\Psi(S, t)dS , \qquad (4.33)$$

where $\Psi(S, t)$ is given by (4.32). Figure 4.3 indicates the range over which the integral needs to be calculated for a call or a put option respectively. Outside the indicated range the value of the options is zero, which is accounted for by choosing the lower boundary of the integral suitably.

The integral can be evaluated in a straightforward manner by changing the variables back to $z = \ln(S/S_0)$, which yields

$$c = \frac{S_0 e^{-\rho t}}{\sqrt{2\pi\sigma^2 t}} \int_{\ln(K/S_0)}^\infty \exp\left[-\frac{(z - \hat{\rho}t)^2}{2\sigma^2 t}\right] \left(e^z - \frac{K}{S_0}\right) dz . \qquad (4.34)$$

The first of the integrals with the term e^z can be integrated by completing the square in the exponent and changing the integration variable to

$$y = \frac{z - \hat{\rho}t - \sigma^2 t}{\sqrt{2\sigma^2 t}} . \qquad (4.35)$$

Additionally, using a similar substitution in the second integral, we arrive at

$$
c = \frac{S_0 e^{-\rho t + \hat{\rho} t + \sigma^2/2}}{\sqrt{\pi}} \int\limits_{\frac{\ln(K/S_0) - \hat{\rho} t - \sigma^2 t}{\sqrt{2\sigma^2 t}}}^{\infty} e^{-y^2} dy - \frac{K e^{-\rho t}}{\sqrt{\pi}} \int\limits_{\frac{\ln(K/S_0) - \hat{\rho} t}{\sqrt{2\sigma^2 t}}}^{\infty} e^{-y^2} dy , \qquad (4.36)
$$

where we transformed all difficult terms in the exponent of the integrand into boundaries of the integral. The integrals are now given in terms of complementary error functions erfc [5], defined by

$$
\mathrm{erfc}(z) = \frac{2}{\sqrt{\pi}} \int\limits_{z}^{\infty} e^{-y^2} dy = 1 - \mathrm{erf}(z) , \qquad (4.37)
$$

where $\mathrm{erf}(z)$ is the normal error function. The exponent in front of the first integral is zero by virtue of $\hat{\rho} = \rho - \sigma^2/2$ and we finally obtain

$$
c = \frac{S_0}{2} \mathrm{erfc} \left[\frac{\ln(K/S_0) - (\rho + \sigma^2/2)t}{\sqrt{2\sigma^2 t}} \right] \qquad (4.38)
$$
$$
- \frac{K e^{-\rho t}}{2} \mathrm{erfc} \left[\frac{\ln(K/S_0) - (\rho - \sigma^2/2)t}{\sqrt{2\sigma^2 t}} \right] .
$$

In order to recover the commonly known way of presenting the well-known option pricing formula we express the complementary error function erfc by the cumulative normal distribution function $N(z)$ which is given by

$$
N(z) = \frac{1}{\sqrt{2\pi}} \int\limits_{-\infty}^{z} e^{-y^2/2} dy = \frac{1}{2} \mathrm{erfc} \left[\frac{-z}{\sqrt{2}} \right] . \qquad (4.39)
$$

The second equality shows its relation to the complementary error function erfc. Replacing the error functions in (4.38) leads to the commonly used expression

$$
c = S_0 N \left[\frac{\ln(S_0/K) + (\rho + \sigma^2/2)t}{\sigma \sqrt{t}} \right] - K e^{-\rho t} N \left[\frac{\ln(S_0/K) + (\rho - \sigma^2/2)t}{\sigma \sqrt{t}} \right] .
$$
$$
\qquad (4.40)
$$

This expression allows us to assign a price to a call option c with strike price K for the underlying asset S with current value S_0 under the assumption that the volatility of the underlying share is σ and its growth rate is ρ. Note, however, that we are free to choose any value of ρ. We might even pick one that causes the current option price c to be very high. Nobody would then buy the call option, because it is not considered fair. The question therefore remains of how to select ρ in such a way that the option price is fair. This is the question that is addressed by the Black-Scholes equation that we discuss in detail in the next chapter.

This caveat notwithstanding, we can determine the price of the European put option p in the same way we used for the call option above by integrating over the shaded region on the left-hand side in Fig. 4.3. This leads to

$$p = e^{-\rho t} \int_{-\infty}^{K} (K - S)\Psi(S, t)dS , \qquad (4.41)$$

where $K-S$ is the pay-off function of the put option for $S < K$. The integral, in the same way as before, weighs the pay-off with probability $\Psi(S, t)dS$ and then back-propagates this expectation value to the initial time with the discount factor $e^{-\rho t}$. Evaluating the integrals leads to the well-known form

$$\begin{aligned}
p &= Ke^{-\rho t} N\left[\frac{\ln(K/S_0) - \rho t + \sigma^2 t/2}{\sigma \sqrt{t}}\right] \\
&\quad - S_0 N\left[\frac{\ln(K/S_0) - \rho t - \sigma^2 t/2}{\sigma \sqrt{t}}\right] \\
&= Ke^{-\rho t} N\left[-\frac{\ln(S_0/K) + (\rho - \sigma^2/2)t}{\sigma \sqrt{t}}\right] \\
&\quad - S_0 N\left[-\frac{\ln(S_0/K) + (\rho + \sigma^2/2)t}{\sigma \sqrt{t}}\right] ,
\end{aligned} \qquad (4.42)$$

where, in the second form, the cumulative distribution function N has the same arguments as in (4.40). These arguments are commonly used in the financial literature and are denoted by

$$d_1 = \frac{\ln(S_0/K) + (\rho + \sigma^2/2)t}{\sigma \sqrt{t}}$$

$$d_2 = \frac{\ln(S_0/K) + (\rho - \sigma^2/2)t}{\sigma \sqrt{t}} = d_1 - \sigma \sqrt{t} . \qquad (4.43)$$

Note that the derivation did not use any assumptions of risk-balancing a portfolio. All we did was calculating the expectation value of the expected pay-off if the average value grows with rate ρ and the distributions of stocks, shares or options are log-normal.

The prices of put and call options with the same strike price K are linked by the so-called *put-call parity*, which follows directly from the pricing formulas for the respective options. To prove this we write the definition of the put option and use the property of the cumulative distribution function, namely that $N(-z) = 1 - N(z)$. We find

$$p = Ke^{-\rho t} N(-d_2) - S_0 N(-d_1)$$
$$= Ke^{-\rho t} [1 - N(d_2)] - S_0 [1 - N(d_1)] \tag{4.44}$$
$$= Ke^{-\rho t} - S_0 + S_0 N(d_1) - Ke^{-\rho t} N(d_2)$$
$$= Ke^{-\rho t} - S_0 + c \, ,$$

which can be rewritten as

$$c + Ke^{-\rho t} = p + S_0 \, . \tag{4.45}$$

The two sides of this equation can be interpreted as the values of two portfolios at the time $t = 0$ the options are sold. The portfolio on the right hand side of equation (4.45) consists of a put option and shares S_0. The one on the left hand side consists of a call option c and a cash value $Ke^{-\rho t}$ invested with rate ρ. But why should the bank give you rate ρ and what is a *fair price* for the call option c and put option p? For simplicity we assume that we deposit some amount $Ke^{-r_f t}$ at the bank at the risk-free rate r_f. In the next chapter we will see that there is an intricate relation between this assumption and fair pricing of options.

We can now ask what the values of the two portfolios is at the strike time $t = T$. First consider the portfolio on the right-hand side. If the stock price exceeds the strike price we forfeit the put option and keep the stock which has a value S_T. If the stock price is below the strike price we exercise the put option and make a profit of $K - S_T$ and the value of the stock is S_T such that the value of this portfolio is $K - S_T + S_T = K$. We now compare this outcome with that of the portfolio described by the left-hand side in (4.45). If the stock value S_T at time T is above the strike price $S_T > K$ we exercise the option which constitutes a value of $S_T - K$. Jointly with the bank deposit which has grown to value K the total value of this portfolio is thus $S_T - K + K = S_T$. If the stock value is below the strike price $S_T < K$ we forfeit the call option but still have the value K in the bank which constitutes the total value of our portfolio.

Summarizing, we find that both portfolios have the value S_T if $S_T > K$ and K in case $S_T < K$. Thus the value of both portfolios is equal in all cases, just as the (4.45) indicates. We note, however, that there is a deeper link between the risk-free rate r_f at which we deposited a sum of cash and the fair prices for the options. In the preceding chapter we always assumed that the growth rate of the stock was given by some rate ρ, possibly derived from CAPM. So the valuation of the options in this chapter is not necessarily fair, but depends on guessing a reasonable value for ρ.

In the next chapter we will resolve this puzzle by determining a differential equation for the temporal evolution of option prices and building portfolios that are (quasi) risk-free, and imply fair prices for the options. But before doing so that let us briefly digress on the calculation of expectation values.

4.7   Digression on Expectation Values

In (4.33) we calculate the present value of a call option c by back-propagating
the payoff function $\max(S - K, 0)$. A complementary view is based on the notion
of averaging the payoff function, weighted by the probability distribution function
$\Psi(S, t)$, which is the same as the expectation value of the payoff function in the
"state" $\Psi(S, t)$.

We note that this is conceptually analogous to calculating the expectation value
$\langle|\phi|H|\phi\rangle$ of an operator H in the state $|\phi\rangle$ in quantum mechanics. Let us assume
that H is the Hamiltonian. Expanding both H and ϕ into eigenfunctions ψ_i of H we
have $H = \sum_i E_i |\psi_i\rangle\langle\psi_i|$ and $|\phi\rangle = \sum_j c_k |\psi_j\rangle$ with, in general complex, expansion
coefficients c_j. Here E_i are the eigenvalues of H and c_j the expansion coefficients
of $|\phi\rangle$. For the expectation value we find

$$\langle|\phi|H|\phi\rangle = \sum_k c_k^* \langle\psi_k| \sum_i E_i |\psi_i\rangle\langle\psi_i| \sum_j c_j |\psi_j\rangle \qquad (4.46)$$

$$= \sum_k \sum_i \sum_j c_k^* c_j \langle|\psi_k|\psi_i\rangle\langle|\psi_i|\psi_j\rangle = \sum_i |c_i|^2 E_i$$

where we used that the eigenfunctions $\langle|\psi_i|\psi_j\rangle = \delta_{ij}$ are orthogonal. We observe
that the expectation value $\langle|\phi|H|\phi\rangle$ can be calculated as the eigenvalues E_i weighed
by the probability $|c_i|^2$ of finding the state $|\phi\rangle$ in eigenstate $|\psi_j\rangle$.

Calculating the odds of in a game of throwing dice is conceptually similar of
betting on stocks exceeding a certain strike price. Let us work out what is more
likely, either rolling seven eyes or rolling a double-digit value—10, 11, or 12. Let
us assume that the payoff for rolling seven eyes is $V_1 = 13\,\$$ and for rolling double-
digit eyes it is $V_2 = 10\,\$$. We can work out the expected payoff $\langle V \rangle$ by weighing the
payoff with the probabilities of rolling either outcome. Since each dice has six faces,
the probability of rolling a specific one is $1/6$. Since the dices roll independently
the probability of any throw with two dices is $1/36$. Now we only have to calculate
the number of possible combinations that result in seven to calculate the probability
p_1. There are six different combinations: $1 + 6, 2 + 5, \ldots, 6 + 1$, which results in
a probability of $p_1 = 6/36 = 1/6$ to roll seven eyes. Rolling double-digit eyes is
also possible with 6 different combinations: $4 + 6, 5 + 5, 6 + 4$ all result in 10 eyes,
and $5 + 6, 6 + 5$, and $6 + 6$ result in 11 or 12 eyes. Even for these combinations the
probability is $6/36$ or $p_2 = 1/6$. If I bet on rolling seven eyes I receive $13\,\$$ if I win
and have to pay $10\,\$$ to my opponent if I lose. Therefore the expectation value of
my payoff is given by $\langle V \rangle = p_1 V_1 - p_2 V_2 = 3/6\,\$$. Thus, on average I will have an
advantage of half a dollar per game.

The notion of propagating a source function to a different point—the payoff
function back in time in the financial context—plays a central role in optics, where
we use *Huygen's principle* [6] to find the intensity on a distant screen cause by light
shining through an aperture. Here the aperture function that is unity where light shines
through and zero otherwise plays the role of the source. In optics, the *point-spread*

function $e^{2\pi i r/\lambda}/r$ plays the role of the Green's function. Here λ is the wavelength of the light and r is the distance between the source point and the observation point on the screen.

After pointing out the analogies in physics, let's get straight back to one of the core topics of finance, the Black-Scholes equation.

Exercises

1. Use a binomial tree with two layers to calculate the value of a call option that, after one year, has a strike price K. The annual growth rate is $\rho = 0.05$/year, and the annual volatility is $\sigma = 0.3/\sqrt{\text{year}}$.

2. Verify that, in three dimensions, the Green's function $G(r) = 1/4\pi r$ satifies $\Delta G(r) = \delta(r)$.

3. Determine the Green's function for a damped harmonic oscillator, described by $\ddot{x} + 2\alpha\dot{x} + \omega^2 x = v_0\delta(t)$. Here ω is the oscillation frequency, α is the damping constant, and v_0 is an initial change in the velocity, instantaneously applied to the oscillator at $t = 0$. The solution to this equation is the Green's function. It is equivalent to the impulse response of an instantaneous change in velocity $\Delta\dot{x} = v_0$.

4. Equation 4.15 can be used to describe the dependence of the temperature $T \rightarrow \psi$, where $D \rightarrow \sigma^2/2$ is the heat diffusion constant of the material. At $t = 0$ heat, which causes an instantaneous temperature rise ΔT_0, is injected at a distance d from the insulated end of a semi-infinite slab with heat diffusion constant D. Calculate the temperature at the insulated end as a function of time. Hint: think of "image heat loads" and that $\partial T/\partial x = 0$ at the insulated end.

5. Show that $\psi(z, \tau)$ from (4.31) (a) is normalized and (b) satisfies (4.30).

6. What is the probability that a stock with an annual return $\rho = 0.1$/year and an annual volatility of $\sigma = 0.3/\sqrt{\text{year}}$ at least doubles its value in 2 years?

7. What is the probability that the stock from Exercise 6 has less than half its value after two years?

8. You invent a new option O that has a payoff function that grows linearly with the stock value S in the range from $(1 - u)K$ to K and decreases linearly from K to $(1 + u)K$, where u is in the range $0.05 < u < 0.2$. (a) Sketch the payoff function. (b) Calculate the price of the option for a stock with annual return $\rho = 0.1$/year and volatility $\sigma = 0.3/\sqrt{\text{year}}$.

9. Huygen's principle tells us that the image of an aperture can be calculated by summing all emanating spherical waves with the point-spread function $e^{2\pi i r/\lambda}/r$, where λ is the wavelength of the light. Considering only one dimension, calculate the image of an aperture of width a on a screen at a distance $d \gg a$ from the aperture. Hint: you can assume that r in the denominator of the point-spread function is approximately equal to d.

References

1. J.C. Hull, *Options, Futures, and Other Derivatives*, 8th edn. (Pearson, Boston, 2012)
2. L. Bachelier, Theorie de la speculation. Annales Scientifiquies de l'Ecole Normale Superieure **17**, 21 (1900)
3. A. Einstein, Über die von der molekularkinetischen Theorie der Wärme geforderte Bewegung von in ruhenden Flüssigkeiten suspendierten Teilchen. Annalen der Physik **322**, 549 (1905)
4. C.W. Gardiner, *Handbook of Stochastic Methods*, 2nd edn. (Springer, Berlin, 1985)
5. M. Abramowitz, I. Stegun, *Handbook of Mathematical Functions* (Dover, New York, 1972)
6. E. Hecht, *Optics*, 2nd edn. (Addison-Wesley, Reading, 1987)

Chapter 5
Black-Scholes Differential Equation

Abstract After deriving the Black-Scholes equation for a call option from the requirement to make a portfolio risk-free, the equation is solved using a number of variable substitutions, which transforms it into a diffusion equation. Using the latter's Green's function is then used to value European call options. The resemblance of the solution found in this chapter to that in Chapter 4 stimulates the discussion of martingale processes. In order to better understand the mechanics of using options for hedging, a MATLAB simulation for the temporal evolution of stocks, options and bank deposits is presented.

The Black-Scholes differential equation contributed significantly to the rapid expansion of the derivative market in the 1970s, because it enabled traders to set a fair price on a large number of options and other financial products. Consequently, the trust in these products increased and boosted their attractivity. Later, the Black-Scholes equation was also blamed to have caused market crashes [1], because traders relied on the existence of risk-free portfolios, which is a corner stone of the theory. But the traders missed that some of the prerequisites were not valid any more. For example, the unlimited liquidity, the availability of funds at the risk-free rate, is no longer true. Moreover, the underlying stochastic dynamics is not necessarily Gaussian in times of financial distress. In Chap. 9 we will address some of the criticism, but first let us discuss the standard Black-Scholes theory and calculate a fair price for standard, often called "vanilla" call and put options.

5.1 Derivation

Let us therefore consider the temporal variation of option prices and assume that the option $c(S, t)$ depends on the stock value S and on the time t. Thus c is a derivative of the underlying asset S, which is a stochastic variable itself and obeys the Langevin equation from (4.7). Thus, also c is a stochastic variable and we need to derive a stochastic differential equation for it in order to analyze how it evolves in time. To first order in the temporal increment dt, we write the Taylor expansion of c

$$dc = \frac{\partial c}{\partial t}dt + \frac{\partial c}{\partial S}dS + \frac{1}{2}\frac{\partial^2 c}{\partial S^2}dS^2 \,, \tag{5.1}$$

where, as before, we keep the terms up to second order in the stochastic variable S. Similar to what we did in Sect. 4.5 we now insert dS from (4.7) and keep terms up to first order in dt

$$\begin{aligned} dc &= \frac{\partial c}{\partial t}dt + \frac{\partial c}{\partial S}[\rho S dt + \sigma S dW] + \frac{1}{2}\frac{\partial^2 c}{\partial S^2}\sigma^2 S^2 dt \\ &= \left[\frac{\partial c}{\partial t} + \rho S\frac{\partial c}{\partial S} + \frac{1}{2}\frac{\partial^2 c}{\partial S^2}\sigma^2 S^2\right]dt + \sigma S\frac{\partial c}{\partial S}dW \,, \end{aligned} \tag{5.2}$$

which describes the stochastic differential equation of a quantity, here c, that depends on another stochastic variable, here S. Note that we used the substitutions already used in Sect. 4.4, in particular $dW^2 = dt$. Moreover, ρ is the expected, but unknown, growth rate of stock S and σ is its volatility.

We now consider a portfolio Π, consisting of one option c and a number of shares S. We try to construct it in such a way that the portfolio is risk-free in the sense that the term with the Wiener-process dW vanishes. This is accomplished by acquiring a fraction $\Delta = \partial c/\partial S$ shares and selling one option c. The portfolio Π is thus given by

$$\Pi = -c + \frac{\partial c}{\partial S}S \,. \tag{5.3}$$

Its value will change in time according to

$$\begin{aligned} d\Pi &= -dc + \frac{\partial c}{\partial S}dS \\ &= -\left[\frac{\partial c}{\partial t} + \rho S\frac{\partial c}{\partial S} + \frac{1}{2}\frac{\partial^2 c}{\partial S^2}\sigma^2 S^2\right]dt \\ &\quad -\sigma S\frac{\partial c}{\partial S}dW + \frac{\partial c}{\partial S}[\rho S dt + \sigma S dW] \\ &= -\left[\frac{\partial c}{\partial t} + \frac{1}{2}\sigma^2 S^2\frac{\partial^2 c}{\partial S^2}\right]dt \,. \end{aligned} \tag{5.4}$$

And here the magic happens: both the random, stochastic part, proportional to dW, and the unknown growth rate ρ cancel and do not appear in the final line of (5.4), which therefore is no longer a stochastic, but an ordinary partial differential equation. Thus, in a risk-neutral environment, the value of the portfolio will grow at the risk-free rate r_f and we have

$$d\Pi = r_f \Pi dt = r_f\left[-c + \frac{\partial c}{\partial S}S\right]dt \,. \tag{5.5}$$

Equating the two expressions for $d\Pi$ from (5.4) and (5.5) and dividing by dt, we obtain the differential equation that is linked to the names of its main contributors: F. Black, M. Scholes, and R. Merton

$$-\frac{\partial c}{\partial t} = \frac{1}{2}\sigma^2 S^2 \frac{\partial^2 c}{\partial S^2} + r_f S \frac{\partial c}{\partial S} - r_f c \,. \tag{5.6}$$

It links the temporal evolution of the option c to that of the underlying asset S. It is interesting that the temporal evolution of the option c is only determined by the volatility σ and by the risk-free rate r_f, rather than the assumed growth rate ρ of the stock S.

We point out that (5.6) was derived by deliberately constructing the portfolio Π, defined in (5.3), to be risk free. This way of assembling a hedged portfolio of one option and $\Delta = \partial c/\partial S$ shares S is usually referred to as Δ-*hedging*. We find that the temporal evolution of the option c, given by (5.6), is a direct consequence of the definition of Π in (5.3). Let us proceed to determine the value of c at a given time before maturity. To do so, we need to integrate the partial differential equation from (5.6), and this is the topic of the next section.

5.2 The Solution

We now continue to solve this equation for a call option with the boundary condition $c = \max(S - K, 0)$ at time T. Inspired by [2], we note that this partial differential equation is very similar to the diffusion equation, except for the additional term $-r_f t$ and that time runs in the "wrong" direction due to the minus sign in front of the temporal derivative. Since we are interested in the evolution of the option price towards maturity at time T we introduce the new variable $\tau = T - t$.

$$c(S, t) = e^{-r_f \tau} g(S, \tau) \,, \tag{5.7}$$

which transforms (5.6) into

$$\frac{\partial g}{\partial \tau} = \frac{1}{2}\sigma^2 S^2 \frac{\partial^2 g}{\partial S^2} + r_f S \frac{\partial g}{\partial S} \,. \tag{5.8}$$

Moreover, the powers of S in front of the derivatives indicate that it is beneficial to introduce $z = \ln S/K$. Here we chose to normalize the stock value S to the strike price K, because it is the only variable that has units of monetary value in the problem. After some algebra we find

$$\frac{\partial g(z, \tau)}{\partial \tau} = \left(r_f - \frac{1}{2}\sigma^2\right)\frac{\partial g(z, \tau)}{\partial z} + \frac{1}{2}\sigma^2 \frac{\partial^2 g(z, \tau)}{\partial z^2} \,. \tag{5.9}$$

Note that, except for the sign of $\hat{r} = r_f - \sigma^2/2$, this is the same equation as the Fokker-Planck equation for the distribution function of stock prices in (4.30). This is not surprising, because the definition of the portfolio Π in (5.3) links the temporal evolution of the derivative c to that of the underlying stock S. Finally, by introducing the substitution $x = z + \hat{r}\tau$, we arrive at

$$\frac{\partial g(x, \tau)}{\partial \tau} = \frac{\sigma^2}{2} \frac{\partial^2 g(x, \tau)}{\partial x^2} , \tag{5.10}$$

which we already encountered in (4.15) in Sect. 4.3. There we found that its fundamental solution, the Green's function, is given by (4.18). Adding the factor $e^{-r_f \tau}$ from (5.7) gives us the Green's function for the Black-Scholes equation

$$G(x, \tau) = \frac{1}{\sqrt{2\pi\sigma^2\tau}} \exp\left[-\frac{x^2}{2\sigma^2\tau} - r_f\tau\right] . \tag{5.11}$$

After expressing $x = z + \hat{r}\tau = \ln(S/K) + \hat{r}\tau$ in terms of the original variables, this Green's function solves (5.6), but it does not satisfy the boundary conditions, namely to reproduce the payoff function at $\tau = 0$. On the other hand, the Black-Scholes equation is a linear differential equation, such that linear combinations of solutions are also solutions. As discussed in Sect. 4.3 is $G(x, \tau)$ the fundamental solution that starts from a point source at $\tau = 0$. From there it diffuses in a Gaussian fashion with the width increasing in time according to $\sigma\sqrt{\tau}$. Furthermore, τ is running backwards in time and $\tau = 0$ corresponds to maturity of the option at $t = T$. The problem of finding the value of the option c at an earlier time t is mapped onto an initial value diffusion problem, where the the payoff function $\max(S - K, 0)$ takes the role of the initial distribution. We rewrite the payoff function using the variables $x' = \log(S/K)$ and τ and then use the Green's function to propagate it backwards in time, and finally translate back to variables S and t.

The distribution of "heat sources" at location x' that correspond to $\max(S - K, 0)$ at $t = T$ is thus given by the linear superposition of fundamental solutions, weighed with the payoff function

$$\max(Ke^{x'} - K, 0) = K \max(e^{x'} - 1, 0) \quad \text{at} \quad \tau = T - t = 0. \tag{5.12}$$

This is equivalent to $K(e^{x'} - 1)$ for $x' \geq 0$ and 0 for $x' < 0$. To find the distribution after some time τ, we integrate over the contributions of all source points x' with their respective strengths, given in (5.12), and weighted them with the Green's function that acts as a propagator from "source" point x' to "observation" point x

$$c(x, \tau) = \int_0^\infty K(e^{x'} - 1)G(x' - x, \tau)dx' . \tag{5.13}$$

Inserting $G(x, \tau)$ from (5.11), we find

$$c(x, \tau) = \frac{Ke^{-r_f\tau}}{\sqrt{2\pi\sigma^2\tau}} \left\{ e^{x+\sigma^2\tau/2} \int_0^\infty \exp\left[-\frac{(x'-x-\sigma^2\tau)^2}{2\sigma^2\tau}\right] dx' \right.$$
$$\left. - \int_0^\infty \exp\left[-\frac{(x'-x)^2}{2\sigma^2\tau}\right] dx' \right\}, \tag{5.14}$$

where we completed the square in the exponent of the first integral. After shifting the integration variable to simplify the exponent and substituting $y' = x'/\sigma\sqrt{\tau}$, we express the integrals in terms of cumulative distribution functions $N(x)$, already defined in (4.39)

$$c(x, \tau) = Ke^{x+\sigma^2\tau/2-r_f\tau}\left[1 - N\left(-\frac{x+\sigma^2\tau}{\sigma\sqrt{\tau}}\right)\right] - Ke^{-r_f\tau}\left[1 - N\left(-\frac{x}{\sigma\sqrt{\tau}}\right)\right]$$
$$= Ke^{x+\sigma^2\tau/2-r_f\tau}N\left(\frac{x+\sigma^2\tau}{\sigma\sqrt{\tau}}\right) - Ke^{-r_f\tau}N\left(\frac{x}{\sigma\sqrt{\tau}}\right). \tag{5.15}$$

Now we can substitute back the original variables $x = z + \hat{r}\tau$ and $z = \ln S/K$. Using $\hat{r} = r_f - \sigma^2/2$ to simplify the previous equation, we arrive at

$$c(S, \tau) = SN\left[\frac{\ln(S/K) + (r_f + \sigma^2/2)\tau}{\sigma\sqrt{\tau}}\right] \tag{5.16}$$
$$- Ke^{-r_f\tau}N\left[\frac{\ln(S/K) + (r_f - \sigma^2/2)\tau}{\sigma\sqrt{\tau}}\right].$$

After replacing $\tau = T - t$ we find the price of the call option $c(S, t)$

$$c(S, t) = SN\left[\frac{\ln(S/K) + (r_f + \sigma^2/2)(T - t)}{\sigma\sqrt{T - t}}\right] \tag{5.17}$$
$$- Ke^{-r_f(T-t)}N\left[\frac{\ln(S/K) + (r_f - \sigma^2/2)(T - t)}{\sigma\sqrt{T - t}}\right],$$

which almost coincides with the result found in Sect. 4.6 in (4.40). Only the growth rate ρ is replaced by the risk-free rate r_f in (4.40).

It is remarkable that the different initial assumptions about the option price used in the derivation in Sect. 4.6 and in this section lead to similar results. In the former case we calculated the expectation value of the future profit back-propagated to today's value using the expected growth rate of the asset. In this section we derived a differential equation that describes a hedged portfolio (5.3) that is risk-free and therefore grows with the risk-free rate r_f. It links the temporal evolution of the stocks S to that of the option c. The solution to the differential equation with the boundary conditions, pertinent to the type of option, resulted in the same option price, including its temporal evolution towards maturity $(T - t)$, albeit with ρ replaced by r_f.

5.3 Risk-Neutrality and Martingales

In this section we illustrate how the two approaches, from Sects. 4.6 and 5.2 are related. Consider two stochastic processes, one for the share price S that was already used in Sect. 4.6 and a second one for a stochastic process X for which the growth rate ρ is replaced by r_f

$$dS = \rho S dt + \sigma S dW(t)$$
$$dX = r_f X dt + \sigma X dW(t) \tag{5.18}$$

where dW is a Wiener-process. We assume that both processes have the same initial conditions $S_0 = X_0$ and the same volatility but have different growth rates ρ and r_f. Therefore, the processes are clearly different. If we use the first process with S we arrive at a price for the call option in (4.40) and if we use the second process with X we would arrive at (5.17), which was initially derived by creating a risk-free hedged portfolio. Note that the growth rate ρ depends on the expectation of the option writer or some market analyst and is therefore somewhat arbitrary.

We can then compare call options c_S priced according to process S to options c_X based on process X. To be specific, we consider a call option with strike price K above the present share value S. This makes the option c_S more expensive than c_X. Imagine an option writer selling an option with price c_S and at the same time purchases a less expensive option c_X, based on process X which is hedged with the underlying asset and thus generates a risk-free profit. In this case the package based on c_X produces risk-free profit on top of the difference between the option prices c_X and c_S which is already in the pocket of the option writer and therefore also risk free. The total profit of the option writer is risk-free and higher than the risk-free rate. But this contradicts the basic assumption that systematic *arbitrage* or a risk-free profit above the risk-free rate is not possible. We conclude that writing options based on expected growth rates will give an unfair advantage to the option writer. In practice nobody would purchase option c_S.

We find that, at least for simple options, we can use the hypothetical process X with the same volatility as the underlying share but with risk-free growth rate r_f to calculate the fair, no-arbitrage, price of the option by integrating the expectation values of the payoff function using the log-normal distribution function (4.32) but with $\hat{\rho} = \rho - \sigma^2/2$ replaced by $r_f - \sigma^2/2$. This modified distribution function we denote by $\Psi_{r_f}(S, t)$. Calculating expectation values using this modified distribution function is called *risk-neutral* valuation and allows us to calculate option prices fairly.

Central in the discussion of efficient markets is that arbitrage—unfair advantage—to make a risk-free profit above the risk-free rate, is prohibited. Yet, there are arbitrageurs who exploit minute imbalances to make just such a profit. This is similar to the physical law of conservation of energy and yet, there are vacuum fluctuations where energy conservation is violated, provided it happens rapid enough, limited by Heisenberg's uncertainty principle. In the same way arbitrage is prohibited in an efficient market, but can be briefly violated. In that sense, to paraphrase Wilmott's [3] dictum about the no-arbitrage law: "There is no such thing as a free lunch, just a quick exchange of snacks."

The notion of a fair process leads to the concept of a *martingale,* which describes a stochastic process in which the outcome is un-biassed in the sense that the expectation value of tomorrows share value, knowing the share values on all previous days, is the value of today. In other words, the share value can go up or down with equal probability. For a stochastic variable x this can be written as

$$E(x_{n+1}|x_n, \ldots x_0) = x_n , \tag{5.19}$$

where the left hand side denotes the expectation value of x_{n+1} conditioned on previous values x_j with $j \leq n$. Thus the best guess of tomorrows value is equal to todays value. If the x were to denote share prices and the n labels the days, we compare value at different days and in that case we need to take discounting into account, but the discussion above should have made clear that we need to discount with the risk-free rate r_f such that the martingale condition now reads

$$E(e^{-r_f T} S_T | S_0) = S_0 \tag{5.20}$$

with the interpretation that tomorrows share value S_T, discounted to today, conditioned on today's value S_0, is S_0. In a fair game we cannot foresee the next roll of the dice so-to-speak. Evaluating expectation values of any function $g(S)$ is then done by weighted averaging with Ψ_{r_f} such that we obtain

$$E(g(S)) = \langle g(S) \rangle = \int g(S)\Psi_{r_f}(S, t)dS , \tag{5.21}$$

which constitutes a method to evaluate expectation values in a risk-neutral way using the weighting function Ψ_{r_f}, where $\Psi_{r_f}dS$ is the *martingale measure.*

5.4 Dynamic Hedging

We now take a closer look at the mechanics of dynamic or Δ–hedging, namely the continuous adjustment of the ratio of shares to options in a risk-free portfolio. First consider the left plot in Fig. 5.1, which shows the price of a European call option as a function of the share price for different times until maturity. At maturity, the price of the option is obviously the payoff function, shown as the solid black line. At earlier times the price of the option is higher than the payoff function, which reflects the expectation that there is a finite probability that the option will end up *in the money,* which means it will create a profit for the option holder, because the share price S at maturity exceeds the strike price K. If the share price is below the strike price—*out of the money*—the option price is low, but not zero, because there is still a chance that an upward motion of the share price can make the option profitable at maturity.

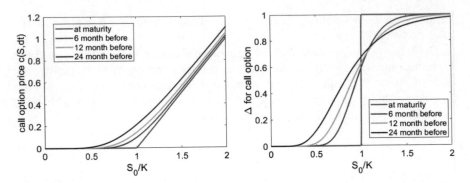

Fig. 5.1   The price (left) and delta (right) of a European call option at maturity, 6, 12, and 24 month before maturity. The annual growth rate and volatility is assumed to be 5% and 30%, respectively

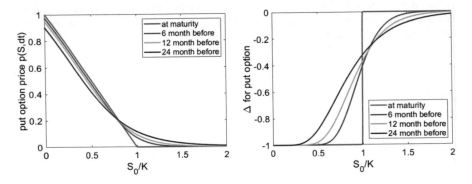

Fig. 5.2   The price (left) and delta (right) of a European put option at maturity, 6, 12, and 24 month before maturity. The annual growth rate and volatility is assumed to be 5% and 30%, respectively

The plot on the right-hand side in Fig. 5.1 shows $\Delta = \partial c / \partial S$, the hedging ratio according to (5.3). It is near zero, if the share price S is much below the strike price K reflecting that there is a finite, but small chance that the option ends up in the money. Therefore a small number of shares is required to hedge the small probability that the option writer has to hand over the shares to the option owner at maturity. As the share price gets closer and eventually exceeds the strike price the hedging ratio Δ approaches unity, indicating a high probability that the owner will exercise the option at maturity. Thus, the option writer better acquires shares to cover her commitment towards the option buyer. The fraction Δ of shares in the portfolio thus reflects the probability to deliver the shares at maturity. In Fig. 5.2 we show the corresponding plots for a European put option but will not further discuss the details here.

Note that the Δ of call and put options can be calculated explicitly by differentiating c with respect to S in (5.17). The equivalent expression for the put option is given by the corresponding derivative of p in (4.42), where we have to replace ρ by the risk-free rate r_f. Explicitly performing the calculations, we find

$$\Delta_c = N(d_1) \qquad \text{for a European call option}$$
$$\Delta_p = N(d_1) - 1 \qquad \text{for a European put option} \qquad (5.22)$$

Here $N(x)$ is the normal distribution from (4.39) and d_1 is defined in (4.43).

We now illustrate how dynamic hedging works in detail by writing a simulation of the process. See Appendices B.2 and B.3 for the MATLAB code. We assume that I do not have any capital initially, but sell one option and receive the option price c_0 in cash that I deposit in a bank at the risk-free rate r_f. Then I calculate the initial hedge $S\Delta = S\partial c/\partial S$, borrow that amount at the risk-free rate and purchase $S\Delta$ shares. During the remaining days until maturity, at the end of every trading day d I calculate Δ_d using the Black-Scholes formula with the current share price S_d and the remaining time until maturity and use that information to calculate the updated hedge value $\Delta_d S_d$. Depending on whether this value is larger or smaller than the hedge value from the previous day, I either buy additional shares or sell from my portfolio. Furthermore, I have to pay interest on the money borrowed from the bank to pay for the shares. This procedure of re-calculating the hedge value every day keeps the portfolio risk-free. As maturity approaches, Δ either approaches zero or unity as is obvious from the plot on the right-hand side in Fig. 5.1. If the option is in the money, Δ will approach unity and my position to provide shares is covered by the hedge. Then I can hand over the shares in the hedge to the owner of the option but only receive the strike price for the shares. That money I use to pay off the loan plus accumulated interest from the bank. See the upper plot in Fig. 5.3 for the temporal evolution of the share price, shown as the solid black line. The hedge is shown as the dot-dashed red line and the money borrowed from the bank is displayed as a dashed blue line. The cost for me, the writer of the option, is the borrowed money minus the strike price, but that is covered by the initially received price of the option c_0, that was invested at the risk-free rate r_f.

The lower plot in Fig. 5.3 shows the situation, where, at maturity, the option is out of the money. In this case Δ will approach zero and I will not need a hedge, because the owner of the option will forfeit. We find that the initial hedge, shown as a dot-dashed red line, is reduced to zero as time progresses towards maturity and it becomes more and more obvious that the option will not be exercised. Therefore Δ approaches zero and I can sell off the shares in the hedge. At maturity no shares are present and therefore all money was returned to the bank and I can cover the accumulated interest from the initially received money for the option c_0.

Note that in both cases discussed, I, the writer of the option, make a profit. This is mostly due to the fact that the shares increased without fluctuating, whereas in the calculation of the option price and the Δ an annual volatility of 30% was assumed. In reality I will always need to increase the hedge when the share price goes up and sell shares to reduce the hedge when it goes down and therefore need to buy shares at a high price and sell at a lower price. This will reduce my profits, but is, on average, covered by the initially calculated option price c_0.

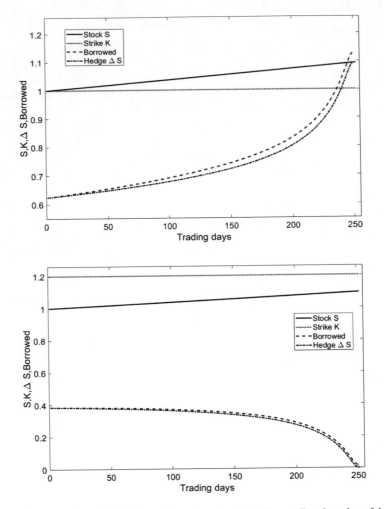

Fig. 5.3 The share price (solid black) and the strike price (dotted) as well as the value of the hedge (dot-dashed in red) and the borrowed amount (dashed in blue) as a function of trading days until maturity on day 252, the end of one trading year. For simplicity the volatility is set to zero and the shares rise by 9% over the year, the risk-free rate is 5% and the strike price is equal to the initial share price (top) and 20% above it (bottom)

5.5 Other Examples

One method to calculate the price of financial derivatives such as options is based on using the Green's function from (5.11) and convoluting it with the payoff function. This is what we did for the call option in (5.13). Inspired by the close relation of the Black-Scholes to the diffusion equation, we can interpret the pay-off function as an initial heat distribution that diffuses with time and observe the "heat" arriving at

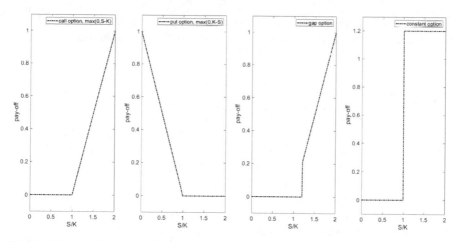

Fig. 5.4   The payoff functions of call (left), put (left center), gap-option (right center) and a constant option (right) as a function of the ratio of share and strike price S/K

another location as a function of time. The second method is based on evaluating the expectation value of the pay-off function over the expected log-normal distribution of the final share value given by (4.32) and back-propagating the expectation value back to the initial time. Here we have to keep in mind to use the risk-free rate instead of some assumed growth rate ρ, in order to ensure a fair valuation of our derivative in accordance with the discussion in Sect. 5.3.

It is illustrative to show the pay-off functions for call, put and two exotic options in Fig. 5.4. The call and put options were covered before. A gap option, shown in the center right pays off the difference between the share and strike price, but only if a second strike level has been surpassed. The payoff function on the right pays 20% more than the strike price, provided the latter is exceeded. Such an option closely resembles a gamble; if the strike price is exceeded at all, a bonus of 20% over the strike price is paid out. Since this option can become very expensive for the option writer, we can expect it to be rather expensive. We refer to [3] for the discussion of a large number of further options and the methods of how to value them.

A *forward contract*, as discussed in Chap. 2, can be valued directly, without using the differential equation. The agreement to deliver a share or other asset at a later time T for the agreed-upon price K has, at maturity the value $S_T - K$ which can be either positive or negative. At an earlier time t its value is given by the back-propagated value

$$f = e^{-r_f t} (S_T - K) = S - e^{-r_f t} K \qquad (5.23)$$

where S is today's value of the share. It is straightforward to verify that the forward contract f actually satisfies the Black-Scholes differential equation, just calculate all derivatives and insert into (5.6) with c replaced by f. The fact that the forward contract satisfies the Black-Scholes differential equation implies that it can be traded without providing any chance for a zero-risk profit. In other words, there are no possibilities of arbitrage. Trade commissions that supervise and authorize the use of new financial instruments at exchanges check this to ensure that all trade is fair.

Exercises

1. Derive the Black-Scholes equation for the portfolio $\hat{\Pi} = -p + S\frac{\partial p}{\partial S}$.
2. Calculate the fair price of the option shown on the right-most plot in Fig. 5.4.
3. You have the bright idea to issue an option that pays twice the present stock value S_0, if the value of the stock S_1 after one year lies in the range $K_1 < S_1 < K_2$. You want to keep K_1 and K_2 general, in order to be able to offer variants of the option to different customers. (a) Sketch the pay-off function. (b) Determine a fair price for the options as a function of K_1 and K_2.
4. Calculate a fair price for an option with the payoff function defined by $\max(S^2/K - K, 0)$.
5. Use your own words to explain the concept of a Martingale measure.
6. Explain why the valuation of options with the Martingale measure from (5.21) yields the same price as the calculation using the Black-Scholes equation from Sect. 5.2.
7. Calculate $N'(z) = dN/dz$ and show that $N(z) + N(-z) = 1$.
8. Show that the forward contract f, given by (5.23), satisfies (5.6).

References

1. I. Steward, *The Mathematical Equation That Caused the Banks to Crash*, The Guardian (2012). Available online at https://www.theguardian.com/science/2012/feb/12/black-scholes-equation-credit-crunch
2. D. Silverman, *Solution of the Black Scholes Equation Using the Green's Function of the Diffusion Equation*, unpublished note (UC Irvine, 1999)
3. P. Wilmott, S. Howison, J. Dewynne, *The Mathematics of Financial Derivatives* (Cambridge University Press, Cambridge, 2005)

Chapter 6
The Greeks and Risk Management

Abstract This chapter introduces the Greeks as derivatives of the option price with respect to the variables and parameters of the Black-Scholes equation. Their use in hedging risks and assessing the resilience of a hedge with respect to parameter variations, is discussed. A discussion of the volatility smile and the concept of value-at-risk follows. Finally, combinations of simple options, such as bull-spreads or straddles, are introduced and their use to adapt one's investment to one's expectations are briefly touched upon.

After covering the derivation of the pricing formulas for the options we now proceed and investigate properties of the solutions, especially their robustness. Then we discuss the use of options in order to create a portfolio insurance as well as other financial instruments to bet on some expected behavior of the market.

6.1 The Greeks

Usually, options are used to hedge the risk of shares in a portfolio. The number of shares that are needed to balance one option is given by the condition that the portfolio Π, given by (5.3), is zero. In that case any variation in the share value S is balanced by a corresponding opposite variation in the value of the option. The proportionality constant

$$\Delta = \frac{\partial c}{\partial S} \qquad (6.1)$$

is called—to everybodies surprise—*Delta* and is the first member of the quantities to characterize the stability of a portfolio named *the Greeks*.

Electronic supplementary material The online version of this chapter (https://doi.org/10.1007/978-3-030-63643-2_6) contains supplementary material, which is available to authorized users.

Delta can be calculated from the valuation equation for the corresponding option, for example (4.40) or (5.17), by differentiation, which is how we derived (5.22). The delta is specific to the type of option, the underlying share price, and the time until maturity of the option. If we calculate the option value, or its delta, again at a later time, the value may have changed, because both the share value and the time to maturity have changed. This implies that the hedge is compromised, unless the fraction Δ of shares and options is repeatedly balanced. In practice this re-balancing is typically done on a daily basis. The reason is that new *information* about the share value S is added from one day to the next and updating the hedge takes this additional information into account.

Apart from the dependence on time t and stock price S, the option price also depends on the risk-free rate r_f and the volatility σ. In Chaps. 4 and 5 we assumed the latter to be constant, while in real life they vary. Their dependence on external influences is perceived as an additional risk and we therefore investigate how much they actually influence the value of an option, or, more generally, of a portfolio.

We first consider the variation of Δ with that of the underlying shares, which is addressed by the quantity *Gamma*, denoted by Γ, and defined by the rate of change, the derivative, of Δ

$$\Gamma = \frac{\partial \Delta}{\partial S} = \frac{\partial^2 c}{\partial S^2} . \tag{6.2}$$

Γ indicates how quickly the hedging needs to be re-balanced. Here we defined Γ for the option c alone, but often it is also calculated for a portfolio Π. The dependence of Γ on the stock price is shown in Fig. 6.1 for 6, 12, and 24 month before maturity. We observe that it is peaked near the strike price where $S/K \approx 1$ and the peaking becomes more pronounced the closer we approach maturity. This reflects the fact that a share price close to the strike price can cause large changes in the hedging ratio Δ. If the share price meanders around the strike price, hedging becomes very

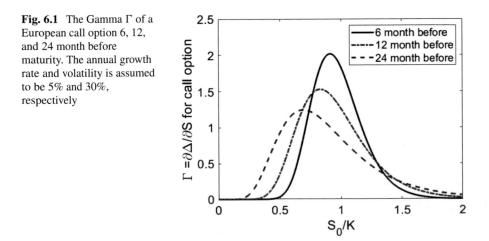

Fig. 6.1 The Gamma Γ of a European call option 6, 12, and 24 month before maturity. The annual growth rate and volatility is assumed to be 5% and 30%, respectively

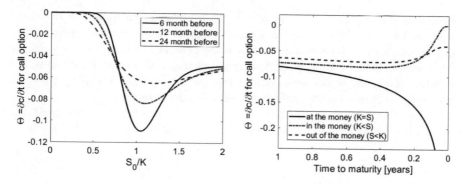

Fig. 6.2 The Theta Θ (left) of a European call option 6, 12, and 24 month before maturity and the temporal evolution of Θ towards maturity for cases where the share price S is close to the strike price, much below or much above. The annual growth rate and volatility is assumed to be 5% and 30%, respectively

difficult. If, on the other hand, the share price differs significantly from the strike price, the hedge varies very little.

The derivative with respect to time is called *Theta,* denoted by Θ, and defined by

$$\Theta(c) = \frac{\partial c}{\partial t} \quad \text{or} \quad \Theta(\Pi) = \frac{\partial \Pi}{\partial t} , \tag{6.3}$$

where we distinguish between the Θ of the option c or of a portfolio Π. In practice Θ is mostly monitored. We show the dependence of Θ on the ratio of share to strike price in the plot on the left-hand side in Fig. 6.2. Similar to Γ, the time derivative Θ varies most, if the share price is close to the strike price, again indicating that hedging is difficult is the share price is close to the strike price. On the right-hand side in Fig. 6.2 we show how Θ varies with time as maturity is approached. If the option is clearly in or out of the money, shown as dashed curves in red and blue, the variability is moderate, but if the option is at the money, shown as solid black curve, Θ can become very negative as maturity approaches.

The interest rate r_f is also an, albeit slowly varying quantity, often in response to the change of discount rates by central banks. The variation of a financial instrument, for example a portfolio Π, with the risk-free rate r_f is called *Rho,* denoted by ρ and defined by

$$\rho(\Pi) = \frac{\partial \Pi}{\partial r_f} . \tag{6.4}$$

Furthermore, the variation of a portfolio Π with the volatility σ is called *Vega,* denoted by $\mathcal{V}$ and defined by

$$\mathcal{V}(\Pi) = \frac{\partial \Pi}{\partial \sigma} . \tag{6.5}$$

Note that Vega is not a character of the Greek alphabet, but rather an oversized letter ν. The volatility σ was also assumed constant in the previous chapters, but in practice depends dynamically on the trading activities of the underlying shares. For example, at times of uncertainty, with impending strikes or political unrest, the volatility increases. Under such circumstances it is beneficial to make a portfolio insensitive to changes of the volatility σ, namely minimizing $\mathcal{V}$.

A portfolio Π consisting of shares and other derivatives is a derivative itself and as such it must obey the Black-Scholes differential equation, given by (5.6), thus

$$-\frac{\partial \Pi}{\partial t} = \frac{1}{2}\sigma^2 S^2 \frac{\partial^2 \Pi}{\partial S^2} + r_f S \frac{\partial \Pi}{\partial S} - r_f \Pi . \tag{6.6}$$

After substituting the definitions of the Greeks we obtain

$$-\Theta = \frac{1}{2}\sigma^2 S^2 \Gamma + r_f S \Delta - r_f \Pi \tag{6.7}$$

In case the portfolio is perfectly hedged, or risk-balanced with $\Delta = 0$ we have

$$\Theta = r_f \Pi - \frac{1}{2}\sigma^2 S^2 \Gamma . \tag{6.8}$$

Furthermore, if the risk-free rate is small we can ignore the term with r_f. If, at the same time, Θ is large we see that Θ and Γ are anti-correlated and we can monitor one as a proxy for the other [1]. This anti-correlation is also visible by comparing Γ in Fig. 6.1 and Θ on the left plot in Fig. 6.2. One is the very closely the negative of the other.

6.2 Volatility Smile

The option pricing formula derived in the previous sections are simply guidelines to value options. In practice the sellers of options can set a price to anything and the market, or more accurately, the buyer decides, whether to accept the price or to buy somewhere else. Following the market crash of 1987, option sellers realized that some of the options that had a very low probability of ever being exercised, all of a sudden were actually exercised and became very expensive for the option seller. The reason was that the shares dropped to very low values in the course of the crash and, for example, put options became extremely expensive. The low selling price of the option was due to the assumptions that large market variations are extremely rare in a system described by fluctuation that are distributed according to a Gaussian model. The tails of the Gaussian, which describe large fluctuations, become exponentially small. This caused the unlikely events to be under-valued.

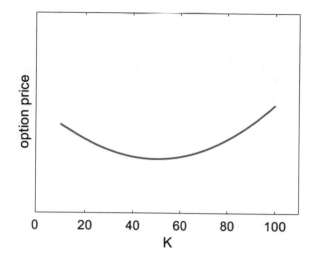

Fig. 6.3 The selling price of an option as a function of the strike price K. In this case the present value of the underlying share is $S = 50$.

The market, or the collective wisdom of its participants, adjusted to the unexpected preponderance of large fluctuations by increasing the price for those options with a strike price K much different from the current value, and the increase was larger the larger the absolute difference. When plotting the price of the option as a function of the strike price we observe a plot similar to the one shown in Fig. 6.3. There we assumed the present value of the underlying share is $S = 50$ and we observe that the price of the option increases as the strike price K deviates from 50. For obvious reasons this behavior is called the *volatility smile,* but is basically a heuristic method to take the possibility of extreme events into account.

6.3 Value at Risk

An important quantity that permits to estimate the resilience of an investment to large losses is *value at risk,* or *VaR.* It describes the value, measured in Euro or Dollar that might be lost with a 1% probability within the next few, typically 10 days. The probability level and time horizon, here 10 days, can vary from case to case. The common theme, however, is best described by the converse statement: we have a 99% chance that we do not incur losses exceeding the VaR value within the next 10 days.

Nowadays banks are required to calculate the VaR of their investments and are also required to hold three or more times the VaR value in liquid, rapidly accessible reserves in order to cover these unforeseen events that are responsible for extreme market movements. These rules are laid out by the *Basel Committee on Banking Supervision* [2].

Calculating the VaR for investments that follow a normally distributed random walk is straightforward, we just need to find the volatility for the ten-day time horizon, that can be derived from the annual or daily volatility and scaling with the root of the number of days. As the second ingredient we need to know that the tail of a Gaussian distribution beyond 2.33 standard deviations contains 1% of the area. Thus, for an initial investment of S_0 we have a 1% chance that the share price wanders beyond $S_0 - 2.33 S_0 \sigma \, (10 \, \text{days})$. The value at risk in that case is

$$\text{VaR} = 2.33 S_0 \sigma \, (10 \, \text{days}) \tag{6.9}$$

and is easily estimated, even for complex portfolios, in which case the volatility for the portfolio as a whole is used.

An alternative method uses historic data [1] and considers a number N of consecutive trading days, calculating gains and losses for each day-to-day variation and ranking the losses. The $0.01 \times N$ worst loss may serve as an estimate for the 1% chance to exceed that loss.

It is apparent that the underlying quantity for all the previously discussed methods are the daily varying share prices and in particular the volatility σ. In Chap. 7 we pay closer attention to the characterization and modeling of such time series and how to extract quantities, such as the volatility, directly from the raw data. But before doing so, we will address methods to combine options and shares to build portfolios that allow us to tailor our risk.

6.4 Tailoring Risk to One's Desire

In this section we will discuss different linear combinations of options that address certain expectations of the portfolio owner. To illustrate this, we introduce the profit diagram, which is closely related to the payoff diagram of options encountered in previous chapters, but has a subjective twist because the profit of one partner in a trade is the loss of the other partner. In this sense their profit axes are inverted with respect to each other. This is illustrated in Fig. 6.4, where we display the payoff function of a call option on the left and the profit functions for the writer and the owner of the option, respectively. The profit functions are shifted vertically because the writer of the option makes an initial profit by selling the option and the purchaser and later owner of the option has to pay for it, thus incurring a negative profit.

We first discuss *portfolio insurance* with a put option. Our portfolio $\Pi = S + p$ consists of one share S and one put option p with a strike price below the current value of the shares. Note that we do not hedge in this scenario, we simply buy one put option for each of the initially acquired shares at the same time. The put option with the lower strike price will allow us to limit losses for the shares, should the share price drop below the strike price K. In that case we can sell at K instead of some even lower price to which the shares may have dropped. In the other case where the share value increases, we had just incurred a small cost when purchasing the put option.

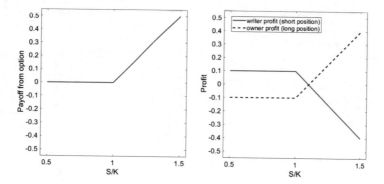

Fig. 6.4 The payoff function of a call option (left) and the profit functions for the writer and owner of the option (right). Note that the profit functions are shifted vertically by the purchase price of the option

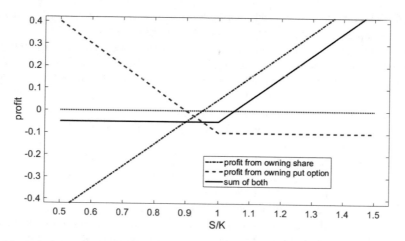

Fig. 6.5 The profit diagram for a portfolio insurance scheme using a put option to limit losses in case of a dramatic share price drop

A plot of the profit of the different constituents of the portfolio as a function of the share price in Fig. 6.5 illustrates this. The dot-dashed red line denotes the profit from owning the share; if it grows, our profit increases; if it drops, we lose. The dashed blue line is the profit line for the put option. If the share price is large, we incur a small loss from the purchasing price of the option, but if the share price drops, the payoff from the put option increases and covers the losses we incur from owning the shares. The solid black line shows the sum of both contributions, which is the value of our portfolio. Note that the construction of one share plus one put option is indeed an insurance for uncontrolled losses. It limits our losses, but it costs a little money up front, just as any normal insurance does. Note that a put option with a much lower strike price than the current share price has a rather low initial cost, because it has a

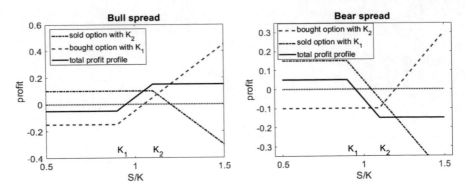

Fig. 6.6 Profit diagram of bull-spread (left) and bear-spread (right)

small initial probability, as per Black-Scholes theory, to be exercised at maturity. It thus is inexpensive, but pays off big-time in times of market crashes.

It should be obvious that the financial constructions are equivalent to betting on an expected performance of some share. If we wish to bet on a rising market, called a *bull market*, we can sell a call option with a high strike price K_2 and buy another with a smaller strike price $K_1 < K_2$. This is called a *bull-spread* and its profit diagram is shown on the left-hand side in Fig. 6.6. The profit lines of the two options are shown as dashed lines in red and magenta and the sum of both is shown as the solid black line. Note that the cash flow of selling and purchasing the two options nearly balances and requires little extra funds. The profit, on the other hand, shown by the solid black line, is distinctly favoring higher share values, hence this financial instrument implements betting on a rising market.

The converse mechanism, namely betting on a falling market, also called a *bear market*, is achieved by a *bear-spread* where a call option with a high strike price is purchased and one with lower strike price sold. The corresponding profit diagram is shown on the right-hand side in Fig. 6.6.

In a similar fashion one can bet on a stable market by creating a *Butterfly spread* where two options with an intermediate strike price K_2 are sold and one option is bought with strike price $K_1 < K_2$ below K_2 and one with K_3 above $K_2 < K_3$. The option with K_2 is at the money and likely to be rather profitable to sell, but the two options are far away from the present share price and rather inexpensive. Provided the market does not move much, the option seller will be able to keep the difference as profit.

The complementary strategy, namely betting on a varying market, is accomplished by purchasing one call and one put option with the same strike price. This is called a *straddle* and the profit diagram is shown in the top left plot in Fig. 6.7. We see that the put and call options rise if the share prise S is significantly different from the strike price. Note, however, that the options are purchased at the money and are likely to be expensive. If the speculator is reasonably certain that the market is very volatile and the stock market varies significantly, the straddle can be "pulled apart"

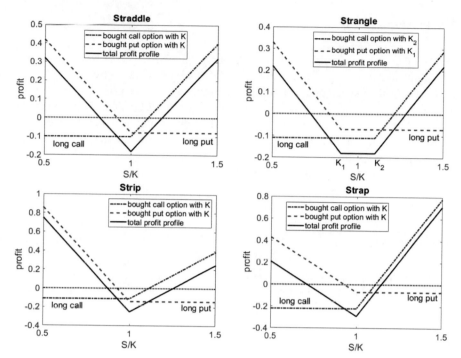

Fig. 6.7 Profit diagrams of straddle (top left) and strangle (top right) and of strip (bottom left) and strap (bottom right)

by choosing a strike price below the current share value for the put option and above for the call option. This construction is called a *Strangle* and is shown in the top-right plot in Fig. 6.7. Choosing the strike prices of the options different from the share price is likely to reduce the cost of initially purchasing the options.

Closely related financial constructs are *Strip* and *Strap*. The former is a close variant of the straddle, but instead of one put and one call option, the speculator purchases two put and one call options. In this case the betting is on a volatile market, but a fall is deemed more likely than an increase in the market. A *Strap* is based on purchasing two call and one put option, which is essentially a bet on a volatile market with a higher probability of a move upwards. The profit diagrams for both strip and strap are shown in the bottom row of Fig. 6.7.

Now it should be obvious that the underling assets, the shares, are moving more or less randomly from one point in time to the next and therefore they constitute a time series of data. And we are interested in extracting information from them; examples are the average growth rate ρ or the volatility σ. Extracting such information from data is often based on methods using regression models, which are commonly encountered in physics by their colloquial name "linear fit." These models are the topic of the next two chapters.

Exercises

1. Calculate Δ and Γ for the call option given in (5.17).
2. Calculate Δ for the put option given in (4.42) with ρ replaced by r_f.
3. Compare Γ for a call and a put option with the same strike price.
4. Calculate the value-at-risk of the two shares in the file stocks.dat from Exercise 2 in Chap. 3. Express the VaR as a fraction of the investment! First do so for one stock at a time, then for a portfolio where the investment divided evenly among the two stocks.
5. Sketch the profit diagram of a butterfly spread.

References

1. J.C. Hull, *Options, Futures, and Other Derivatives*, 8th edn. (Pearson, Boston, 2012)
2. Web site of the Basel Committee on Banking Supervision. Available online at https://www.bis.org/bcbs/

Chapter 7
Regression Models and Hypothesis Testing

Abstract This chapter covers the basics of adapting regression models, also known as linear fits in physics, to find the parameters that best explain data in a model and then estimate the error bars of the parameters. The analysis of the model's reliability stimulates a discussion of χ^2 and t-distributions and their role in testing hypotheses regarding the parameters; for example, whether a parameter can be omitted from the fit. A more elaborate method, based on the F-test, follows. The chapter closes with a discussion of parsimony as a guiding principle when building models.

Both in physical and social sciences as well as in economics we often try to describe the behavior of observable parameters—the measurements—in terms of model parameters. Frequently the dependence of the measurements on the parameters of a model—the fit parameters—is linear; the archetypical problem of this type is the fit of data points to a straight line. Let us therefore assume that we are given a list of n data points (s_i, y_i) for $i = 1, \ldots, n$ that come from varying a variable s and observing another quantity y. Figure 7.1 shows two examples. Visual inspection indicates that the data points cluster around a straight line that we can describe by two parameters a and b, given by $y_i = as_i + b$. Our task is thus to find the slope a and the intercept b. This is most easily accomplished by writing one copy of the equation for each of the n measurements, which defines one row in the following matrix-valued equation

$$\begin{pmatrix} \vdots \\ y_i \\ \vdots \end{pmatrix} = \begin{pmatrix} \vdots & \vdots \\ s_i & 1 \\ \vdots & \vdots \end{pmatrix} \begin{pmatrix} a \\ b \end{pmatrix}. \tag{7.1}$$

Electronic supplementary material The online version of this chapter (https://doi.org/10.1007/978-3-030-63643-2_7) contains supplementary material, which is available to authorized users.

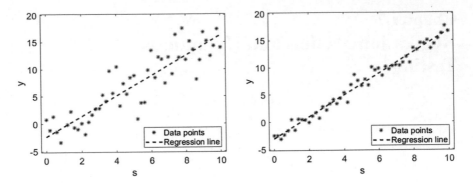

Fig. 7.1 Two examples with data points scattered around a linear trend (asterisks, crosses) and dashed regression lines

Here the n measurements y_i are assembled into a column vector and the values of the independent variable s_i are assembled, together with n times a one, in a $n \times 2$-matrix. The unknown parameters a and b are assembled in a second column vector. Note that the equation has the form of $\mathbf{y} = A\mathbf{x}$, where $\mathbf{y}$ is the column vector with the y_i, the matrix A with the s_i, and a column vector $\mathbf{x}$ with a and b. Our task is now to find values for a and b such that the line approximates the data points as good as possible (in some sense, defined below). Moreover, we want to assess the error bars on the fit parameters a and b. The larger scatter of the data points on the on the left-hand plot in Fig. 7.1 will likely reduce our confidence in the fitted value of a and b, compared to the values derived from the data points on the right-hand plot. The scatter of the data points y_i is commonly described by adding a vector $\Delta\mathbf{y}$ to $\mathbf{y} = A\mathbf{x} + \Delta\mathbf{y}$ and the assumption that the components Δy_i of $\Delta\mathbf{y}$ are sampled from Gaussian distributions with standard deviation σ_i, the error bars of the measurements y_i. Furthermore, increasing the number of fit parameters, for example, by fitting a third-order polynomial through the data points, will likely make the fit better. One might ask, however, whether a small improvement is really worth having to deal with an additional fit parameter? This fit parameter may even have large error bars associated. We will address questions, such as this one, throughout this chapter.

Before considering a number of examples we need to elaborate how to actually find the model parameters such as the slope or intercept in a straight-line fit.

7.1 Regression and Linear Fitting

In (7.1) we only try to determine two fit parameters, but we can easily generalize the method to include many, say m, fit parameters x_j with $j = 1, \ldots, m$. Since we assume the dependence of the measurements y_i on the model parameters x_j to be linear, the corresponding equation is matrix valued and in general will have the following form

$$\begin{pmatrix} y_1 \\ y_2 \\ \vdots \\ y_n \end{pmatrix} = \begin{pmatrix} A_{11} & A_{12} & \dots & A_{1m} \\ A_{21} & A_{22} & \dots & A_{2m} \\ \vdots & \vdots & \ddots & \vdots \\ A_{n1} & A_{n2} & \dots & A_{nm} \end{pmatrix} \begin{pmatrix} x_1 \\ x_2 \\ \vdots \\ x_m \end{pmatrix} + \begin{pmatrix} \Delta y_1 \\ \Delta y_2 \\ \vdots \\ \Delta y_n \end{pmatrix}, \quad (7.2)$$

which can be also written as $\mathbf{y} = A\mathbf{x} + \Delta\mathbf{y}$. The information about the dynamics of the system, which relates the n measurements y_i to the m fit parameters x_j, is encoded in the known coefficients of the matrix A_{ij} with $i = 1, \dots, n$ and $j = 1, \dots, m$. The measurement errors are described by Δy_i with rms values σ_i.

Systems of equations, defined by (7.2), are over-determined linear systems if n, the number of observations or measurements, is larger than the number of parameters m. Such systems can be easily solved by standard linear-algebra methods in the least squares sense. For this we minimize the so-called χ^2, which is the squared difference between the observations and the model

$$\chi^2 = \sum_{i=1}^n \left(\frac{y_i - \sum_{j=1}^m A_{ij} x_j}{\sigma_i} \right)^2. \quad (7.3)$$

Here we introduce the error bar or standard deviation σ_i of the observation y_i to assign a weight to each measurement. If an error bar of a measurement is large, that measurement contributes less to the χ^2 and will play a less important role in the determination of the fit parameters x_j. If all error bars are equal $\sigma_i = \sigma$ for all i the system is called *homoscedastic* and if they differ, it is called *heteroskedastic*.[1]

If we consider a homoscedastic system and ignore the σ_i for the moment, we see that (7.3) can be written as a matrix equation $\chi^2 = (\mathbf{y}^t - \mathbf{x}^t A^t)(\mathbf{y} - A\mathbf{x})$, where $\mathbf{x}$ and $\mathbf{y}$ are column vectors, albeit of different dimensionalities n and m. Minimizing the χ^2 with respect to the fit parameters $\mathbf{x}$, we find $0 = \partial\chi^2/\partial\mathbf{x} = -2A^t(\mathbf{y} - A\mathbf{x})$, where $\partial/\partial\mathbf{x}$ denotes the gradient with respect to the components of $\mathbf{x}$. Solving for $\mathbf{x}$ we obtain

$$\mathbf{x} = (A^t A)^{-1} A^t \mathbf{y}, \quad (7.4)$$

where $(A^t A)^{-1} A^t$ is the well-known pseudo-inverse of the matrix A.

For heteroscedastic systems, we introduce the diagonal matrix Λ that contains the inverse of the error bars on its diagonal $\Lambda_{ii} = 1/\sigma_i$, which transforms (7.3) into $\chi^2 = (\Lambda\mathbf{y}^t - \mathbf{x}^t A^t \Lambda^t)(\Lambda\mathbf{y} - \Lambda A\mathbf{x})$ such that both $\mathbf{y}$ and A are prepended with Λ. Performing these substitutions in (7.4), we obtain

$$\mathbf{x} = (A^t \Lambda^2 A)^{-1} A^t \Lambda^2 \mathbf{y} = J\mathbf{y}. \quad (7.5)$$

---

[1]Greek: homo = equal, hetero = unequal, skedasis = dispersion or spread.

Here we introduce the matrix J that maps $\mathbf{y}$ onto $\mathbf{x}$ to explicitely show that the fit parameters $\mathbf{x}$ depend linearly on the measurements $\mathbf{y}$. We denote the matrix by the symbol J to remind us that it has the same function as a Jacobian in a coordinate transformation. In our case the transformation is linear, and consequently the Jacobian has constant matrix elements.

The error bars and the covariance matrix of the fit parameters x_j can be calculated by standard error propagation techniques from the covariance matrix of the measurements y_i, which we is given by $C_{ij}(\mathbf{y}) = \langle \Delta y_i \Delta y_j \rangle$. The analysis is based on the realization that the covariance matrix $C_{ij}(\mathbf{y})$ is defined through the second moments of the deviations Δy_i. On its diagonal $C_{ij}(\mathbf{y})$ contains the squared error bars of the individual measurements σ_i^2 and the off-diagonal elements carry information about correlations among the measurements. If, on the other hand, the measurements are uncorrelated, as we assume them to be, all off-diagonal elements are zero. In our example $C_{ij}(y)$ is simply the square of the inverse of the matrix Λ, namely $C(\mathbf{y}) = (\Lambda^2)^{-1}$, which has $\sigma_1^2, \ldots, \sigma_n^2$ on its diagonal. From (7.5) follows that small deviations of the measurements Δy_i lead to small deviations of the model parameters $\Delta \mathbf{x} = J \Delta \mathbf{y}$. This observation allows us to calculate the covariance matrix $C_{kl}(\mathbf{x}) = \langle \Delta x_k \Delta x_l \rangle$. A little algebra then leads to

$$C(\mathbf{x}) = JC(\mathbf{y})J^t , \tag{7.6}$$

which describes how covariance matrices transfrom under a change of variables given by (7.5). Furthermore, inserting the definition of J from (7.5) in the previous equation, we find after some additional algebra

$$C(\mathbf{x}) = (A^t \Lambda^2 A)^{-1} , \tag{7.7}$$

which allows us to find the covariance matrix $C(\mathbf{x})$ of the fit parameters x_j by simple matrix operations from the error bars of the initial measurements that are buried in Λ and the system matrix A that initially defined the problem in (7.2).

In order to illustrate the methodology, let us consider a few examples of regression models taken from several disciplines.

7.2   Examples

Here we illustrate how we can use regression methods to extract information from data provided in tabular form, either to obtain quantitative numbers or to test some model.

Our first example addresses the question whether education actually pays off. Do extra years of education increase the earnings and if so, at which rate. Moreover, can we find a quantitative description of how well education protects from unemployment? We base our analysis on information for 2018 from the US Bureau of

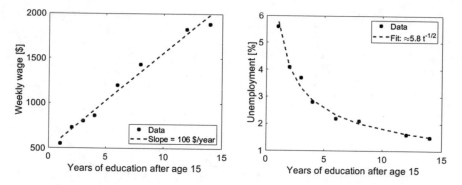

Fig. 7.2 The median weekly earnings (left) and the unemployment rate (right) as a function educational attainment, here specified as the years of education past the age of 15. The data are based on [1] from the US bureau of labor statistics

Labor Statistics [1], which provides data on the median weekly earnings w and the unemployment rate u based on the highest level of educational attainment. Here we characterize the attainment by the additional years t of education past the age of fifteen; we assume a high-school diploma completed after two years, a bachelor after six, a master after eight, a PhD after twelve, and a professional degree after fourteen years. The left plot in Fig. 7.2 shows the weekly wage w as a function of t as black asterisks and a straight-line fit of $w = at + b$ as a dashed red line. The slope a of the line is $a = 106$ \$/year of education, such that each additional year, on average, increases the weekly earnings by 106 \$. So, financially, education pays off!

The right plot in Fig. 7.2 shows the unemployment rate u as a function of t. The data points do not follow a straight line, but plotting $\log(u)$ versus $\log(t)$ reveals a linear dependence. Fitting a straight line to $\log(u) = a \log(t) + b$ provides fit parameters $a \approx -0.5$ and $b \approx 1.76$. The dashed red line on the right plot is based on these parameters. We thus find that the unemployment rate u scales as $1/\sqrt{t}$. Education even protects against unemployment, at least to some extent.

The second example uses data about the number of infected persons and fatalities from the corona epidemic, provided by the Johns-Hopkins University [2]. We ask ourselves whether the reported number of infected people I and reported number of deaths D are consistent. In the basic SIR model [3] of epidemics the rate of recovered and subsequently immune patients dR/dt is proportional to I. Analogously, we assume here that the rate of change of fatalities dD/dt is also proportional to the number infected I, or $dD/dt = \alpha I$ with a proportionality constant α that may differ from country to country, depending on the health system and the accounting of the infected and the corona-related fatalities. Note that the reported numbers of infected persons in the data from [2] do not reflect those having recovered from the infection. We therefore expect the model to work only during the first few weeks of the epidemic, while we can neglect the recovery rate. With these considerations in mind we process the data by integrating the model equation once to obtain $D = \alpha \int I \, dt$. The data files from [2] contain day-by-day values of D and I. Therefore we store the values

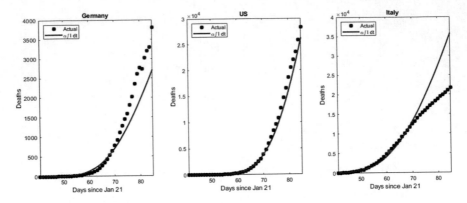

Fig. 7.3 The number of Corona-virus related deaths (black dots) and the simple model (red line), discussed in the text, for Germany, the US, and Italy

in arrays and access the number of fatalities on day i by D_i and likewise the number of infected by I_i. The integral K_i of I_i is then given by $K_i = \sum_{j \leq i} I_j$ and we then try to determine α from $D_i = K_i \alpha$, which is of the type specified by (7.2) with K_i corresponding to a single-column matrix A and I_i to y_i. We then find α, the model parameter, from (7.4).

Figure 7.3 shows the number of reported deaths from March 1 until April 16, 2020 in three countries. The solid red line shows the integral of the infected persons K_i, scaled with the country-specific factor α, which was determined by fitting the range up to 14 days before the end. We observe that the regression line for Germany underestimates the fatalities in the last 14 days, which could be explained by many unaccounted infected persons. The fit and the data for the US show good agreement between model and data, indicating that the assumptions in the model are approximately valid. The data for Italy show less actual fatalities than the model predicts. One could hypothesise that this might be attributed to a significant number of recovered persons such that the underlying assumptions of our simple model are not valid towards the end of the range. As a disclaimer, we need to stress that this example is intended to illustrate regression methods, not to develop policies of how to deal with an epidemic.

Our third example is motivated by the fact that systems in equilibrium, if slightly perturbed, perform harmonic oscillations. Thus, the dynamics is goverend by $\ddot{x} + \omega^2 x = 0$ with amplitude x and frequency ω. We now want to determine the frequency ω and the initial conditions x_0 and $\dot{x}_0$ at time $t = 0$ from a sequence of measurements of the amplitude x_n at times $t = n\Delta t$. This can be visualized as a stroboscopic recording of the oscillation at discrete points in time. To do so, we first realize that $x = x_0 \cos(\omega t) + (\dot{x}_0/\omega) \sin(\omega t)$ solves the differential equation and satifies the initial conditions. Likewise, we find that the position x_{n+1} after $(n + 1)\Delta t$ is related to the position x_n after $n\Delta t$ by $x_{n+1} = x_n \cos(\omega \Delta t) + (\dot{x}_n/\omega) \sin(\omega \Delta t)$ and we find the velocity $\dot{x}_{n+1} = -\omega x_n \sin(\omega \Delta t) + \dot{x}_n \cos(\omega \Delta t)$ by differentiation. Both equations can we written as a matrix-valued equation

$$\begin{pmatrix} x_{n+1} \\ \dot{x}_{n+1} \end{pmatrix} = \begin{pmatrix} \cos(\omega\Delta t) & (1/\omega)\sin(\omega\Delta t) \\ -\omega\sin(\omega\Delta t) & \cos(\omega\Delta t) \end{pmatrix} \begin{pmatrix} x_n \\ \dot{x}_n \end{pmatrix}. \qquad (7.8)$$

Inverting this equation, we find $(1/\omega)\sin(\omega\Delta t)\dot{x}_{n+1} = x_{n+1}\cos(\omega\Delta t) - x_n$. Moreover, using (7.8) shifted by one time step, results in $x_{n+2} = x_{n+1}\cos(\omega\Delta t) + (\dot{x}_{n+1}/\omega)\sin(\omega\Delta t)$. Inserting the former into the latter equation yields

$$x_{n+2} + x_n = 2\cos(\omega\Delta t)x_{n+1}. \qquad (7.9)$$

This is a linear equation in the positions x_n, which allows us to determine $\cos(\omega\Delta t)$ and thereby ω by casting the equation in the form of (7.2) with $x_{n+2} + x_n$ populating the vector $\mathbf{y}$ on the left-hand side and filling the single-column matrix A with $2x_{n+1}$. Here $\cos(\omega\Delta t)$ corresponds to a single fit parameter, called $\mathbf{x}$ in (7.2). As opposed to Fourier methods, which require many consecutive samples to achieve a high frequency resolution, here we only need a minimum of three positions x_n from consecutive periods in order to determine $\cos(\omega\Delta t)$. Using more data points makes the problem over-determined and improves the accuracy.

Now that we have determined the frequency ω, we proceed to determine the initial conditions x_0 and $\dot{x}_0$ from subsequent readings of $x_n = x_0\cos(n\omega\Delta t) + (\dot{x}_0/\omega)\sin(n\omega\Delta t)$. Furthermore, we assume that each measurement x_n is known to be within its error bars $\pm\sigma$. This allows us to write the problem as a matrix-valued equation

$$\begin{pmatrix} \vdots \\ \frac{x_n}{\sigma} \\ \vdots \end{pmatrix} = \begin{pmatrix} \vdots & \vdots \\ \frac{\cos(n\omega\Delta t)}{\sigma} & \frac{(1/\omega)\sin(n\omega\Delta t)}{\sigma} \\ \vdots & \vdots \end{pmatrix} \begin{pmatrix} x_0 \\ \dot{x}_0 \end{pmatrix}, \qquad (7.10)$$

where we explicitly absorbed the matrix Λ with the error bars into (7.10), which is of the type of (7.2) and can be solved by the methods discussed in Sect. 7.1 yielding the initial conditions x_0 and $\dot{x}_0$ of the harmonic oscillation. The error bars of the fit parameters can be determined with (7.7).

Going beyond fitting for one or two parameters we consider a fit to a polynomial of order p to n data points (t_i, y_i) for $i = 1, \ldots, n$

$$y_i = a_p t_i^n + \cdots + a_1 t_i + a_0, \qquad (7.11)$$

which is straightforward to convert to an equation similar to (7.2). Here the matrix A is of size $n \times (p+1)$ and the vector $\mathbf{x}$ contains the $p+1$ coefficients a_p.

In Chap. 8, where we will analyze time series, we will use regression methods, similar to those discussed in this chapter to determine models for the temporal behavior of dynamics systems. In Sect. 11.1 we will encounter macroeconomic models that are described through difference equations, relating economical quantities, such as company output, profit, and investment rate, from one time period to the next. The validity of these models can be corroborated or rejected by analyzing the consistency of the models when comparing them with published data for companies.

The above manipulations allow us to calculate the fit parameters and their error bars, but we still do not know how reliable the calculations are, because we could have a small χ^2 and just had misjudged the initial measurement errors σ_i. Therefore we will first introduce a parameter R^2 that characterizes how well the fitted model actually explains the observed data y_i and then calculate the distribution of χ^2 we can expect and can judge how likely the obtained χ^2 actually is.

7.3 Goodness-of-Fit R^2

The goodness-of-fit parameter $R^2 = \text{SSE}/\text{SST}$ compares the spread of the measurement values y_i, the *total sum of squares* (SST), around their mean $\bar{y}$

$$\text{SST} = \sum_{i=1}^{n}(y_i - \bar{y})^2 \quad \text{with} \quad \bar{y} = \frac{1}{n}\sum_{i=1}^{n} y_i \tag{7.12}$$

to the *explained sum of squares* (SSE), defined as the spread of the values $\hat{y}_i$ predicted by the fitted model, to the mean $\bar{y}$

$$\text{SSE} = \sum_{i=1}^{n}(\hat{y}_i - \bar{y})^2 \quad \text{with} \quad \hat{y}_i = \sum_{i} A_{ij}x_j \ . \tag{7.13}$$

Here $\hat{y}_i$ are the values estimated with the model A_{ij} and the fit parameters x_j. R^2 thus characterizes how well the fitted model can explain the variation around the mean $\bar{y}$.

Let us now inspect how SST and SSE are related to the χ^2 that we introduced in (7.3). We therefore insert and subtract the $\hat{y}_i$ in the definition of SST

$$\text{SST} = \sum_{i=1}^{n}(y_i - \hat{y}_i + \hat{y}_i - \bar{y})^2$$

$$= \sum_{i=1}^{n}(y_i - \hat{y}_i)^2 + \sum_{i=1}^{n}(\hat{y}_i - \bar{y})^2 + 2\sum_{i=1}^{n}(y_i - \hat{y}_i)(\hat{y}_i - \bar{y}) \tag{7.14}$$

$$= \text{SSR} + \text{SSE} + 0$$

where the first term is the sum of squared differences of the measured values and the model predictions $y_i - \hat{y}_i = y_i - \sum_{j} A_{ij}x_j$ and is called the *sum of squared residuals* (SSR). We recognize it as $\sigma^2\chi^2$ from (7.3) if all the error bars are equal and have magnitude σ. The second term we recognize as the SSE and the last term is zero, which can be shown by inserting $\hat{y}_i = \sum_{i} A_{ij}x_j$ and (7.4).

Using (7.14) to write SSE=SST−SSR, the Goodness-of-fit R^2 can be written using either SSE or SSR

$$R^2 = \frac{SSE}{SST} = 1 - \frac{SSR}{SST}.$$ (7.15)

We see that the smaller the SSR, or equivalently the χ^2 are, the closer R^2 approaches unity, where $R^2 = 1$ describes a perfect fit of the model to the data.

Calculating R^2 is rather straightforward; first calculate the sample variance SST from the average and variance of the measurement or sample values. Then perform the fit procedure and determine the sum of squared residuals between the measurement samples and the corresponding fitted values SSR. Finally calculate R^2 from (7.15).

We already noted that the SSR is closely related to the χ^2 of the fit. In the following section we derive the probability distribution function of the χ^2 values that we can expect. It will help us to assess our confidence in the fitted parameters.

7.4 χ^2-Distribution

The quantity we minimize to find a regression model is the χ^2, which is given by

$$\chi^2 = \sum_{i=1}^{n} \left(\frac{y_i - f_i(\mathbf{x})}{\sigma_i} \right)^2 = \sum_{i=1}^{n} z_i^2 ,$$ (7.16)

where y_i are the measurement values with error bars σ_i and $f_i(\mathbf{x})$ is a model function with fit parameters x_j. In Sect. 7.1 we used a linear dependence $f_i(\mathbf{x}) = \sum_j A_{ij} x_j$. If we have estimated the error bars σ_i correctly, all individual factors $z_i = (y_i - f_i(\mathbf{x}))/\sigma_i$ in the sum should be of order unity and normally (Gaussian) distributed. Thus it is natural to ask what the distribution function of n squares of normally distributed random variables is, namely the probability distribution function $\psi_n(q)$ such that we have the probability $\psi_n(q = \chi^2)dq$ of finding a value of $q = \chi^2$ within the interval $[q - dq/2, q + dq/2]$.

We can calculate this distribution by assuming that the individual constituents z_i of the sum above are normally distributed random variables and we need to find the distribution of the $q = \sum_{i=1}^{n} z_i^2$. We start by writing the product of n independent and normalized Gaussian distribution functions for the z_i

$$1 = \frac{1}{(2\pi)^{n/2}} \int\limits_{-\infty}^{\infty} dz_1 \int\limits_{-\infty}^{\infty} dz_2 \cdots \int\limits_{-\infty}^{\infty} dz_n \, e^{-(z_1^2 + z_2^2 + \cdots z_n^2)/2}$$

$$= \frac{1}{(2\pi)^{n/2}} \int\limits_{0}^{\infty} dr \, r^{n-1} \int d\Omega \, e^{-r^2/2} ,$$ (7.17)

where we introduce spherical coordinates with $r^2 = z_1^2 + z_2^2 + \cdots z_n^2$ in the second equation. Since the integrand does not depend on angular variables we know that $\int d\Omega = S_n$ is the surface area of an $n-$dimensional sphere. After substituting $q = r^2$ we arrive at

$$1 = \frac{S_n}{2(2\pi)^{n/2}} \int\limits_0^\infty q^{n/2-1} e^{-q/2} dq = \int\limits_0^\infty \psi_n(q) dq \, , \tag{7.18}$$

which implicitly defines the probability distribution $\psi_n(q)$ for the $q = \chi^2$.

The previous equation still depends on the unknown surface area S_n. We can, however, determine S_n from the requirement that the distribution function is normalized. Substituting $p = q/2$ we find

$$1 = \frac{S_n}{2(2\pi)^{n/2}} 2^{n/2-1} \int\limits_0^\infty dp p^{n/2-1} e^{-p} = \frac{S_n}{2(2\pi)^{n/2}} 2^{n/2-1} \Gamma(n/2) \tag{7.19}$$

where $\Gamma(z)$ is the Gamma function [4], defined as

$$\Gamma(z) = \int\limits_0^\infty t^{z-1} e^{-t} dt \, . \tag{7.20}$$

Solving for S_n, we find

$$S_n = \frac{2\pi^{n/2}}{\Gamma(n/2)} \, , \tag{7.21}$$

which we insert into (7.18) to obtain the following expression for the probability distribution function

$$\psi_n(q) dq = \frac{1}{2^{n/2} \Gamma(n/2)} q^{n/2-1} e^{-q/2} dq \, . \tag{7.22}$$

Figure 7.4 displays $\psi_n(q)$ for $n = 2, 5$, and 10.

The probability of finding a χ^2 smaller than a given limit ξ is given by the integral from zero to ξ of the probability distribution function

$$Q_n(\xi) = \int\limits_0^\xi \psi_n(q) dq = \int\limits_0^\xi \frac{1}{2^{n/2} \Gamma(n/2)} q^{n/2-1} e^{-q/2} dq = P(n/2, \xi/2) \tag{7.23}$$

where $P(a, x)$ is the incomplete gamma function [4]. Now we are in a position to answer the question whether a χ^2 that arises from a fitting procedure or regression analysis is actually probable. In particular, the probability of finding a value of χ^2

Fig. 7.4 The χ^2-probability distribution function $\psi_n(q)$ from (7.22) for $n = 2, 5,$ and 10.

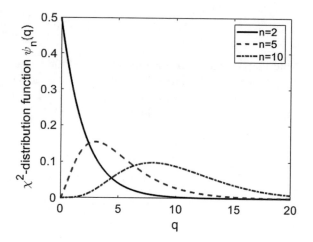

smaller than $\hat{\chi}^2$ is given by $P(n/2, \hat{\chi}^2/2)$. Conversely, finding a value that is even larger than $\tilde{\chi}^2$, is given by $1 - P(n/2, \tilde{\chi}^2/2)$.

The χ^2-distribution plays an important role in testing hypothesis, but also in assessing the reliability of estimates from a small number of samples. The latter is the topic of the next section.

7.5 Student's t-Distribution

Assume that you are responsible for the quality control of the base materials for your favorite beer. For example, you need to work out how many bags of barley you have to test in order to assess the average quality of the entire delivery, which could be quite large, say, a hundred bags. This was the task that William Gosset—"Student" was his pen name—faced during the early years of the 20th century. He worked for the Guinness brewery in Dublin and had to ensure the quality of the delivered barley. We follow his lead and work out how well we can estimate the true mean μ and variance σ^2 of a Gaussian distribution from a small number of samples n.

Since all we have are the n samples x_i, we start by calculating their average $\bar{X}_n$ and the variance S_n^2. These quantities are given by

$$\bar{X}_n = \frac{1}{n} \sum_{i=1}^{n} x_i \quad \text{and} \quad S_n^2 = \frac{1}{n-1} \sum_{i=1}^{n} (x_i - \bar{X}_n)^2 , \quad (7.24)$$

where the x_i are the samples. In the following analysis we assume the samples x_i to be drawn from a Gaussian distribution $N(x; \mu, \sigma)$, defined by

$$N(x; \mu, \sigma) = \frac{1}{\sqrt{2\pi}\sigma} \exp\left[-\frac{(x-\mu)^2}{2\sigma^2}\right]. \tag{7.25}$$

Note that $\bar{X}_n$ and S_n^2 are commonly called the *sample average* and the *unbiased sample variance*, respectively.[2] We first need to understand how fast the estimates $\bar{X}_n$ and S_n^2 converge towards the true values μ and σ^2 as we increase n. We therefore calculate the expectation value of the squared difference between the estimate and the true value

$$\left\langle (\bar{X}_n - \mu)^2 \right\rangle = \left\langle \left(\frac{1}{n}\sum_i x_i - \mu\right)^2 \right\rangle = \left\langle \left(\frac{1}{n}\sum_i (x_i - \mu)\right)^2 \right\rangle$$

$$= \frac{1}{n^2}\sum_i \sum_j \langle (x_i - \mu)(x_j - \mu) \rangle \tag{7.26}$$

$$= \frac{1}{n^2}\sum_i \langle (x_i - \mu)^2 \rangle = \frac{1}{n^2}\sum_i \sigma^2 = \frac{\sigma^2}{n},$$

where we exploited the fact that the different x_i are statistically independent and therefore $\langle (x_i - \mu)(x_j - \mu) \rangle = \langle (x_i - \mu)^2 \rangle \delta_{ij} = \sigma^2 \delta_{ij}$. Finally we see that the estimate $\bar{X}_n$ on average approaches the true mean μ with $\sigma/\sqrt{n}$ as we increase the number of samples n.

Next we need to address the reliability of an estimate based on a small number n of samples. To achieve this goal we introduce a test-statistic t, which is given by the deviation of the estimate X_n from the real mean μ, but divided by the estimated standard deviation $S_n/\sqrt{n}$.

$$t = \frac{\bar{X}_n - \mu}{S_n/\sqrt{n}} = \frac{(\bar{X}_n - \mu)/(\sigma/\sqrt{n})}{\sqrt{S_n^2/\sigma^2}}. \tag{7.27}$$

In the second equality in (7.27) we divide numerator and denominator by σ to visualize that the numerator $x = (\bar{X}_n - \mu)/(\sigma/\sqrt{n})$ indeed stems from a normal distribution with unit variance, given by

$$\psi_x(x) = \frac{1}{\sqrt{2\pi}} e^{-x^2/2}. \tag{7.28}$$

Moreover, the denominator $y = \sqrt{S_n^2/\sigma^2}$ stems from the square root of a χ^2-distribution, the latter given by (7.22). Thus $t = x/y$ is a random variable, defined by the ratio of a Gaussian random variable x and a random variable y, which is derived

---

[2]Defining the sample variance with n in the denominator yields a value that is too small and is called *biassed*, because using $\bar{X}_n$, which is derived from the same samples as S_n^2, is closer to the samples x_i than the "true" mean μ. This is compensated by dividing by $n-1$ instead, which leads to the *unbiased sample variance*, defined in (7.24).

from the root of numbers drawn from a χ^2-distribution. Note that originally we start from n samples or independent measurements and one degree of freedom was used to determine the average $\bar{X}_n$. Therefore the estimate of the variance S_n only contains information of $\nu = n - 1$ degrees of freedom. The χ^2–distribution that we need to consider is therefore one for $\nu = n - 1$ degrees of freedom.

On our way to calculate the distribution $\Phi_\nu(t)$ of the test-statistic t we first determine the distribution function of the square root of the χ^2 variable divided by $\sqrt{\nu}$. We introduce $y = \sqrt{q/\nu}$ and change variables from q to y in (7.22)

$$\phi_y(y) = \int_0^\infty \frac{1}{2^{\nu/2}\Gamma(\nu/2)} q^{\nu/2-1} e^{-q/2} \delta\left(y - \sqrt{q/\nu}\right) dq$$

$$= \frac{1}{2^{\nu/2}\Gamma(\nu/2)} \int_0^\infty q^{\nu/2-1} e^{-q/2} \frac{\delta(q - \nu y^2)}{1/(2\sqrt{\nu q})} dq \qquad (7.29)$$

$$= \frac{\nu^{\nu/2}}{2^{\nu/2-1}\Gamma(\nu/2)} y^{\nu-1} e^{-\nu y^2/2} .$$

In appendix A we follow [5] and show that the sum of random variables and the sum of squares of random variables are statistically independent and therefore we can calculate the probability distribution function $\Phi_\nu(t)$ of $t = x/y$ from the product of the two distribution functions in the following way

$$\Phi_\nu(t) = \int_0^\infty dy\, \phi_y(y) \int_{-\infty}^\infty dx\, \psi_x(x) \delta(t - x/y)$$

$$= \int_0^\infty \phi_y(y) \psi_x(yt) |y| dy . \qquad (7.30)$$

Inserting the distribution functions $\psi_x(x)$ from (7.28) and $\phi_y(y)$ from (7.29) and rearranging terms we find

$$\Phi_\nu(t) = \frac{1}{\sqrt{2\pi}} \frac{\nu^{\nu/2}}{2^{\nu/2-1}\Gamma(\nu/2)} \int_0^\infty y^\nu e^{-(\nu+t^2)y^2/2} dy . \qquad (7.31)$$

The substitution $z = (\nu + t^2)y^2/2$ and some rearrangements lead to

$$\Phi_\nu(t) = \frac{1}{\sqrt{\pi}} \frac{\nu^{\nu/2}}{\Gamma(\nu/2)} \frac{1}{(\nu + t^2)^{(\nu+1)/2}} \int_0^\infty z^{(\nu-1)/2} e^{-z} dz . \qquad (7.32)$$

The integral is given by the gamma function, defined in (7.20), as $\Gamma((\nu + 1)/2)$, such that we finally arrive at

$$\Phi_\nu(t) = \frac{1}{\sqrt{\nu\pi}} \frac{\Gamma(\frac{\nu+1}{2})}{\Gamma(\frac{\nu}{2})} \left(\frac{1}{1 + t^2/\nu}\right)^{\frac{\nu+1}{2}}, \tag{7.33}$$

which is the commonly found form of Student's t-distribution. Note, that it is expressed in terms of ν, the number of degrees of freedom, rather than the number of samples $n = \nu + 1$.

It is instructive to consider the limiting cases of the distribution. In particular, for $n = \nu + 1 = 2$ we recover the Lorentz, Breit-Wigner, or Cauchy-distribution and for infinitely many degrees of freedom it can be shown that we recover the Gaussian distribution. The distributions for other values of ν lie between these extremes. Figure 7.5 shows $\Phi_\nu(t)$ for $\nu = 1, 2$, and 100.

In order to assess the reliability of testing only a few samples we need to calculate the probability that our test statistic $t = (\bar{X}_n - \mu)/(S_n/\sqrt{n})$ lies within the central range of the distribution. If we specify this central range to lie within $\pm\hat{t}$ this probability is given by the integral

$$A(\hat{t}, \nu) = \int_{-\hat{t}}^{\hat{t}} \Phi_\nu(t)dt = 1 - I_x\left(\frac{\nu}{2}, \frac{1}{2}\right) \tag{7.34}$$

where $I_x(a, b)$ is the (regularized) incomplete beta function [4] and $x = \nu/(\nu + \hat{t}^2)$. The second equality follows from the substitution $s = t^2$ and the definition of the incomplete beta function as an integral with the same integrand [4]. On the left-hand side in Fig. 7.6 we show $A(\hat{t}, \nu)$ for $\nu = 1, 2, 5$, and 100 as a function of $\hat{t}$. The solid black line corresponds to the case $\nu = 100$, where $\Phi_\nu(t)$ approaches a Gaussian, which contains 95% of the distribution within two standard deviations and corresponds to $\hat{t} \approx 2$. The range containing a percentage, say 95%, of the distribution, defines the 95% *confidence level*. When taking fewer samples $n = \nu + 1$ the probability to find a value of t within $\pm\hat{t}$ is reduced compared to the Gaussian. For example, if we only test three samples ($\nu = 2$) the 95% confidence level is only reached at $|\hat{t}| \approx 4.3$. For convenience, we show the values $|\hat{t}|$ where $A(\hat{t}, \nu)$ as a function of ν reaches 95, 90, and 80% on the right-hand plot in Fig. 7.6. The solid black curve corresponds to the 95% confidence level and indeed we find the point mentioned above at $\nu = 2$ and $|\hat{t}| \approx 4.3$ on it. Furthermore, for large values of ν it approaches $\hat{t} \approx 2$ as expected for Gaussian distributions. Below the curve for the 95% level, we find the curves for 90 and 80%. They correspond to a smaller area around the center of the distribution function and consequently a smaller probability of finding a value of t closer to the center. We point out that for $\nu > 10$ the curves rapidly approach their asymptotic values, which agree with those of a Gaussian. This is the reason for the common method to use two standard deviations to specify the 95% confidence level.

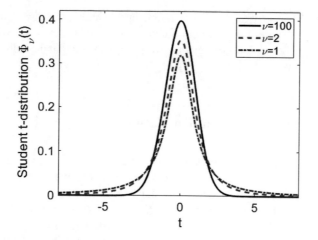

Fig. 7.5 The probability distribution function $\Phi_\nu(t)$ from (7.33) for $\nu = 1, 2$, and 100.

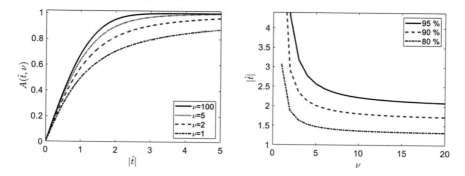

Fig. 7.6 The fraction $A(\hat{t}, \nu)$ for the distribution $\Phi_\nu(t)$ for $\nu = 1, 2, 5$, and 100 of contained within $\pm\hat{t}$ (left) and the value of $|\hat{t}|$ that contains 95, 90, and 80% of $A(\hat{t}, \nu)$ as a function of ν (right)

Let us come back to William Gosset at Guinness and ensure the quality of the delivery of a hundred bags of barley. Since we are lazy, we only test $n = 3$ randomly selected bags that have quality indicators $x = 16, 12$, and 17. So, what is the range of values x that encloses the true mean μ with 90% confidence? To find out, we use (7.24) and calculate $\bar{X}_3 = 15$ and $S_3 \approx 2.6$ from the three samples. From the right-hand plot in Fig. 7.6 we find $|\hat{t}| \approx 2.92$ for $\nu = n - 1 = 2$ degrees of freedom at the desired 90% confidence level. From (7.27) we now determine the corresponding range of μ to be $\mu = \bar{X}_3 \pm |\hat{t}| S_3/\sqrt{3} \approx 15 \pm 4.5$, which is a rather large range. We therefore decide to take a fourth sample, which turns out to be $x = 18$. Repeating the above calculation, we now find the 90% range to be $\mu \approx 15.8 \pm 3.1$. Maybe we should take a fifth sample, but that is left as an exercise.

In this section, we determined the range that contains the "true" mean μ of a distribution with a certain level of confidence. In the next section we will address a related problem: the validation or rejection of a hypothesis about the value a parameter is expected to have.

7.6   Hypothesis Testing and p-Values

In a regression analysis we might wonder whether we really need to include a certain fit parameter $\hat{X}$ or whether the model works equally well when omitting it. We can address this problem by testing the hypothesis $\hat{X} = 0$. In particular, we reject the hypothesis $\hat{X} = 0$ if the test-statistics $t = X/\sigma(X)$ is very large and lies in the tails of the distribution. Here X is the value of a fitted parameter from (7.5) and $\sigma(X)$ is its error bar, extracted from (7.7). In particular, if we find a value of t that lies in the tails containing 10% probability, as shown by the red areas in Fig. 7.7, we say that "the hypothesis $\hat{X} = 0$ is rejected at the 10% level."

Once we have determined the test-statistics t from the regression analysis we can calculate the probability—the p-value—of finding an even more extreme value as

$$p = \int_{t}^{\infty} \Phi_{v}(t')dt' \, , \tag{7.35}$$

where we assumed that t was positive. If it is negative we have to integrate from $-\infty$ to t instead. The parameter v is the number of degrees of freedom $v = n - m$ of the regression, where n and m are defined in the context of (7.2).

Fig. 7.7 Student's t-distribution for one degree of freedom $v = n - m = 1$ with the tails, containing 5% each, indicated by the red area. The tails begin at $|\hat{t}| = 6.31$. If the actually observed t lies in the tail region, the hypothesis $\hat{X} = 0$ is rejected at the 10% level

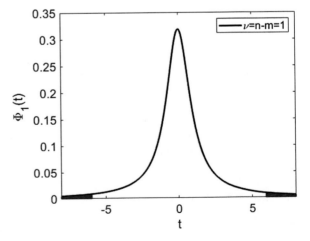

For an example let us return to the question posed in the introduction to this chapter whether all fit parameters are really necessary. Therefore we consider fitting data to a straight line as shown in Fig. 7.1. It appears reasonable to fit the linear dependence $y_i = a + bs_i$ to the dataset but we might even consider a second-order polynomial $y_i = a + bs_i + cs_i^2$, which will follow the data points even closer and result in a smaller χ^2, because we have the additional parameter c to approximate the dataset. But since we expect the data to lie on the straight line we state the hypothesis that $c = 0$. We test it by fitting the second-order polynomial to the dataset, then determine the error bars $\sigma(c)$ of the fit parameter c using (7.7), and finally calculate the test statistics $t = c/\sigma(c)$. If we have many degrees of freedom $\nu = n - m \gg 1$, we can approximate $\Phi_\nu(t)$ by a Gaussian and check whether t is larger than 2, which would indicate that our hypothesis $c = 0$ is rejected at the 10% level. Conversely, if t is smaller, we corroborate the hypothesis that c is consistent with zero and we might as well omit it from the fit.

The discussed method works well to test hypotheses about individual fit parameters, but occasionally we have to figure out whether we can omit a larger number of fit parameters at the same time. This is the topic of the following section.

7.7 F-Test

In the previous section we used the t-statistic to determine whether one coefficient in a regression model is compatible with zero and can be omitted. This works well if there is only a single obsolete coefficient, but might fail if several of the coefficients are significantly correlated. In that case the error bars for each of the coefficients is large and leads to the conclusion that the coefficient can be omitted, but the real origin of the problem is that the fitting procedure fails to work out whether to assign the uncertainty to one or the other(s) of the correlated coefficients. This leads to serious mis-interpretations of fit results.

One way to solve the dilemma with several potentially correlated coefficients and to determine which ones to omit is the F-test. In this test the contribution of a group of p fit parameters to the χ_p^2 is determined. We define this by the squared difference of the N "measurement" values y_i and the regression model, normalized to the measurement error σ_i, thus $s_i = (y_i - \sum_{j=1}^{p} A_{ij}x_j)/\sigma_i$. For χ_p^2 we then obtain

$$\chi_p^2 = \sum_{i=1}^{N} s_i^2 = \sum_{i=1}^{N} \left(\frac{y_i - \sum_{j=1}^{p} A_{ij}x_j}{\sigma_i} \right)^2 . \tag{7.36}$$

In the next step we increase the number of fit parameters to $q > p$ and test whether χ_q^2 is significantly smaller than χ_p^2. In order to quantify this improvement, we introduce the F-statistic

$$f = \frac{(\chi_p^2 - \chi_q^2)/(q - p)}{\chi_q^2/(N - q)} . \tag{7.37}$$

Here the denominator is the χ_q^2 per degree of freedom of the "larger" fit, whereas the numerator is the difference $\chi_p^2 - \chi_q^2$ per additional degree of freedom, which come from the additional $q - p$ fit parameters. The F-value thus measures the relative reduction of the χ^2 when increasing the number of fit parameters. We point out that the denominator depends on $N - q$ squared random number s_i and the numerator on the $q - p$ additional s_i. In Appendix A we motivate that the random numbers in the numerator and those in the denominator are independent, which allows us to derive the distribution of the F-statistic as the ratio of two independent χ^2-distributions; one with $n = q - p$ degrees of freedom, the other with $m = N - q$ degrees of freedom. For easy reference, we again display the χ^2-distribution function, already shown in (7.22)

$$\psi_n(x) = \frac{1}{2^{n/2}\Gamma(n/2)} x^{n/2-1} e^{-x/2} .$$

We will use it once to describe the χ^2–variable x with $n = q - p$ degrees of freedom that appears in the numerator of (7.37) and once for the χ^2-variable y for the $m = N - q$ degrees of freedom in the denominator. Using these variables the F–statistic f is given by

$$f = \frac{x/n}{y/m} = \frac{m}{n}\frac{x}{y} \quad \text{with} \quad m = N - q \quad \text{and} \quad n = q - p . \tag{7.38}$$

Our next task is to calculate the distribution function $\Psi(f)$ for the f from

$$\Psi_{n,m}(f) = \int_0^\infty dy\,\psi_m(y) \int_0^\infty dx\,\psi_n(x)\delta(f - mx/ny)$$

$$= \frac{n}{m} \int_0^\infty dy\,\psi_m(y)\psi_n(nyf/m)y , \tag{7.39}$$

where we used that the χ^2–variable y is positive and that $\partial f/\partial x = m/ny$. Again, the delta function collects all values of x and y that result in $f = mx/ny$. Inserting the definition of ψ_m and ψ_n we obtain

$$\Psi_{n,m}(f) = \frac{n/m}{2^{m/2+n/2}\Gamma(m/2)\Gamma(n/2)} \left(\frac{nf}{m}\right)^{n/2-1} \int_0^\infty y^{(m+n)/2-1} e^{-y(1+nf/m)/2} dy$$

$$= \frac{\Gamma((m+n)/2)}{\Gamma(m/2)\Gamma(n/2)} \left(\frac{n}{m}\right)^{n/2} \frac{f^{n/2-1}}{(1 + nf/m)^{(n+m)/2}} , \tag{7.40}$$

Fig. 7.8 The F−probability distribution function $\Psi_{n,m}(f)$ from (7.42) for a few combinations of n and m.

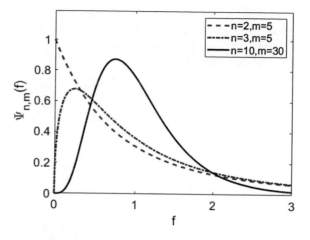

where we substitute $z = (1 + nf/m)y/2$ in the second step and then recognize the remaining integral as a representation of a Gamma-function with argument $(n + m)/2$. Note that the specific combination of Gamma-functions can be expressed as a Beta-function $B(p, q)$ with [4]

$$B(p, q) = \frac{\Gamma(p)\Gamma(q)}{\Gamma(p + q)} , \qquad (7.41)$$

which allows us to write the F−distribution in the form

$$\Psi_{n,m}(f) = \frac{1}{B(n/2, m/2)} \left(\frac{n}{m}\right)^{n/2} \frac{f^{n/2-1}}{(1 + nf/m)^{(n+m)/2}} \qquad (7.42)$$

that clearly shows that it depends only on the number of degrees of freedom $m = N - q$ and $n = q - p$. We can now use it to assess whether we need to include n additional fit parameters in a model and whether this added complexity is worth the effort.

Figure 7.8 shows the F−distribution function for a few values of n and m. In the first two cases we have a small number of degrees of freedom $N - q = m = 5$ where we fit q parameters to N data points. We then compare what distributions of our test-characteristic f we can expect, if we add n additional fit parameters The dashed blue line corresponds to a situation, where we add $n = 2$ additional fit parameters. We see that the distribution function is peaked near zero, which indicates that small values of f are very likely. Adding a third fit parameter ($n = 3$) causes $\Psi_{3,5}(f)$ to assume the shape indicated by the dot-dashed red line. Now very small values near zero are less likely and the distribution shows a peak. The solid black line in Fig. 7.8 illustrates a case where we add $n = 10$ additional fit parameters to a fit that originally had $N - q = m = 30$ degrees of freedom. We observe that the peak of the distribution moves towards $f = 1$ but is rather broad and shows significant tails.

In order to intuitively asses whether a found f-statistics is likely or not, we calculate the probability that its value is even smaller or larger, depending whether we find a particularly small or large value of f. The probability can be expressed in terms of the cumulative distribution function of $\Psi_{n,m}(f)$ by

$$\int_0^f \Psi_{n,m}(f')df' = \frac{1}{B\left(\frac{n}{2}, \frac{m}{2}\right)} \int_0^{\frac{n}{m}f} \frac{t^{\frac{n}{2}-1}dt}{(1+t)^{\frac{n+m}{2}}} \tag{7.43}$$

$$= \frac{1}{B\left(\frac{n}{2}, \frac{m}{2}\right)} \int_0^{\frac{nf}{m+nf}} u^{\frac{n}{2}-1}(1-u)^{\frac{m}{2}-1}du = I_{\hat{x}}\left(\frac{n}{2}, \frac{m}{2}\right)$$

with $\hat{x} = nf/(m+nf)$ and where we used the substitution $t = u/(1-u)$ in the second equality. $I_{\hat{x}}(a, b)$ is the (regularized) incomplete beta function [4], we already encountered in (7.34), and $B(a, b)$ is the beta function [4]. Now, to assess the probability, the p-value $p(f)$, of finding an even smaller value f is given by $p(f) = I_{\hat{x}}(n/2, m/2)$. Note that $\hat{x}$ depends on f. Conversely, the probability of finding a larger value, which is relevant on the on the right-hand side of the maximum, is given by $p(f) = 1 - I_{\hat{x}}(n/2, m/2)$.

Rejecting hypotheses works in the same way as discussed above in Sect. 7.6. If an F-value f, computed from data, lies in the tails of the distribution and exceeds the value for a 10% tail-fraction, we say that the hypothesis is rejected at the 10% level.

7.8 Parsimony

In philosophy, a *razor* is a criterion to remove explanations that are rather improbable. The image of shaving off something unwanted, for example, one's beard, comes to mind. One well-known example is *Occam's razor*, which, translated from Latin, reads "plurality should not be posited without necessity." Today, it is often rephrased as "the simpler solution is probably the correct one." Applied to the our regression analysis and fitting parameters to models it guides us to seek models with as few fit parameters as possible, which is also called the principle of *parsimony*. It helps us to avoid adding unnecessary parameters to a model that may lead to *over-fitting* which causes the model parameters to be overly affected, or over-constrained, by the noise in the data. Such a model then works very well with the existing data set with its particular noise spectrum, but its predictive power to explain new data is limited.

A classical example of over-fitting is the fit of a polynomial of degree $n-1$ to n data points. With many data points the polynomial is of very high order. It perfectly fits the data set and makes the χ^2 of the fit to zero, but outside the range of the original data set large excursions typically occur. Any additional data point is unlikely to be well-described by this highly over-constrained polynomial.

A further reason to avoid too many fit parameters is that groups of parameters are highly correlated and, again, this degeneracy is heavily affected by noise. It is therefore advisable to construct a model with the least number of fit parameters—to be parsimonious, in other words. Using fewer parameters therefore makes models more *robust* against the spurious influence of noise.

Now that we have the methods to fit parameters to models and assess their validity and robustness, we can take a closer look at time series and extract useful information from the raw data.

Exercises

1. In an experiment you recorded the parameter y as you changed another parameter s in Table 7.1. (a) You expect a linear dependence between s and y and therefore set up (7.1), use the methods from Sect. 7.1, and determine the slope a and intercept b of a straight-line fit to the data. If you assume that all error bars are $\sigma = 1$, what values do you obtain for a and b and what are the error bars of these fit parameters. (b) You realize that you were less careful when recording the data points in the range $-1 < s < 4$. You therefore double the error bars for the corresponding data points, as indicated by the last row in the table, and redo the fit. What values do you obtain for a and b and what are their error bars?
2. Calculate the R-value of the data and fit from Exercise 1a.
3. Use (7.5) and (7.6) to prove that (7.7) is correct.
4. If x is a Gaussian random variable, calculate the probability distribution functions of (a) $y = x - a$, (b) $y = bx$, and (c) $y = cx^2$.
5. You know that the data in the file ex7_5.dat, available from the book's web page, comes from a process that can be fitted by a polynomial of third order. (a) Plot the data. (b) Find the coefficients of a third-order polynomial in a regression analysis. (c) Estimate the error bars σ_y of the y–values (the "measurements") from the rms deviation of your fit-polynomial to the data points. Note that is a very heuristic approach to estimate error bars and can be criticised! (d) Calculate the covariance matrix, based on your estimate of the error bars σ_y, and deduce the error bars of the polynomial coefficients. Is there a coefficient that is so small

Table 7.1 Parameter s and measurement value y for Exercise 1. The last row shows the error bars for (b)

s	-2	-1	0	1	2	3	4	5
y	-7.0	-3.5	-3.3	0.1	1.6	0.3	1.5	5.5
σ	1	2	2	2	2	2	2	1

as to be consistent with zero. (e) Determine its F-value $\hat{f}$ and the probability to find an F-value that is even larger than $\hat{f}$.

6. Lets pursue the quality control at Guinness from the end of Sect. 7.5 and assume that we test a fifth bag of barley with the test result $x = 13$. What is the 90% confidence range now? If we were to be satisfied with an 80% confidence level, what would that range be?

References

1. US Bureau of Labor Statistics data on educational attainment (2018). https://www.bls.gov/emp/tables/unemployment-earnings-education.htm
2. Novel Coronavirus COVID-19 (2019-nCoV) Data Repository by Johns Hopkins CSSE (2019). https://github.com/CSSEGISandData/COVID-19
3. H. Hethcote, The mathematics of infectious diseases. SIAM Rev. **42**, 599 (2000)
4. M. Abramowitz, I. Stegun, *Handbook of Mathematical Functions* (Dover, New York, 1972)
5. G. Casella, R.L. Berger, *Statistical Inference*, 2nd edn. (Brooks/Cole, 2008)

Chapter 8
Time Series

Abstract This chapter introduces the Box-Jenkins methodology to analyze time-series data, which first removes trend and seasonality, before trying to explain the residual time series through moving average or auto-regressive models. In order to determine the order of a process, autocorrelation and partial autocorrelation functions are introduced. They are then used to construct a suitable model for the dynamics. All concepts are illustrated with environmental data. Later in the chapter, the basic ideas behind forecasting applications are developed. A discussion of various types of models, ARIMA, EWMA, and GARCH among them, concludes the presentation.

Raw data in physics, finance, and other fields are often produced at a constant rate. Examples are share prices that are updated daily, hourly, or even by the fraction of a second and we seek to extract information from the data. We might either try to

- determine a model to *characterize the dynamics* of the system;
- predict or *forecast* the next data points;
- deduce inherent characteristics of the data, for example its *volatility*.

In the following sections, we will address these points one at a time. Inspired by the discussion from the NIST web site [1, 2], we base our discussion on measurements of the CO_2 concentration from the Mauna Loa observatory in Hawaii. Time series of the averaged monthly measurements from 1958 until today are available in the file co2_mm_mlo.txt from [3]. We reformat the data in this text file, extract data for the period between 1995 and 2008, and generate a file that contains three columns: the date in decimal form, the averaged monthly CO_2 concentration expressed in μmole/mole of dry air, and the month as a number (Jan $= 1,\dots,$ Dec $= 12$). We show the first few lines to illustrate that format.

Electronic supplementary material The online version of this chapter
(https://doi.org/10.1007/978-3-030-63643-2_8) contains supplementary material, which is
available to authorized users.

Fig. 8.1 The CO_2
concentration at Mauna Loa
from 1995 until 2008 [3] and
a linear fit to the data

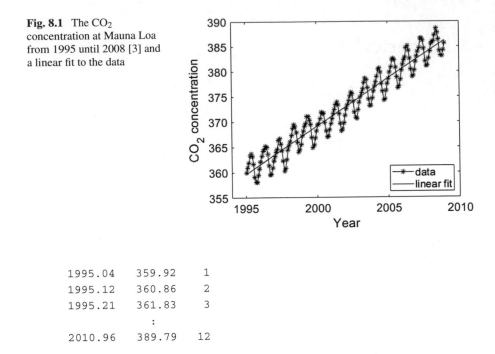

```
1995.04      359.92      1
1995.12      360.86      2
1995.21      361.83      3
               :
2010.96      389.79     12
```

We can obtain a first impression of the data by simply plotting the data points, as
shown in Fig. 8.1. We observe a periodic—seasonal—variation on top of a linearly
increasing base line—a trend—with some residual fluctuations remaining. Our task
is to first describe the obvious variations (trend and possibly seasonality) and then
build a model that accounts for the residual fluctuations in the sense that we want
construct a dynamic model or a (digital) filter (FIR or IIR, finite or infinite impulse
response) that transforms white noise to the residual fluctuation pattern. But first
things first: let us remove trend and seasonality.

8.1   Trend and Seasonality

In our sample data the trend appears to be linear, but in other cases it could be quadratic
or have any other functional dependence on time or sample number. Since it appears
to be linear here, we fit a straight line of the type $y = p_1 + p_2 x$ with intercept p_1
and slope p_2 to the raw data, which is shown as the straight line in Fig. 8.1. We then
subtract that line from the raw data and show the result in Fig. 8.2, where the trend
is removed and the seasonal variations are the most prominent remaining feature.

The seasonal variations become apparent when plotting the CO_2 concentration
after trend removal as a function of the month, which is what we show in Fig. 8.3,
where there are several data points at each month coming from the different years.
The sinusoidal oscillation is clearly visible and that months of different years cluster

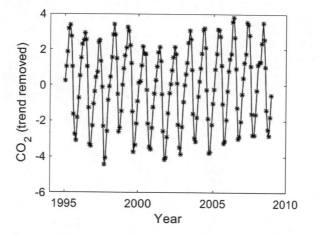

Fig. 8.2 The CO_2 concentration with trend removed

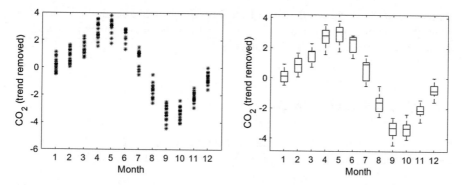

Fig. 8.3 The CO_2 concentration with trend removed, plotted versus the month number. Left are the raw data points and right a box-plot

around the same value. The clustering can be made very obvious by employing a *box-plot*, shown for the same data on the right-hand side in Fig. 8.3. For each month, it displays the median as a red line and the central 50% of the data points (25–75% percentile) as a box and the extreme values are shown as short horizontal markers. It permits us to judge the quality of the periodicity and whether the month-to-month variations are significant compared to the intra-month spread, which it clearly is.

One way to remove most of the seasonality is to difference the data with the lag of the seasonality, provided it is known. In this procedure a new time series s_k is produced by subtracting the data of the original series at point $k - 12$ from that at point k, or $s_k = x_k - x_{k-12}$, where x_k is the original time series. The resulting time series after differencing is shown in Fig. 8.4. We see that the seasonality is removed.

We observe that, after removing trend and seasonality, the resulting time series still does not resemble white noise. Thus, our task is to construct a filter that transforms white noise to the time series shown in Fig. 8.4. If we succeed in doing so, we have

Fig. 8.4 The CO_2
concentration with trend
removed and differenced
with lag 12 to remove
seasonality

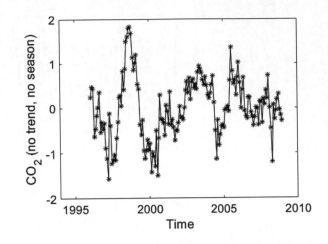

characterized the system to the best of our knowledge, because all the information
about the dynamics is encoded in the filter coefficients and all we can know about
the white noise that excites the system, is its rms amplitude.

8.2   MA, AR, and ARMA

Let us start by specifying the white noise through its time series ε_i. This describes a
series of shocks ε_j that excite a system. The statistics of the shocks is given by

$$\langle \varepsilon_i \rangle = 0 \quad \text{and} \quad \langle \varepsilon_i \varepsilon_j \rangle = \sigma^2 \delta_{ij} , \tag{8.1}$$

where the angle brackets denote average over the ensemble of shocks or over time.
This basically means that the shocks have zero mean, variance σ^2, and are uncorrelated from shock to shock. We assume that a stream of these shocks excites our
system, which we model as a process that extracts particular features from the white
noise and is modeled as a filter.

The simplest model is a *moving average* or MA process, which multiplies a number
of $n + 1$ consecutive shocks ε_i with fixed numbers a_j and outputs their weighted sum

$$y_i = \sum_{j=0}^{n} a_j \varepsilon_{i-j} . \tag{8.2}$$

Note that the output y_i is the scalar product of the filter coefficients and the last
$n + 1$ shocks. We can visualize this as the filter coefficients providing a window
through which we observe the sequence of shocks ε_i. As an example, consider the
MA process with all $n + 1$ coefficients equal to $a_k = 1/(n + 1)$, which calculates the

moving average of $n + 1$ consecutive shocks. The notation in (8.2) makes it easy to understand why the process is named "moving average." In the literature, however, a notation is commonly used where the first coefficient is set to unity and the remaining coefficients are denoted by θ with their sign reversed. A general MA(q) process is therefore characterized by

$$y_i = \varepsilon_i - \theta_1 \varepsilon_{i-1} - \cdots - \theta_q \varepsilon_{i-q} \tag{8.3}$$

with the coefficients θ_k related to the coefficients a_j through $a_0 = 1$ and $a_j = -\theta_j$ for $j > 0$.

We point out that in electrical engineering (8.2) is called a finite-impulse response (FIR) filter, provided that the random shocks ε_j are replaced by a sampled input signal x_j. The output signal y_i is then given by the weighted average of the most recent input samples x_j, where the filter coefficients a_j are the weights applied to the respective samples. The name "finite-impulse response" describes the feature, that a particular input samples x_j requires $n + 1$ time steps to trickle through the filter. For example, an input signal with a single non-zero value $x_k = 1$ will produce the filter coefficients as output signal y_i.

A slightly more complex filter is an infinite-impulse response (IIR) filter, which is based on feeding a fraction of the output signal back to the input. Its functional dependence is given by

$$y_i = \sum_{j=1}^{n} b_j y_{i-j} + a_0 x_i , \tag{8.4}$$

where the coefficients b_k describe the weights of how much of the previous output signals y_{i-j} is fed back to the input. The name "infinite-impulse" describes the feature that combinations of coefficients can cause a single non-zero input sample to result in an output signal with infinite duration.

An example of an IIR filter is the *exponentially weighted moving average* (EWMA) filter, which is given by

$$y_i = \frac{1}{m+1} (m y_{i-1} + x_i) , \tag{8.5}$$

related to the coefficients in (8.4) through $b_1 = m/(m + 1)$ and $a_0 = 1/(m + 1)$. The interpretation is straightforward. The output consists of a heavily weighted previous sample (with weight $m/(m + 1)$) and a weakly weighted new sample update (with weight $1/(m + 1)$) in case m is moderately large. This filter is also a low-pass filter, because it smoothes fluctuations of the input signal x_i over approximately m samples. IIR and FIR filters are widely used in digital data-acquisition applications to clean up noisy signals. But let us return to the discussion of statistical processes that are excited by random shocks ε_j.

In the time-series literature the processes, corresponding to IIR filters from electrical engineering, are called *autoregressive* (AR) processes, which describe the feeding back of the output to the input. Again the nomenclature varies a little bit and we write

$$y_i = \phi_1 y_{i-1} + \cdots + \phi_p y_{i-p} + \varepsilon_i = \sum_{j=1}^{p} \phi_j y_{i-j} + \varepsilon_i \qquad (8.6)$$

to denote an AR(p) process.

It is not surprising that the combined effect of a moving average process and an autoregressive process is called an ARMA(p, q) process and its functional description is the sum of the effect of a MA and an AR process

$$y_i = \phi_1 y_{i-1} + \cdots + \phi_p y_{i-p} + \varepsilon_i - \theta_1 \varepsilon_{i-1} - \cdots - \theta_q \varepsilon_{i-q} , \qquad (8.7)$$

where we use the conventionally used variables θ_j and ϕ_j to denote the coefficients describing this ARMA(p, q) process. Here we only consider *stationary* processes in the sense that the coefficients θ_j and ϕ_k are constant for the entire duration of the time series that we record.

Let us now return to the CO_2 time series y_i, shown in Fig. 8.4, and investigate the suitability of the processes discussed in this section to characterize that time series. We therefore need to to determine the coefficients θ_j and ϕ_j that generate the y_i from shocks ε_i, as defined in (8.1). But first we need to find out which orders of processes p and q are suitable. This is facilitated by the autocorrelation function (ACF) and partial autocorrelation (PACF) function.

8.3 Auto-Covariance and Autocorrelation

We first consider the MA(q) process from (8.3) and calculate the *auto-covariances* γ_j, which are the averaged sums of the time series y_i, weighted by the time series shifted by j samples y_{i-j}, where we assume that undefined samples are zero

$$\gamma_j = \langle y_i y_{i-j} \rangle , \qquad (8.8)$$

where the angle brackets denote an average over different realizations of the noise process ε_i. Note that the auto-covariances γ_j are symmetric in the sense that $\gamma_j = \gamma_{-j}$.

In order to explore how these quantities can be useful, let us calculate the autocorrelation γ_j for an MA(q) process given by (8.3) with $q = 1$, thus $y_i = \varepsilon_i - \theta_1 \varepsilon_{i-1}$. For γ_0 we then obtain

$$\begin{aligned} \gamma_0 &= \langle (\varepsilon_i - \theta_1 \varepsilon_{i-1})^2 \rangle \\ &= \langle \varepsilon_i^2 - 2\theta_1 \varepsilon_i \varepsilon_{i-1} + \theta_1^2 \varepsilon_{i-1}^2 \rangle \\ &= \sigma^2 (1 + \theta_1^2) , \end{aligned} \qquad (8.9)$$

where we use (8.1) to evaluate the average over the shocks. For the other auto-covariances γ_j we find

$$
\begin{aligned}
\gamma_1 &= \langle (\varepsilon_i - \theta_1 \varepsilon_{i-1})(\varepsilon_{i-1} - \theta_1 \varepsilon_{i-2}) \rangle = -\theta_1 \sigma^2 \\
\gamma_2 &= \langle (\varepsilon_i - \theta_1 \varepsilon_{i-1})(\varepsilon_{i-2} - \theta_1 \varepsilon_{i-3}) \rangle = 0
\end{aligned}
\tag{8.10}
$$

and all higher γ_j with $j > 1$ are zero as well. This is noteworthy, because it will help us to estimate and characterize the order of the process when calculating estimates of the covariances from a given time series.

The *autocorrelations* (ACF) ρ_j are given by the auto-covariances γ_j, divided by γ_0

$$
\rho_j = \frac{\gamma_j}{\gamma_0} ,
\tag{8.11}
$$

which makes the autocorrelations dimensionless. They neither depend on the physical units of the process nor the numerical magnitude of the time series values y_i.

Generalizing to higher order MA(q)-processes we find that γ_0 and the γ_j of higher-order can be calculated in the same way as before. They are given by

$$
\begin{aligned}
\gamma_0 &= \langle (\varepsilon_i - \theta_1 \varepsilon_{i-1} - \cdots - \theta_q \varepsilon_{i-q})^2 \rangle \\
&= \sigma^2 (1 + \theta_1^2 + \cdots + \theta_q^2) \\
\gamma_j &= \sigma^2 (-\theta_j + \theta_1 \theta_{j+1} + \cdots + \theta_{q-j} \theta_q) \quad \text{for } 1 < j < q \\
\gamma_j &= 0 \qquad\qquad\qquad\qquad\qquad\qquad\quad \text{for } j > q.
\end{aligned}
\tag{8.12}
$$

Again, we see that all coefficients for $j > q$ vanish. The ACF ρ_j are defined in the same way as for the $q = 1$ case by normalizing the γ_j by γ_0 as in (8.11). In practice we calculate estimates for the γ_j from the time series data y_i such as from Fig. 8.4, plot them and inspect whether some distinct peaks show up and whether the covariances or correlations vanish beyond a cutoff, which we can interpret as the order q of the process.

Let us briefly discuss what "vanish" or "small" means in this context with a very heuristic and qualitative discussion. The auto-covariances are calculated from the sum of n products $y_i y_{i-j}$ and we can intuitively understand that an increasing number of data points n will improve the precision to which the auto-covariances can be calculated. Since the random component of the y_i are uncorrelated, we can assume that the product of two such parameters are uncorrelated. The random component of coefficient γ_j will therefore grow like $\sqrt{n}$ while the magnitude of γ_j and also γ_0 will grow linearly with n. From these two dependencies on the length of the time series n we can intuitively understand that the random component of the autocorrelations ρ_j, will decrease as $1/\sqrt{n}$, where this indicates the standard error of the random component. See Quenouille [4] for a more detailed discussion of this dependency. But we can use this information to specify what we mean by "small," namely that the coefficient ρ_j is small compared to two times the standard error or $\pm 2/\sqrt{n}$, which indicates the band that describes the 95% confidence level in the sense discussed in Sect. 7.5.

After considering the MA processes and how to interpret the magnitude of the coefficients in autocorrelation plots, we need to investigate the auto-covariances and autocorrelations of the $AR(p)$ processes from (8.6). As before, we first consider the simplest process with $p = 1$, which is given by $y_i = \phi_1 y_{i-1} + \varepsilon_i$. The auto-covariances are given by (8.8) and can be calculated by considering a system at rest $y_j = 0$ for $j < 0$ before the sequence of shocks starts to excite the system. After the first iteration we have $y_1 = \varepsilon_1$ and after the second iteration we have $y_2 = \varepsilon_2 + \phi_1 \varepsilon_1$ and after the third $y_3 = \varepsilon_3 + \phi_1 \varepsilon_2 + \phi_1^2 \varepsilon_1$. Obviously, the value of y_i depends on the contemporary and all previous shocks ε_{i-j} with $i \geq 0$ where each shock is weighted by ϕ_1^j. Generalizing this we can write

$$y_i = \varepsilon_i + \phi_1 \varepsilon_{i-1} + \phi_1^2 \varepsilon_{i-2} + \cdots = \sum_{j=0}^{\infty} \phi_1^j \varepsilon_{i-j} \,. \tag{8.13}$$

In this form we can calculate the auto-covariances to be

$$\begin{aligned} \gamma_0 &= \langle y_i y_i \rangle = \sigma^2 (1 + \phi_1^2 + \phi_1^4 + \cdots) \\ \gamma_j &= \langle y_i y_{i-j} \rangle = \sigma^2 \phi_1^j (1 + \phi_1^2 + \phi_1^4 + \cdots) \end{aligned} \tag{8.14}$$

and, provided that $|\phi_1| < 1$, we can sum the series and arrive at

$$\gamma_0 = \frac{\sigma^2}{1 - \phi_1^2} \quad \text{and} \quad \gamma_j = \frac{\sigma^2 \phi_1^j}{1 - \phi_1^2} \,. \tag{8.15}$$

The autocorrelations ρ_j then turn out to be

$$\rho_j = \frac{\gamma_j}{\gamma_0} = \phi_1^j \,, \tag{8.16}$$

which implies that the autocorrelations ρ_j exponentially decay from $j = 0$ towards larger j, which is the signature of an $AR(1)$ process.

Returning to the data shown in Fig. 8.4, we display the ACF for the CO_2 data after removing trend and seasonality in Fig. 8.5. The red lines denote the 95% confidence level around zero. We see that a large number of autocorrelations are highly relevant and no cutoff is visible. On the other hand there appears to be an exponential decay of the ρ_j as a function of the shift j which suggests that the underlying process is of the AR type.

The large number of non-negligible autocorrelations makes it difficult to identify the relevant parameters ϕ_j directly. To remedy this deficiency, we discuss a method to determine the most important parameters in the following section.

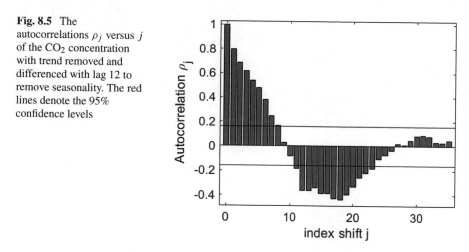

Fig. 8.5 The autocorrelations ρ_j versus j of the CO_2 concentration with trend removed and differenced with lag 12 to remove seasonality. The red lines denote the 95% confidence levels

8.4 Partial Autocorrelation Function

First, we determine relations between the experimentally determinable autocorrelations ρ_j and the sought-after parameters ϕ_k. Such a relation for AR(p) processes can be found by inserting the definition of the AR-process from (8.6) into the definition of the auto-covariances. First we consider γ_0

$$\gamma_0 = \langle y_i y_i \rangle = \phi_1 \gamma_1 + \cdots + \phi_p \gamma_p + \sigma^2 \quad \text{for } j = 0 , \tag{8.17}$$

where the term σ^2 comes from $\langle \varepsilon_i y_i \rangle = \langle \varepsilon_i \left(\varepsilon_i + \sum_{j=1}^{p} \phi_j y_{i-j} \right) \rangle = \sigma^2$. For γ_j with $j > 0$ we calculate

$$\begin{aligned} \gamma_j &= \langle (\phi_1 y_{i-1} + \cdots + \phi_p y_{i-p} + \varepsilon_i) y_{i-j} \rangle \\ &= \phi_1 \langle y_{i-1} y_{i-j} \rangle + \cdots + \phi_p \langle y_{i-p} y_{i-j} \rangle + \langle \varepsilon_i y_{i-j} \rangle \\ &= \phi_1 \gamma_{j-1} + \cdots + \phi_p \gamma_{j-p} \quad \text{for } j > 0, \end{aligned} \tag{8.18}$$

where $\langle \varepsilon_i y_{i-j} \rangle = 0$, because the shocks at later time i are uncorrelated to the time series value y_{i-j} that precede it.

Dividing the equations for γ_j by γ_0, we obtain the relation for the autocorrelations ρ_j

$$\rho_j = \phi_1 \rho_{j-1} + \phi_2 \rho_{j-2} + \cdots + \phi_p \rho_{j-p} . \tag{8.19}$$

with $\rho_0 = 1$ and $\rho_j = \rho_{-j}$. These equations are called the *Yule-Walker equations*. They show that the autocorrelations ρ_j obey the same recursion relation as the original time series, but without the random shocks ε_k.

If we write (8.19) for consecutive j, we obtain a system of equations for the AR(p) coefficients ϕ_j

$$\rho_1 = \phi_1 + \phi_2\rho_1 + \cdots + \phi_p\rho_{p-1}$$
$$\rho_2 = \phi_1\rho_1 + \phi_2 + \cdots + \phi_p\rho_{p-2}$$
$$\vdots \tag{8.20}$$
$$\rho_p = \phi_1\rho_{p-1} + \phi_2\rho_{p-2} + \cdots + \phi_p .$$

This set of equations can be cast into a matrix-valued equation with the result

$$
\begin{pmatrix} \rho_1 \\ \rho_2 \\ \vdots \\ \rho_{p-1} \\ \rho_p \end{pmatrix}
=
\begin{pmatrix}
1 & \rho_1 & \cdots & \rho_{p-1} \\
\rho_1 & 1 & \cdots & \rho_{p-2} \\
\vdots & \vdots & \ddots & \vdots \\
\rho_{p-2} & \rho_{p-3} & \cdots & \rho_1 \\
\rho_{p-1} & \rho_{p-2} & \cdots & 1
\end{pmatrix}
\begin{pmatrix} \phi_1 \\ \phi_2 \\ \vdots \\ \phi_{p-1} \\ \phi_p \end{pmatrix}
\tag{8.21}
$$

and, if the matrix is non-degenerate, we can determine the coefficients ϕ_j by inverting the matrix. We find

$$
\begin{pmatrix} \phi_1 \\ \phi_2 \\ \vdots \\ \phi_{p-1} \\ \phi_p \end{pmatrix}
=
\begin{pmatrix}
1 & \rho_1 & \cdots & \rho_{p-1} \\
\rho_1 & 1 & \cdots & \rho_{p-2} \\
\vdots & \vdots & \ddots & \vdots \\
\rho_{p-2} & \rho_{p-3} & \cdots & \rho_1 \\
\rho_{p-1} & \rho_{p-2} & \cdots & 1
\end{pmatrix}^{-1}
\begin{pmatrix} \rho_1 \\ \rho_2 \\ \vdots \\ \rho_{p-1} \\ \rho_p \end{pmatrix},
\tag{8.22}
$$

provided that we know the order p of the AR(p)-process. But we do not know this a-priori. However, after initially calculating the estimates of the autocorrelations, we can set up (8.22) for increasing $p = 1, 2, \ldots$, which results in a sequence of linear systems of order $p = 1, 2, \ldots$. At each step, we record the *coefficient of the highest order* p, which we call ϕ_{pp}. This coefficient is small if all information is already accounted for by the lower-order coefficients. Conversely, if it is large, it carries new information. By redoing the fit with progressively increasing order p and always saving the highest-order coefficient ϕ_{pp}, we progressively check whether the new highest coefficient carries new information. Moreover, if the order of the process that generated the time-series from which the autocorrelations are derived, is $\hat{p}$, all ϕ_{pp} with $p > \hat{p}$ should be zero or at least small. The coefficients ϕ_{pp} are called *partial auto correlation function* (PACF). Since they are small or zero beyond the value of $\hat{p}$, characteristic for the underlying process, they provide a cutoff that helps us to identify the largest p needed to adequately describe the process dynamics.

In Fig. 8.6 we display the PACF derived from the autocorrelations depicted in Fig. 8.5 for our CO_2 data with 95% confidence level around zero indicated as red lines. Here we find that the coefficients for $p = 1$ and 12 exceed the 95% confidence level significantly and those at $p = 7, 9$, and 13 only slightly. In the spirit of building a parsimonious model we dare to neglect the latter, because the lowest order model with $p = 1$ contains most of the dynamics and the $p = 12$ component is a residual seasonal

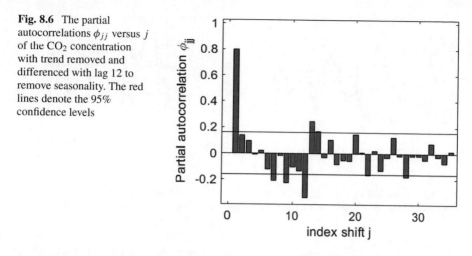

Fig. 8.6 The partial autocorrelations ϕ_{jj} versus j of the CO_2 concentration with trend removed and differenced with lag 12 to remove seasonality. The red lines denote the 95% confidence levels

component that escaped our earlier attempt to remove seasonality. Summarizing, our best guess for the dynamics is

$$y_i = \phi_1 y_{i-1} + \phi_{12} y_{i-12} + \varepsilon_i \tag{8.23}$$

and we will determine the parameters ϕ_1 and ϕ_{12} in the next section.

8.5 Determining the Model Coefficients

Once we know that the underlying process is auto-regressive and what coefficients are relevant we can set up a system of equations from which we then determine the coefficients ϕ_j. We simply write down copies of (8.23) for $i = 13, 14, \ldots$ up to the last data point n and assemble them in matrix form

$$\begin{pmatrix} y_{13} \\ y_{14} \\ \vdots \\ y_n \end{pmatrix} = \begin{pmatrix} y_{12} & y_1 \\ y_{13} & y_2 \\ \vdots & \vdots \\ y_{n-1} & y_{n-12} \end{pmatrix} \begin{pmatrix} \phi_1 \\ \phi_{12} \end{pmatrix}, \tag{8.24}$$

which is of the same form as (7.2) in Chap. 7. We solve it in the least square sense using the pseudo-inverse from (7.4)

$$\begin{pmatrix} \phi_1 \\ \phi_{12} \end{pmatrix} = \left(A^t A\right)^{-1} A^t \begin{pmatrix} y_{13} \\ y_{14} \\ \vdots \\ y_n \end{pmatrix}, \tag{8.25}$$

where A is the matrix with the two columns of shifted data points from (8.24).

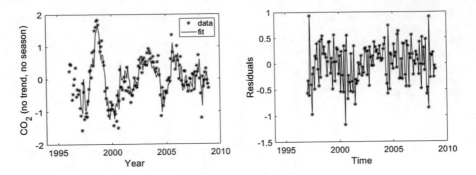

Fig. 8.7 On the left we show the data from Fig. 8.4 (asterisks) with the fitted time series (solid red line) and on the right-hand side we show the remaining differences between data and fit

We test how well our model works by showing both the approximation and the original data on the left-hand side in Fig. 8.7. We conclude that the overall trend of the data is reproduced quite well, considering the small number of fit parameters. The fit residuals are shown in the plot on the right-hand side in Fig. 8.7. We conclude that the procedure of first removing trend, then seasonality and finally identifying a particular model with the help of the ACF and PACF yields a parsimonious, yet satisfactory model based on two parameters, only.

In our example the analysis of ACF and PACF pointed to an AR model and fitting it was straightforward using (8.24) because the coefficients of an auto-regressive process depend on the known time series data y_i only. If the analysis had pointed to a MA model this had not been possible, because the fit parameters are given by de-convoluting the output and input data. But the input data are the shocks, which we, unfortunately, do not know.

8.6 Box-Jenkins

The procedure to analyze the time series used earlier in this chapter loosely follows the procedure that is commonly called the Box-Jenkins procedure [5]. The procedure is based on the following steps:

- Plot the data and verify that it is stationary, remove trend, if necessary. If all values are positive, consider using the logarithm of the values.
- Determine seasonality or periodicity from spectral analysis (Fourier transform, FFT), autocorrelation or other information, for example a-priori knowledge.
- Difference time series to remove seasonality and make it stationary.
- Determine the order of the ARMA process from autocorrelation and partial auto-correlation plot of the remaining time series.
- Determine the coefficients of the model.

- Check whether the model describes the data adequately. If the fit is unsatisfactory, possibly use more or different fit parameters.
- Once the model is established we can use it to forecast how the system evolves into the future.

There are standardized software tools to perform all these steps. They use much more sophisticated verification algorithms than those we used in this chapter. An extensive list of available software is maintained on Wikipedia under the heading of *Comparison of statistical packages.*

8.7 Forecasting

In the simplest case we can iterate the parsimonious model calculated in the earlier section (and assuming all future unknown shocks to be equal to their expectation value zero) to determine the forecast result. This works for both MA, AR, or ARMA models. The error bars of the predicted values, however, are easiest to evaluate in case we have an MA model. Therefore we first present a method that converts an $AR(p)$ or $ARMA(p, q)$ model into an $MA(\infty)$ model such that we can later present a unified method of forecasting the error bars. First, we introduce the lag operator $\hat{L}$ that is defined by the following relation on any sampled variable y_i

$$\hat{L} y_i = y_{i-1} , \tag{8.26}$$

which simply changes a sample to the previous one. Repeated application of $\hat{L}$ will produce earlier samples such that $\hat{L}^m y_i = y_{i-m}$. This notation allows us to write the general $ARMA(p, q)$ model from (8.7) as

$$\left(1 - \sum_{j=1}^{p} \phi_i \hat{L}^j\right) y_i = \left(1 - \sum_{j=1}^{q} \theta_j \hat{L}^j\right) \varepsilon_i , \tag{8.27}$$

and, provided that the polynomial in $\hat{L}$ does not have unit roots, we can divide both sides by $\left(1 - \sum_{j=1}^{p} \psi_i \hat{L}^j\right)$ resulting in

$$y_i = \left(1 - \sum_{j=1}^{\infty} \pi_j \hat{L}^j\right) \varepsilon_i \tag{8.28}$$

where the π_j can be calculated recursively, order by order, from the θ_i and the ϕ_k. Since we know all the actual values ε_j up to $j = m$ we can calculate the $n-$sample forecast value

$$\langle y_{m+n} \rangle = \left(1 - \sum_{j=n+1}^{\infty} \pi_j \hat{L}^j \right) \varepsilon_m \ . \tag{8.29}$$

Here we replace earlier, and unknown, values of ε_{m+k} for $k = 1 \ldots n$ by their expectation value, which is zero.

The error bar for the forecast is then based on the uncertainty of the values of the shocks from $k = m + 1$ to $k = m + n$ and the variance (or square of the error bar) $V(y, m + n)$ for y_{m+n} is given by

$$V(y, m + n) = \sigma^2(y_{m+n}) = \left(1 + \sum_{j=1}^{n-1} \pi_j^2 \right) \sigma^2 \tag{8.30}$$

where σ^2 is the variance of the ε_i as defined in (8.1). So, all the information about the error bars for the forecast values is embedded in the coefficients π_i.

All information, available to us in order to generate the forecast y_{n+m}, is a finite number of earlier values y_j for $j \le m$. Therefore, we use these available previous data points in order to predict the forecast $\hat{y}_{i+n}$ and its error bars, where we mark the future estimate $\hat{y}_{i+n}$ with a hat. To do so, we assume that the estimate of a future value $\hat{y}_{i+n}$, which lies n time steps in the future, is a linear function of the last known m samples y_{i-j} with $j = 0, \ldots, m - 1$

$$\hat{y}_{i+n} = \sum_{j=0}^{m-1} \alpha_j y_{i-j} \ . \tag{8.31}$$

This representation can be shown [6] to minimize the mean squared error. Note that the coefficients $\alpha_j = \alpha_j(n)$ depend in the forecasting distance n, though we will omit the argument to make the equations more readable. All information about the forecast is accounted for in the coefficients α_j such that the difference between the left-hand and right-hand side of (8.31) is uncorrelated to any of the earlier values y_{i-k}. This requirement leads to

$$\left\langle \left(\hat{y}_{i+n} - \sum_{j=0}^{m-1} \alpha_j y_{i-j} \right) y_{i-k} \right\rangle = 0 \qquad \text{for } k = 0, \ldots, m - 1 \ , \tag{8.32}$$

where the angle brackets denote ensemble average over many realizations of the time series. The previous equation can be rewritten as

$$\left\langle \hat{y}_{i+n} y_{i-k} \right\rangle = \sum_{j=0}^{m-1} \alpha_j \left\langle y_{i-j} y_{i-k} \right\rangle \qquad \text{for } k = 0, \ldots, m - 1. \tag{8.33}$$

Here we note that $\langle y_{i-j} y_{i-k} \rangle = \gamma_{j-k}$, where γ_k are the auto-covariances encountered in Sect. 8.3. For a given time series y_i, we can estimate the γ_j by calculating $\gamma_j = \frac{1}{n} \sum_{i=1}^{n} y_i y_{i-j}$ with unspecified values of y_i set to zero. The expression on the left-hand side of (8.33) equals $\langle \hat{y}_{i+n} y_{i-k} \rangle = \gamma_{n+k}$. Using these relations and writing the previous equation in component form we arrive at

$$
\begin{pmatrix} \gamma_n \\ \gamma_{n+1} \\ \vdots \\ \gamma_{n+m-1} \end{pmatrix} = \begin{pmatrix} \gamma_0 & \gamma_1 & \cdots & \gamma_{m-1} \\ \gamma_1 & \gamma_0 & \cdots & \gamma_{m-2} \\ \vdots & \vdots & \ddots & \vdots \\ \gamma_{m-1} & \gamma_{m-2} & \cdots & \gamma_0 \end{pmatrix} \begin{pmatrix} \alpha_0 \\ \alpha_1 \\ \vdots \\ \alpha_{m-1} \end{pmatrix} , \tag{8.34}
$$

which we invert in order to solve for the forecast coefficients $\alpha_k(n)$

$$
\begin{pmatrix} \alpha_0(n) \\ \alpha_1(n) \\ \vdots \\ \alpha_{m-1}(n) \end{pmatrix} = \begin{pmatrix} \gamma_0 & \gamma_1 & \cdots & \gamma_{m-1} \\ \gamma_1 & \gamma_0 & \cdots & \gamma_{m-2} \\ \vdots & \vdots & \ddots & \vdots \\ \gamma_{m-1} & \gamma_{m-2} & \cdots & \gamma_0 \end{pmatrix}^{-1} \begin{pmatrix} \gamma_n \\ \gamma_{n+1} \\ \vdots \\ \gamma_{n+m-1} \end{pmatrix} . \tag{8.35}
$$

Here we included the argument n to the coefficients $\alpha_k(n)$ as a reminder that the coefficients are specific to the forecast distance n. Using these coefficients in (8.31) will result in a forecast derived from the previously recorded values y_k. For the convenience in further calculations we introduce the following shorthand notation for the previous equation

$$
\boldsymbol{\alpha}(n) = \Gamma^{-1} \boldsymbol{\gamma}(n) , \tag{8.36}
$$

where $\boldsymbol{\alpha}(n)$ is the column vector as shown in (8.34), $\boldsymbol{\gamma}(n)$ the column vector on the right-hand side. The transpose of the vectors will be denoted by a superscripted letter 't'. Γ denotes the matrix in (8.34).

Now we will address the expected accuracy of the forecast by calculating the expectation value of the quadratic deviation of the forecast $\sigma^2(y_{i+n})$

$$
\sigma(\hat{y}_{i+n})^2 = \left\langle \left(\hat{y}_{i+n} - \sum_{j=0}^{m-1} \alpha_j y_{i-j} \right) \left(\hat{y}_{i+n} - \sum_{j=0}^{m-1} y_{i-k} \alpha_k \right) \right\rangle \tag{8.37}
$$

$$
= \langle \hat{y}_{i+n} \hat{y}_{i+n} \rangle - 2 \sum_{j=0}^{m-1} \alpha_j \langle \hat{y}_{i+n} y_{i-j} \rangle + \sum_{j=0}^{m-1} \sum_{k=0}^{m-1} \alpha_j \alpha_k \langle y_{i-j} y_{i-k} \rangle
$$

$$
= \gamma_0 - 2 \boldsymbol{\alpha}(n)^t \boldsymbol{\gamma}(n) + \boldsymbol{\alpha}(n)^t \Gamma \boldsymbol{\alpha}(n) ,
$$

where we used the abbreviations introduced above and, in particular $\Gamma_{jk} = \langle y_{i-j} y_{i-k} \rangle$. Inserting (8.36), we finally obtain

$$
\sigma(\hat{y}_{i+n})^2 = \gamma_0 - \boldsymbol{\gamma}(n)^t \Gamma^{-1} \boldsymbol{\gamma}(n) \tag{8.38}
$$

and the 95% confidence level bands around the forecast value $\hat{y}_{i+n}$ are given by $\hat{y}_{i+n} \pm 2\sqrt{\sigma(\hat{y}_{i+n})^2}$.

Inspired by reading about algorithmic trading methods in [7] let us explore the use of forecasting with a rather contrived example, where we use the day-to-day returns $r_d = S_{d+1} - S_d$ from stock S_d on day d, to predict the *sign* of the return on the following day $d + 1$. If the sign is positive, we keep the stock; if it is negative, we short-sell it. The hope is to turn a loss caused by a falling stock value into a gain. On the other hand, if the prediction is wrong, we turn the gain from stocks going up into a loss, because we short sell. We calculate the running tally of our portfolio by accumulating the "real" return r_{d+1} with the sign, predicted by the algorithm. To test it, we used data from Apple Inc. (AAPL) from March 27, 2018 until March 27, 2019, retrieved from [8], and show the result in Fig. 8.8. The solid black line shows the evolution of the stock value as it happened during the period. The dot-dashed blue curve is based on using only data from the last day ($m = 1$) to forcast the next, which basically means "if it went up, it will continue to go up." We observe that it mostly tracks the black line with little worse results. Especially after day 200 it misses the rise; likely the algorithm is fooled by small wiggles in the real data. The dashed red line is based on using the last four days ($m = 4$) to predict the following day. This prediction performs rather poor up to day 150, but works reasonably well during the subsequent 50 days, where the stock value falls. We attribute this to the fact that the black line with the real data shows moderately systematic oscillations that this algorithm correctly identified. Varying the horizon m, over which the prediction is based showed varying results, most om them lying in the range given by the two examples from Fig. 8.8. Also, testing with other stocks showed great variations. For stocks with fairly stable changes in their value, $m = 1$ worked best, but we observe that the algorithm is badly fooled by random stock motion. This is understandable, because using a forecasting method implicitly assumes that the underlying dynamics shows systematic behavior. Thus, if we believe in the efficient market hypothesis and that the stock evolution is based on a martingale process, we should be unable to predict tomorrows stock value, based on historic values. We therefore recommend not to use this algorithm with real money.

Most of the models we encountered so far were autoregressive. But there are more.

8.8 Zoo of Models

In Sect. 8.2 we already addressed MA, AR, and ARMA models, which were defined through (8.3), (8.6), and (8.7). In this section we will discuss a number of generalizations.

ARIMA
ARMA models with coefficients θ and ϕ only work satisfactorily if the underlying process is stationary, rather than having a trend or seasonal variations superimposed. We already observed and remedied this in Sect. 8.1, where we discussed the CO_2

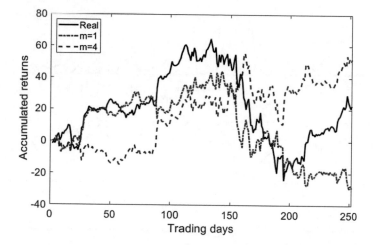

Fig. 8.8 The accumulated returns for Apple stocks as they relay happened (black) and simulated data using the algorithm, discussed in the text, for $m = 1$ (blue) and $m = 4$ (red)

data from Mauna Loa. There, we also found that differencing the data points, as suggested by the Box-Jenkins methodology, made the time series "more stationary." Thus, by adding one or more differencing steps

$$z_i = y_i - y_{i-1} , \tag{8.39}$$

we create a new time series z_i that is closer to being stationary than the original series y_i. This is then called an ARIMA($p, d = 1, q$)-model, if only a single differencing step is prepended, else d can also assume larger values. Since the inverse of differentiation is integration, and, at the end, the differentiation needs to be undone, the model is called an *auto-regressive integrated moving average,* or ARIMA-model. Note that the differentiation stage can be described through the lag-operator $\hat{L}$ from (8.26), which permits us to write

$$z_i = (1 - \hat{L})y_i \tag{8.40}$$

and the second derivative becomes

$$u_i = z_i - z_{i-1} = (1 - \hat{L})z_i = (1 - \hat{L})^2 y_i = y_i - 2y_{i-1} + y_{i-2} \tag{8.41}$$

where we observe that the last expression is indeed the second derivative, because it can be written as $u_i = (y_i - y_{i-1}) - (y_{i-1} - y_{i-2})$.

A wonderful example of using ARIMA models is discussed in [9, 10], where the authors analyze the impact of terrorist attacks on the revenue generated by tourism in Italy in the period from 1971 until 1988. They base their analysis on the data set ITALY.XLS, available from [11]. The file contains three columns with values

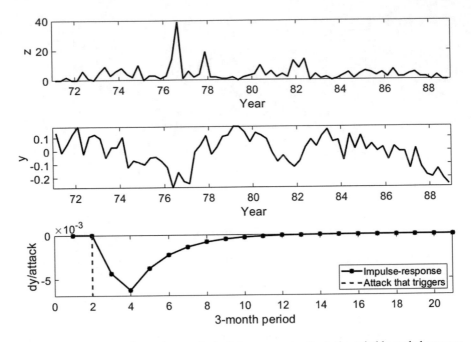

Fig. 8.9 The number of terrorist attacks in Italy per quarter (top), the suitably scaled revenue from tourist activities (middle) and the impulse response of a single terrorist attack on the revenue (bottom). The analysis is based on [9, 10]

for a three-month period: the time t, a quantity y related to the logarithm of the revenue generated by tourism, and z, the number of terrorist attacks during the quarter. We show the terrorist attacks z in the upper graph in Fig. 8.9 and y in the middle graph. Since tourism depends significantly on the time of the year, we first remove seasonality by introducing $y_n^* = y_n - y_{n-4}$ and $z_n^* = z_n - z_{n-4}$ before determining the model

$$y_n^* = ay_{n-1}^* + b_1 z_{n-1}^* + b_2 z_{n-2}^* \tag{8.42}$$

with one auto-regressive coefficient a and two moving-average coefficients b_1 and b_2. In [9] the authors carefully determine which coefficients are useful. We prepared a MATLAB script that first determines $b_1 = -0.0043$ and $b_2 = -0.0036$ from a least-squares fit and then subtracts the direct impact from the time series by calculating $u_n^* = y_n^* - b_1 z_{n-1}^* - b_2 z_{n-2}^*$. In a second step, $a = 0.60$ is determined from a regression analysis of $u_n^* = au_{n-1}^*$. With all coefficients of the model available, we calculate the lost revenue as a consequence of a single terrorist attack in the second quarter $z_2^* = 1$ and iterate (8.42) for a few iterations. Figure 8.9 shows this impulse-response. We see that the attack in the second quarter, shown as the dashed red line, triggers the reduction of the tourist revenue that reaches its maximum impact after two three-month periods and then slowly decays over a period of two to three years.

The authors of [9] then use the area under the curve to estimate the total financial damage that a single terrorist act caused.

But we will move on to other time-series models.

ARFIMA

If the differencing parameter d is a fraction rather than an integer, the model is called *autoregressive fractionally integrated moving average*, or ARFIMA(p, d, q)-model. These models are particularly useful if the underlying process has very long memory, which can be easily seen by writing the differencing steps in terms of the lag operator $\hat{L}$ and formally using the binomial expansion

$$(1 - L)^d = \sum_{i=0}^{d} \binom{d}{k} \hat{L}^k = 1 - dL + \frac{d(d-1)}{2!} \hat{L}^2 - \cdots \tag{8.43}$$

and we find that the series does not terminate for fractional values of d. Therefore, information from *much* earlier samples contributes to the present sample.

EWMA

We already encountered the exponentially weighted moving average method in (8.5). It is essentially an IIR filter that outputs the weighted average of the most-recent output and a new value. The equation is reproduced in the following equation on the left-hand side

$$y_i = \frac{1}{m+1} (m y_{i-1} + x_i) \quad \text{or} \quad \sigma_i^2 = \frac{1}{m+1} \left(m \sigma_{i-1}^2 + u_i^2 \right), \tag{8.44}$$

where we average m times the previous output value y_{i-1} with the new value x_i. In this way, old values are "forgotten" on a time scale of m samples and continuously updated by new values x_i. The equation on the right-hand side is constructed in the same way, but calculates continuously updated values of the volatility σ_i, whose day-to-day variation of the relative return u_j is given by

$$u_j = \frac{S_j - S_{j-1}}{S_{j-1}}. \tag{8.45}$$

Here S_j is, for example, the fluctuating stock value. We illustrate this in Fig. 8.10, which shows the daily returns u (blue) and the corresponding value of σ, averaged with $m = 20$ (red), for Apple Inc. (top) and for Coca-Cola (bottom) from March 2018 until March 2019, downloaded from https://finance.yahoo.com. We find the high-tech stock from Apple to be more volatile than the rather stable stock from Coca-Cola. The latter only shows a one-day glitch near trading day 220 that causes σ to increase before returning towards the average value at a rate, determined by m. Apparently, the thirst for soft-drinks is less volatile than the thirst for high-tech products.

Instead of continuously updating the volatility alone, we can determine variations of the relative covariance σ^{XY} between two sampled variables X_i and Y_i at sample time i. The covariance matrix for two stocks X and Y that appears when calculating

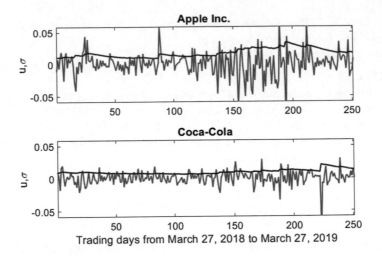

Fig. 8.10 The day-to-day returns (blue) and the values averaged with an exponential filter with $m = 20$ (red) for Apple Inc. and Coca-Cola from March 2018 until March 2019

the optimized portfolios in Chap. 3 may serve as an illustration. To continuously update it, we calculate

$$\sigma_i^{XY} = \frac{1}{m+1} \left[m\sigma_{i-1}^{XY} + \left(\frac{X_i - X_{i-1}}{X_{j-1}} \right) \left(\frac{Y_i - Y_{i-1}}{Y_{i-1}} \right) \right], \qquad (8.46)$$

which gives us an estimate of the covariance matrix that is based on the recent past, as defined by exponentially averaging over the most recent m days.

ARCH(n) and GARCH(n, m)

The generalization of the EWMA models for squared quantities, such as the one on the right-hand side in (8.44), are ARCH and GARCH models. Here ARCH is an acronym for *autoregressive conditional heteroscedasticity* and a model of order n is similar to an MA process or an FIR filter with n coefficients a_j and a constant additional term c, thus

$$\sigma_i^2 = c + \sum_{j=1}^{n} a_j u_{i-j}^2 . \qquad (8.47)$$

Here the sum starts at unity rather than at zero, as in (8.2). The term *heteroscedasticity* refers to the fact that the volatility σ varies, rather than being constant.

The generalization to ARMA models, or IIR filters, that use earlier samples, here σ_{i-k}, in order to calculate the updated sample σ_i is called GARCH, an acronym for *generalized autoregressive conditional heteroscedasticity*. A GARCH(n, m) model is thus written as

$$\sigma_i^2 = c + \sum_{j=1}^{n} a_j u_{i-j}^2 + \sum_{k=1}^{m} b_k \sigma_{i-k}^2 , \qquad (8.48)$$

where n and m are the orders of the moving average and autoregressive part, respectively.

Most of the time the volatility is reasonably constant, but occasionally it increases significantly. One reason for such an increase is political unrest, but others are speculative bubbles and subsequent crashes of a market, the topic of the next chapter.

Exercises

1. Remove the trend and seasonality from the data in file ex8_1.dat. What period do you use? Plot the residuals and then sort them into a histogram with 30 bins. Hint: MATLAB has a built-in function histogram().
2. You suspect that the data in file ex8_2.dat come from an MA(q)–process. Determine the order q of the process.
3. You suspect that the data in file ex8_3.dat come from an AR(p)–process. Determine the order p of the process and discuss which coefficients ϕ_j are significant.
4. Analyze the time series of the CO_2–concentration measured on Mauna Loa from 2009 until 2019, available in the file ex8_2009a.dat from the book's web site. In your analysis

   a. remove the trend and the seasonality from the data;
   b. perform a PACF on the residuals (up to order 4: ϕ_{44}) in order to find the relevant coefficients for an AR(p) model;
   c. determine the model coefficients and display the model together with the data in order to verify that your fitting makes sense.

5. Derive the coefficients π_j for $j = 1, 2$, and 3 that appear in (8.28) from the θ_i and ϕ_k that appear in (8.27).
6. Use an EWMA filter with $m = 1, 3, 10$, and 30 to remove the noise from the time series in file ex8_6.dat, available from the book's web page. Plot both the raw time series and the de-noised copies. Discuss what you observe as you increase m.

References

1. NIST/SEMATECH e-Handbook of Statistical Methods (2016). http://www.itl.nist.gov/div898/handbook/
2. http://www.itl.nist.gov/div898/handbook/pmc/section4/pmc4411.htm

3. NOAA ESRL Global Monitoring Laboratory. 2019, updated annually. Atmospheric Carbon Dioxide Dry Air Mole Fractions from quasi-continuous measurements at Mauna Loa, Hawaii, Barrow, Alaska, American Samoa and South Pole. Compiled by K.W. Thoning, A. Crotwell, and D.R. Kitzis. National Oceanic and Atmospheric Administration (NOAA), Earth System Research Laboratories (ESRL), Global Monitoring Laboratory (GML): Boulder, Colorado, USA. Version 2020-04 at https://doi.org/10.15138/yaf1-bk21
4. M. Quenouille, Approximate tests of correlation in time series. J. R. Stat. Soc. **B11**, 68 (1949)
5. G. Box, G. Jenkins, G. Reinsel, *Time Series Analysis*, 4th edn. (Wiley, Hoboken, 2008)
6. J. Hamilton, *Time Series Analysis* (Princeton University Press, Princeton, 1994)
7. G. Zuckerman, *The Man Who Solved the Market* (Penguin Random House, New York, 2019)
8. https://finance.yahoo.com/quote/AAPL. Accessed 28 Mar 2019
9. W. Enders, T. Sandler, G. Parise, An econometric analysis of the impact of terrorism on tourism. Kyklos **45**, 531 (1992)
10. W. Enders, *Applied Economentric Time Series*, 4th edn. (Wiley, Hoboken, 2015)
11. http://time-series.net/data-sets

Chapter 9
Bubbles, Crashes, Fat Tails and Lévy-Stable Distributions

Abstract After presenting historical bubbles and crashes, this chapter distills a number of pertinent mechanisms behind stock market crashes. One source of the sometimes irrational behavior of investors can be traced to their psychology, which motivates a brief discussion of behavioral economics. The collective behavior of all traders determines the probability distribution function of the daily returns from the stock market, which shows distinctly fat tails. This motivates us to cover power laws, fractals, random walks with increments drawn from fat-tailed distributions, and Levy-stable distributions as their limiting case. This context is used to derive the central limit theorem and question an underlying assumption of the Black-Scholes theory; being based on a (Gaussian) Wiener process. After a short review of extreme-value theory, we introduce Sornette's theory of finite-time divergencies, sometimes visible in time series of stock markets.

Much of the financial and economic theory discussed in previous chapters is based on the notion of markets in equilibrium, a concept heralded as the "efficient market hypothesis." This hypothesis is probably a good approximation to the truth much of the time, but once in a while markets enter a phase where they grow very rapidly—a bubble—and later contract even more rapidly—a crash. On the top panel in Fig. 9.1 we show the historic evolution of the Dow Jones Index [1] from 1900 until 2010 on a logarithmic scale. It rises from about 50 in 1900 to about 10000 in 2010. The curve is neither smooth nor following an exponential growth. Instead, there are periods of steep rises, examples are in the late 1920s, mid 1980s, the 1990s, and the first years of the new millennium. During these times the market was in a bubble that subsequently deflated in a more or less violent crash, visible as a significant drop in the index. The lower panel shows the day-to-day variation of the index $(I(i + 1) - I(i))/I(i)$. We observe that crashes are accompanied by increased volatility or wildly fluctuating day-to-day variations. Crashes are identified by large single-day drops, such as the 30% loss during the crash in October 1987. These patterns of bubbles and crashes are difficult to reconcile with the idea of a market in equilibrium. To better understand this behavior, we will therefore discuss historical examples of crashes, followed by a presentation of the likely mechanisms causing them. Finally, we spend some time to characterize the mathematics of crashes.

© The Author(s), under exclusive license to Springer Nature Switzerland AG 2021
V. Ziemann, *Physics and Finance*, Undergraduate Lecture Notes in Physics,
https://doi.org/10.1007/978-3-030-63643-2_9

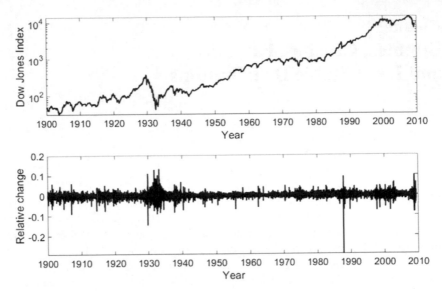

Fig. 9.1 The top graph shows the long term history and the relative variation of the Dow Jones Industrial Index since 1900 until 2010. (data from [1])

But first let us review some of the historical speculative bubbles of stock markets and the subsequent crashes [2].

9.1   Historical Bubbles and Crashes

The first major bubble was the *Tulip mania* in Holland that lasted from 1634 to 1637. At that time Holland had grown rich from the trade of the East-India company and became enchanted with a luxury product introduced from the middle east—tulips. Demand for the precious bulbs accelerated in 1635 and the prospect of ever increasing prices led to speculations to an extent that people mortgaged their houses to invest in tulips. The bubble deflated catastrophically early in 1637 after a buyer of tulips defaulted on his purchase and more and more potential buyers abstained from purchasing, leading to a drastic drop of prices and thus the perceived wealth. People realized that all they had were tulip bulbs, that had lost most of their value and could no longer be used as collateral for loans. After a few month speculation with tulips had stopped.

The *Mississippi bubble* lasted from 1718 until 1720 when the Scottish gambler and financial theoretician John Law through a sequence of lucky (for him) breaks was put in charge to create the first French National Bank in order to amend the desolate finances of the French government. He increased the amount of available money—the liquidity—by introducing "paper money," which he promised, sanctioned by the king, to be freely exchangeable to gold. Based on the success of his method to salvage the state finances, he was granted the monopoly on trading with the newly acquired overseas territories in North America. He started issuing bonds that promised large profits. Soon speculation in these bonds took off and caused a huge bubble, partly fueled by Law being in charge of issuing liquidity in the form of paper money. The bubble crashed early in 1720 when speculators tried to realize their profits by converting Mississippi bonds to gold.

Almost simultaneously the *South sea bubble* developed in England during a prosperous period when the South Sea Trading Company was granted the monopoly to trade with Spanish colonies in the West Indies and South America. The expectation of large profits caused speculation in its shares, despite an on-going war with Spain. The profits were expected to appear once the war ended. The success of the South Sea company caused imitators with less than credible get-rich-quick schemes to appear, a development that was halted by Parliament issuing the "Bubble Act," which required all joint ventures to have a Royal charter. This eliminated many competitors and caused the stock of the South Sea company to rise enormously which triggered a selling avalanche to realize the profits. This in turn caused the bubble to implode, causing a huge destruction of wealth, including part of Isaac Newton's [2].

After the upheaval caused by the first World War, the 1920s are characterized by peace and prosperity, especially in the US. New developments such as automobiles, radios, movies, and air traffic are part of the reason. Simultaneously, the increased flow of people moving from rural to urban areas improved the economic conditions and increased consumption of a large fraction of the public. Many people had money to spend and invested it in the stock market, which appeared to expand without limit. Borrowing money to invest and using one's portfolio as collateral was common, which works well as long as the stock market goes up. In 1929, however, first signs appeared that there is a limit to growth and reduced optimism and doubt caused widespread attempts to realize profits from the stocks. But sellers found few buyers, which caused the value of the stocks to drop. This triggered margin calls from the money lenders, because the stocks did not cover the value of the borrowed money. Thus, the speculative bubble of the 1920s ended in the *great crash of 1929* [3] that led to the great depression of the 1930s.

In the second half of the 1950s the emerging electronics industry following the discovery of the transistor and the start of the space race in 1957 caused a bubble, called the "Tronics boom" that deflated or fizzled in 1962 with a decline of the Dow of about 25% over a few months rather than in a sudden crash.

After having left behind the era of oil crises in the 1970s, the early 1980s were characterized by liberal economics under the recently elected President Reagan. Tax policies favored mergers of companies, which was often funded by massively borrowing money from the public by issuing so-called *junk bonds,* which promised high returns but also carried a high risk. The promise of increased efficiency and increased prospective profits caused euphoric market conditions with many willing investors. In order to safeguard the gains and limit losses new technologies such as automated trading and portfolio insurance with put options were implemented. The bubble grew but started to wobble when people started to realize profits. Then automatic portfolio insurance kicked in, which led to an enormous surge of sell orders that could not be satisfied and eventually the bubble burst in October 1987 leading to the *Black Monday crash of 1987.* Unlike the aftermath of the crash of 1929, recovery from the 1987 crash was swift and the stock market recovered and reached its previous level within two years.

Just as recovery from the 1987 crash was in full swing, the prevailing optimism was further enhanced by the fall of the Berlin Wall, the demise of the Soviet Union, and the end of the cold war leading to a remarkable growth period during the 1990s. Partly fueled by the newly emerging Internet and technology companies, which promised large profits, the stock market grew at a remarkable pace to ever increasing heights of the stock indices. There were a few hiccups along the way, such as Russia's default on foreign debt in 1998, which wiped out some hedge funds, but the overall trend was upwards. This continued until speculators started to become cautious and attempts to sell shares faced difficulties, which then led to the so-called *dot.com crash of 2000.*

In order to alleviate problems with lacking liquidity following the dot.com crash, interest rates were lowered significantly in the US. At the same time political consensus emerged to stimulate the purchase of private homes, even for those previously not eligible for mortgages. But the increased availability of mortgages led to increased demand for housing, which led to higher prices for homes. Using the newly acquired homes as collateral for the mortgage is possible as long as house prices rise. Around 2007, however, the demand for new homes decreased, which lowered the value of homes, such that the home value did no longer cover the mortgages. The hard-pressed home-owners were required to sell below price or default on their mortgage. The latter happened on a large scale, which wiped out several banks, most notably the Lehman brothers.

Note that none of the above crashes were caused by natural disasters, but rather by speculation on a rising asset price. In the following section we will try to summarize some observations about the underlying mechanisms of the bubble-crash sequence.

9.2 Bubble-Crash Mechanisms

Here we distill some observations pertaining to the speculative bubbles and ensuing crashes from the preceeding section.

- The normal valuation of stocks as sum of discounted future earnings (see Sect. 3.6) does not apply. Instead traders speculate on an increasing asset price and hope to sell at higher prices than they bought the asset.
- Bubbles require *liquidity* to invest. Often the increasing stock value is used as security for the *borrowed money,* which is called "leveraged purchase" or "bought on margin." Cheap money and low interest rates often accompany bubbles.
- Speculative bubbles and crashes appear to be inherently linked. First an unfailing trust in an ever-increasing asset value (what A. Greenspan called "irrational exuberance" in 1996, which is also the title of Shiller's book [4] on bubbles) drives the bubble to ever-higher values. Losing faith in the continued increase of the

stock value and starting to sell stocks then precedes a crash. Note, however, that sometimes bubbles fizzle out rather than end in a full-blown crash.

- Often a *new technology* or an equivalent development, such as expected riches from the Americas or the South Seas drives the bubble. The hype for novel Internet stocks in the dot.com bubble or the excitement about increasing values of homes before 2008 are based on the same mechanism.
- Bubbles sometimes follow a dire period, such as a war and then really take off when *optimism* in the future reappears. Think of the "roaring twenties" preceeding the 1929 crash!
- It is remarkable that often no specific exogenous cause for a crash can be identified. It appears to stem from an intrinsic instability of the economic system, resembling a saturated vapor that spontaneously precipitates, stimulated by a random inhomogeneity.

It is obvious that the concepts such as "optimism" or "faith" are difficult to handle in a physics framework and describe mental states that are better handled in a framework using psychology or other behavioral sciences.

9.3 Behavioral Economics

The field of *behavioral economics* considers the driving forces deep seated in our human psyche and how we make decisions and value their outcome. The measure of how we value such an outcome is called *utility*. Historically economists assumed that the agents of trade behave rational, which is also what we assumed in the first few chapters. Such rational agents are so-called *Econs* according to Thaler [5]. Econs calculate the expected utility or value of an action $\langle V \rangle$ by the well-defined expectation value $\langle V \rangle = \sum_i p_i O_i$, where p_i are the probabilities and O_i the outcomes. In contrast, the vast majority of the population behaves non-rational at times and are called *Humans* by Thaler. They calculate the expected utility by applying some non-linear function of the p_i and O_i. As an example, consider the following bet:

- win 80 Euros with certainty, or
- win 100 Euros with a 90% chance and nothing with a 10% chance.

Econs would pick the second bet, because the expected return is 10 Euros higher, but I would pick the first choice. I value the prospect of certainly winning 80 Euros higher than the prospect of actually winning nothing with a 10% chance. Apparently close to certainty the linear calculation of expectation values does not work for us *Humans*.

Likely the most influential scientists, who analyzed the psychology of decision making, are A. Tversky and D. Kahneman. They describe a theory [6] of how *Humans* make decisions based on what outcomes can be expected, hence the name *prospect theory*. Their work is based on a large number of psychological experiments, or bets, similar to the one above, that they evaluated. They found a number of key concepts

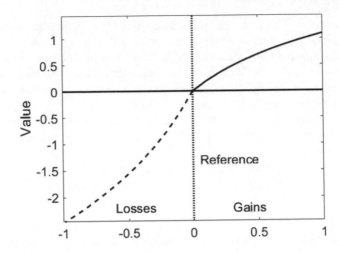

Fig. 9.2 The value function of prospect theory

that make us *Humans* tick. These concepts are concisely represented in Fig. 9.2. The figure shows the experienced value, or utility, as a function of gains or losses with respect to the reference point at the origin.

- Reference point: We *Humans* use the situation just before a decision as reference to judge gains or losses that follow from the decision. Figure 9.2 shows the origin as reference point. In contrast, *Econs* do not care about the reference. For them, only the final result on an absolute scale, such as total wealth, counts.
- Anchoring: Establishing a reference prior to asking a question will affect the answer. In many situations, for example salary negotiations, asking for a lot in the beginning will generate a higher outcome than a modest initial request normally would.
- Loss aversion: Psychological experiments show that losses hurt about two to three times more than a gain of equal magnitude. Therefore, the negative branch in Fig. 9.2, which describes losses, is steeper than the positive branch.
- Non-linear weighting: The relative increase of utility when receiving 20 Euros instead of 10 Euros is felt more positive than receiving 120 instead of 110 Euros. This is indicated in Fig. 9.2 by the curves leveling off at the extremes.

One of the reasons for the "human" behavior are two sub-systems of our brain. System 1 is physically located in the old part of the brain that developed early during the human evolution. It acts very quickly, intuitively and effortlessly, but is guided by rules of thumb, so-called heuristics, that are sometimes fooled. System 2 is physically located in the frontal cortex, a region of the brain that is responsible for analytical reasoning. System 2 is slow and and using it requires much energy in the mental effort. Normally. System 2 supervises System 1, but if the former is busy with other tasks, System 1 makes decisions anyway, and they are sometimes stupid or at least sub-optimal, as Tversky, Kahnemann, and others showed [6].

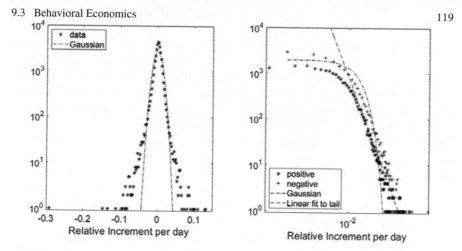

Fig. 9.3 The left graph shows a histogram of the relative day-today changes of the Dow Jones data from the bottom graph in Fig. 9.1 as well as a Gaussian fit with 1.1% standard deviation as the red line. The right graph shows the same data, but separated for the positive (asterisks) and negative (plus signs) as well as the Gaussian fit

Since these psychological mechanisms are common to all of us—including the speculators—we might feel tempted to consider the market participants as a large system of interacting agents, similar to what physicists consider in statistical mechanics. The difficulty is that the interactions in physical systems are rather homogenous (all particles are the same and have the same interaction potential) whereas in economic systems the traders and speculators come in a huge variety. The basic "human" interaction potentials, however, are somewhat restricted by the mechanisms of prospect theory as described above.

Despite being central to the behavior of market participants, we will not pursue the psychological discussion further, but will use some methods borrowed from statistical mechanics and dynamical systems to further investigate the dynamics of stock markets and crashes in particular.

9.4 Fat-Tailed Distributions

A natural place to start is the time series of stock prices and extract information about the crashes which are extreme events in the time series. On the lower panel in Fig. 9.1 we see that the crashes of 1929 and 1987 are accompanied by a large volatility and, in particular, large negative jumps of the day-to-day price variation. If the underlying stochastic process were Gaussian, large jumps were much rarer than actually observed.

This is clearly visible on the left-hand plot in Fig. 9.3. It shows the histogram of all day-to-day changes of the Dow Jones index from 1900 until 2010 on a vertically logarithmic scale as blue asterisks. The red curve is a Gaussian which has the same standard deviation of 1.1% as the raw data points. On the logarithmic scale it appears as an inverted parabola that actually tracks the data points quite well up to about $\pm 5\%$ or about five standard deviations, but beyond that point there is a clear surplus of data points above the Gaussian. Moreover, the deviations on the negative side are somewhat more pronounced. On the right-hand side in Fig. 9.3, we show the same data points on double logarithmic scale, separately for positive and negative deviations, the latter simply flipped to the positive side, as well as the Gaussian. In this representation the Gaussian does not fit the data very well. But at large horizontal deviations the data in double logarithmic representation appears to be linear, which indicates a power law of the form $y \propto x^{-\alpha}$. We refrain from reporting the coefficients because of the limited statistics of the few data points at large deviations. In general, however, power-law distributions appear in many contexts. An example is the distribution of wealthy individuals; there are few extremely rich individuals but very many more individuals of moderate wealth. Plotting a histogram of wealth versus number of individuals in a particular wealth bracket in double logarithmic fashion reveals a straight line, indicating that the distribution follows a power law. This was first observed by V. Pareto, who analyzed the distribution of wealth in Italy in the early 1900s. In the 1960s Mandelbrot [7] and Fama [8] found fat-tailed distributions when analyzing returns of stocks. Other distributions that follow power laws are moderately sized earthquakes when plotting a histogram of the number of quakes versus their magnitude—the well-known Richter law and Zipf's law of word frequency in texts. We refer to [9] for further examples.

So, there is something special about power laws and we will therefore investigate them further.

9.5 Power Laws

We follow [9] and write a generic power law in the form

$$p(x) = Cx^{-\alpha} . \tag{9.1}$$

We point out that the power law behavior will probably be valid only for values of x larger than some minimum value $\hat{x}$. This was the case for the day-to-day variations of the Dow Jones index, which follows a power law only for values larger than $\hat{x} = 0.05$. All distributions we discuss in the following section will cover the region above $\hat{x}$ up to infinity. Since divergences only arise in this region, we ignore the fact that the normalization and all moments differ by an additive constant, depending on whether we take the distribution below $\hat{x}$ into account or not.

The normalization of the distribution $p(x)$ from $\hat{x}$ to infinity is given by

$$1 = \int_{\hat{x}}^{\infty} p(x)dx = \frac{C}{-\alpha + 1}x^{-\alpha+1}\Big|_{\hat{x}}^{\infty} = \frac{C}{\alpha - 1}\hat{x}^{1-\alpha} \tag{9.2}$$

or $C = (\alpha - 1)\hat{x}^{\alpha-1}$, provided $\alpha > 1$ to ensure convergence of the integral.

The moments can then be calculated by standard integrals and are given by

$$\langle x^m \rangle = \int_{\hat{x}}^{\infty} x^m p(x)dx = \frac{\alpha - 1}{\alpha - 1 - m}\hat{x}^m , \tag{9.3}$$

where we expressed the constant C by the one found from normalizing the distribution.

Equation 9.3 already reveals some interesting feature of power laws. In order to have a first moment—the mean—they are required to have $\alpha > 2$, otherwise the right-hand side is negative, despite everything on the left-hand side being positive. Furthermore, the second moment with $m = 2$ only exists for $\alpha > 3$. In the same spirit, for the mth moment $\langle x^m \rangle$ to exists, we must have $\alpha > m + 1$. Conversely, a power law distribution $x^{-\alpha}$ characterized by exponent α only has moments up to order $\alpha - 1$.

As an example, we consider the distribution $\psi_a(x)$, known as Lorentz, Cauchy, or Breit-Wigner distribution in different fields

$$\psi_a(x) = \frac{a}{\pi}\frac{1}{a^2 + x^2} . \tag{9.4}$$

The parameter a is qualitatively related to the width of the distribution. Asymptotically, the distribution behaves as $x^{-\alpha}$ with $\alpha = 2$ and is known not to have a second moment, in correspondence with the general discussion in the previous paragraph.

As an illustration of where power laws can show up we return to the Swedish income distribution and display the integral of the distributions for the years 2010 until 2014 in Fig. 9.4, retrieved from the web site of the Swedish statistical agency [10]. We chose to plot the integral, because the data becomes smoother. The horizontal axis shows the annual income level and on the vertical axis the number of adults above the age of 20 that own *more* than the value on the horizontal axis. The plot on the left-hand side shows the data on a linear scale and on the right-hand side

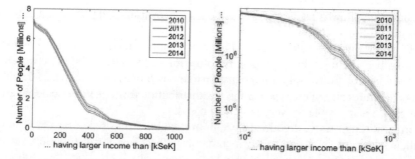

Fig. 9.4 The Swedish income distribution for the years 2010 to 2014. The tail-exponent is between $\alpha + 1 = 3.20$ and 3.25 for all years

using a double-logarithmic scale. It shows the linear dependence for large incomes, consistent with a power-law dependence with a tail exponent $\alpha + 1 = 3.2$ for the integral, such that $\alpha = 2.2$, which is slightly larger than the exponent $\alpha = 2.09$ for the US, reported in [9], and on the same order of magnitude as the exponent for Italy ($\alpha = 2.5$) or Australia ($\alpha = 2.3$), reported in [11]. Note, however, that our determination of the tail exponent is very crude. It should only serve as an example an should not be used for further comparison. The other data from [9] were determined using much more sophisticated methods.

A special feature of power laws is their *scale invariance* in the sense that multiplying the argument will be equal to the same power law multiplied by a constant

$$p(ax) = q(a)p(x) \tag{9.5}$$

which, following [9], can be visualized by considering the scale factor a as a change of units which preserves the overall shape of the distribution. Using this definition of scale invariance we now show that $p(x)$ must be a power law. We first differentiate the left-hand side of (9.5) and then the right-hand side, which leads to

$$\frac{dp(ax)}{da} = xp'(ax) = q'(a)p(x) = \frac{p'(a)}{p(1)}p(x) \xrightarrow{a=1} xp'(x) = \frac{p'(1)}{p(1)}p(x). \tag{9.6}$$

where we used $q(a) = p(a)/p(1)$, which follows from setting $x = 1$ in (9.5). Separating variables and subsequently integrating yields

$$\ln p(x) - \ln p(1) = \frac{p'(1)}{p(1)} \ln x = \ln x^{p'(1)/p(1)} \tag{9.7}$$

where the integration constant $\ln p(1)$ on the left-hand side follows from setting $x = 1$, which causes both sides of the equation to vanish. Rewriting the previous equation gives us

$$p(x) = p(1)x^{-\alpha} \quad \text{with} \quad \alpha = -p'(1)/p(1). \tag{9.8}$$

We have thus shown that the requirement for scale invariance in (9.5) directly implies that $p(x)$ follows a power law.

We point out that scale invariance is a very important concept in our context of crashes, because it implies that there is no natural scale in the day-to-day variations or the volatility. Conversely, Gaussian distributions possess their standard deviation σ as a natural scale, whereas power the law distribution from (9.1) are scale-invariant and can be responsible for very large variations.

In earlier chapters we described the temporal evolution of stock values by a random walk in which the small day-to-day increments with a Gaussian distribution were led to the evolution of the stock values over larger times. Stock values are therefore sums of random numbers. In a later section we will follow this theme further, especially, what happens if the random values are drawn from distributions different from a

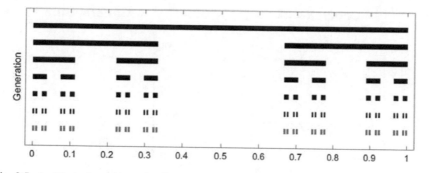

Fig. 9.5 An illustration of how the Cantor set is constructed by repeated removal of the middle third of any remaining interval

Gaussian, such as the fat-tailed power-law distributions. But first, let us explore the concept of scale invariance a bit further as it pertains to the concept of *fractals*.

9.6   Fractals

The notion of self-similarity, or scaling, is closely related to the concept of fractals. They are geometric objects that have a dimension different from an integer. The concept originates from B. Mandelbrot asking himself how the length of the English coastline changes when the size of ruler, used to measure it, changes. With a large ruler we miss all the small bays, while decreasing size of the ruler always increases the measured coastline. In contrast, repeatedly measuring the length of a mathematical line, which we know is one-dimensional, always yields the same length. Decreasing ruler size does not alter the measured length, because the shorter segments are balanced by their larger number.

Another way to determine the geometric dimension of an object on the plane is covering the plane with small squares and counting the number of squares that cover the object. Then we observe how the number of covering squares scales with the size of the squares. If the geometric object is one-dimensional, the number of covering squares grows linearly with the size of the squares. If the geometric object were a circular disk, we would find that the number of squares grows quadratically with the size of the squares. Imagine we had an object with many densely embedded holes; the number of squares to cover the object with holes would be something between one and two. The dimension we find in this way is called *box-counting dimension,* which, in many practical cases, is the same as the *Haussdorff dimension.* In general the number of boxes needed to cover the object N scales with the box size ε as $N(\varepsilon) = \varepsilon^D$ if the object has dimension D. Conversely, we can determine the dimension by counting $N(\varepsilon)$ at scale ε by $D = \ln(N(\varepsilon))/\ln(\varepsilon)$.

In order to develop an intuition about how fractals look or behave we briefly discuss two classical geometric objects with fractal dimension. First, we consider

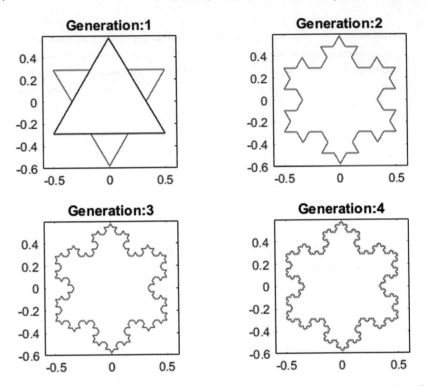

Fig. 9.6 An illustration of how the Koch snowflake is constructed by repeated addition of an 'equilateral detour' at the center part of each interval

the *Cantor set*, which is defined by the transformation of the interval from zero to unity which repeatedly takes out the central third of any interval. This is illustrated in Fig. 9.5, where the top line shows the unit interval. On the line below the central third is removed and on consecutive lines from one level to the next always the central third is removed. We only show the first six iterations, but it is easy to understand that the set of points after infinitely many iterations is not a one-dimensional point, because there are many points (sometimes called Cantor-dust). Neither is it a one-dimensional line, because there are many points missing. Thus we expect the dimension to be somewhere between zero and unity. In fact, we see that changing the size of a covering square by a factor $\varepsilon \sim 3^{-m}$, only half of the points in the line segments remain and they scale by $N \sim 2^{-m}$. The dimension D is then given by $D = \ln(N)/\ln(\varepsilon) = \ln(2^{-m})/\ln(3^{-m}) = \ln(2)/\ln(3) \approx 0.63$, which is indeed between zero and unity. The Cantor set is therefore not a point, but not quite a line yet; it has a fractional dimension of about $D \approx 0.63$.

The second fractal set is *Koch's snowflake*. Figure 9.6 illustrates its construction, which starts from a single triangle (red) and repeatedly replacing the central third of any interval by an "equilateral detour" as can be seen on the top left graph. There, a small blue triangle is added to each side of the red triangle. In the next step, each of

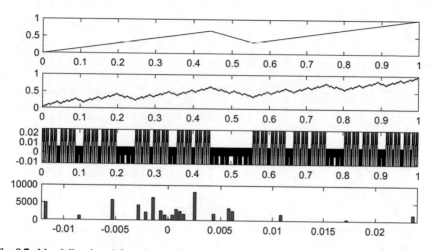

Fig. 9.7 Mandelbrot's uni-fractal map of the unit interval onto itself The top plot shows the first generation map and the second plot shows the map after 10 iterations. The third plot shows the difference between 17 consecutive points and the bottom plot a histogram of the jumps in the third plot

the segments is treated in the same way and then the process is iterated indefinitely. The dimension of Koch's snowflake we expect to be a value between one and two, because the periphery of the snowflake grows as points are added in every iteration. In fact the number of points grows as $N \sim 4^m$ if we increase the scale by a factor of three $\varepsilon \sim 3^m$. For the dimension D we then obtain $D = \ln(4)/\ln(3) \approx 1.26$. We take notice that both the Cantor set and Koch's snowflake are self-similar objects in the sense that zooming into a part of the object looks the same, independent of the magnification we use.

And this self-similarity can serve as a guiding principle to the non-Gaussian noisiness of stock charts. In [12] Mandelbrot describes a method to generate fractals by mimicking the method by which the Koch snowflake was generated. He introduced a map from the unit interval onto itself. An example is shown on the top graph in Fig. 9.7 with parameters taken from [12]. The coordinates of the two points that create the kink in the line are $(4/9, 2/3)$ and $(5/9, 1/3)$. This mapping is scaled to every newly created interval and iterated, resulting in the graph shown in the second plot in Fig. 9.7. The third plot shows the difference between a small number of consecutive data points and is, loosely speaking, the derivative of the second plot. The bottom plot shows a histogram of the differences shown in the third plot. We observe that the second plot has a faint resemblance to the motion of stocks, but still appears to be rather ordered, which is no surprise; it is generated by simply iterating the top plot on each segment.

In order to increase the randomness in the trace, Mandelbrot further suggested to scramble the order of which the segments are placed. Instead of the sequence (long1, short, long2) that we use to denote the sequence of the three segments on the top plot of Fig. 9.7 we use a random permutation such as (long2, long1, short) to place the

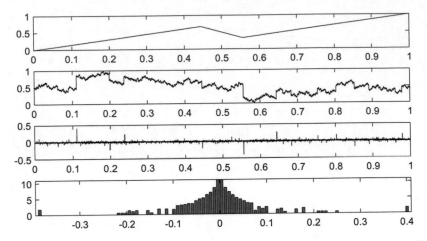

Fig. 9.8 The same type of plots as in Fig. 9.7 but here the three subintervals are randomly scrambled. The differences on the third plot now appear like white noise with a number of large outliers sticking out

three segments for the next iteration. This produces significantly more random traces, which can be seen in the second plot in Fig. 9.8. The third plot shows the increments and they appear to consist of a rather uniform background noise with superimposed large outliers. This is also apparent in the histogram, shown on the bottom plot in Fig. 9.8. Note that here the vertical axis of the histogram is logarithmic in order to make the "fat tails" more noticeable.

We observe that Mandelbrot's algorithm to generate scrambled fractals in fact generates traces and increments that somewhat resemble stock traces and the histogram with the increments shows fat tails. Note that the stretching and folding procedure can also be applied to the horizontal, the time-axis. In that case the resulting curve shows more pronounced periods of tranquility and bursts of activity, very much like trading progresses in time. The resulting traces are called *multi-fractals* as opposed to the uni-fractal traces we discussed above.

A particular feature of the fractal curve generation is the repeated application of a simple "rule" resulting in a self-similar structure, which is what we earlier called "scaling." The structure found on a large scale re-appears on a much reduced scale. We will see in Sect. 9.10 that this self-similarity with discrete scales causes additional large scale features that can be observed in stock charts.

The share values evolving in time, as shown in Fig. 9.1, or the synthetic trace from Fig. 9.8, can be interpreted as adding a random increment to the previous value, much in the spirit of (4.7), only here the increments are drawn from a fat-tailed distribution. This motivates the topic of the next section—adding multiple random numbers.

9.7 Sums of Random Numbers

So, what happens if we add up random numbers drawn from a given distribution? If the underlying distribution is *Gaussian,* we can easily convince ourselves that the resulting distribution is Gaussian as well. For simplicity, consider two independent Gaussian distributions of random numbers x_1 and x_2. Those are typically called *iid,* which is short for *independent identically distributed.* The distribution of the sum $x = x_1 + x_2$ can be calculated by writing the independent distributions as a product, but then constraining their sum with the aid of a delta function. For the distribution $\Psi_2(x)$ of the sum-variable x we find

$$
\Psi_2(x) = \int_{-\infty}^{\infty} dx_1 \int_{-\infty}^{\infty} dx_2 G(x_1, \sigma) G(x_2, \sigma) \delta(x - x_1 - x_2)
$$

$$
= \int_{-\infty}^{\infty} dx_1 G(x_1, \sigma) G(x - x_1, \sigma) , \tag{9.9}
$$

where $G(x_i, \sigma)$ are normalized Gaussians with standard deviation σ or variance σ^2

$$
G(x, \sigma) = \frac{1}{\sqrt{2\pi}\sigma} e^{-x^2/2\sigma^2} . \tag{9.10}
$$

The second equality in (9.9) describes the convolution of two Gaussians and we know that this is just another Gaussian with a variance that is the sum of the variances of the two constituents. Since the variance is the square of the standard deviation σ, we find that

$$
\Psi_2(x) = G(x, \sqrt{2}\sigma) , \tag{9.11}
$$

such that the sum of two random numbers, each of which is distributed according to a Gaussian distribution is again Gaussian, but with a standard deviation that is $\sqrt{2}$ times bigger than the standard deviation σ of the two Gaussians for x_1 and x_2.

The generalization from two to many, say N, i.i.d variables is easy to realize by observing that the convolution of two functions is equivalent to the product of the Fourier transforms, denoted by a tilde and having argument k, of the individual functions and then Fourier-transforming back. In this sense the Fourier-transformation $\tilde{\Psi}_N(k)$ of the convolution of N Gaussians is

$$
\tilde{\Psi}_N(k) = \left[\tilde{G}(k, \sigma) \right]^N , \tag{9.12}
$$

where $\tilde{G}(k, \sigma)$ is the Fourier-transform of the Gaussian $G(x, \sigma)$ from (9.10)

$$\tilde{G}(k, \sigma) = \int\limits_{-\infty}^{\infty} e^{ikx} G(x, \sigma)dx = e^{-k^2\sigma^2/2} . \qquad (9.13)$$

From (9.12) we know $\tilde{\Psi}_N(k)$

$$\tilde{\Psi}_N(k) = e^{-Nk^2\sigma^2/2} = \tilde{G}(k, \sqrt{N}\sigma) , \qquad (9.14)$$

which has the same form as $\tilde{G}(k, \sigma)$, but has σ replaced by $\sqrt{N}\sigma$. Its inverse Fourier transform will yield $\Psi_N(x) = G(x, \sqrt{N}\sigma)$.

Now we add a twist: since we know that the standard deviation grows with $\sqrt{N}$, rather than calculating the distribution function of $x = x_1 + \cdots + x_n$, we calculate the distribution function of $y = (x_1 + \cdots + x_n)/\sqrt{N}$, which means that each random variable x_i is divided by $\sqrt{N}$ and that means that in the definition of the Gaussian in (9.10) the standard deviation σ is replaced by $\sigma\sqrt{N}$ and retracing the following steps we find that the factor $\sqrt{N}$ vanishes from (9.14). In other words, using the scaled variables $y = x/\sqrt{N}$ instead of x the sum of random number sampled from i.i.d Gaussians reproduces the original Gaussian.

Note that we could have used any other distribution $D(x)$, instead of a Gaussian. But before proceeding further we have to briefly introduce the *generating function* of a probability-distribution functions $D(x)$ and its *cumulant expansion*. The Fourier transform $\tilde{D}(k) = \int e^{ikx} D(x)dx$ of $D(x)$ is called the generating function, because the expansion coefficients of its Taylor series are related to the moments $\langle x^m \rangle = \int x^m D(x)dx$ through

$$\tilde{D}(k) = \sum_{m=0}^{\infty} \frac{\langle x^m \rangle}{m!}(ik)^m \quad \text{with} \quad \langle x^m \rangle = (-i)^m \frac{d^m \tilde{D}(k)}{dk^m}\bigg|_{k=0} . \qquad (9.15)$$

Equivalently, we introduce the Taylor-expansion coefficients c_m of the logarithm of the generating function $\log\left(\tilde{D}(k)\right)$. The c_m are called the *cumulants* and they allow us to write $\tilde{D}(k)$ as

$$\tilde{D}(k) = \exp\left[\sum_{m=0}^{\infty} \frac{c_m}{m!}(ik)^m\right] \quad \text{with} \quad c_m = (-i)^m \frac{d^m \log\left(\tilde{D}(k)\right)}{dk^m}\bigg|_{k=0} . \qquad (9.16)$$

The cumulants have the very convenient property that convoluting two distribution functions corresponds to adding their respective cumulants. We already saw this feature for the Gaussians, whose variances are added when convoluting two Gaussians.

Let us now return to calculating the sum of random numbers, drawn from the distribution function $D(x)$ with Fourier-transform $\tilde{D}(k)$. The generating function $\tilde{\Phi}_N(k)$ of the sum of N samples drawn from $D(x)$ is given by the Nth power of $\tilde{D}(k)$

$$\tilde{\Phi}_N(k) = \left[\tilde{D}(k)\right]^N . \tag{9.17}$$

If we now express $\tilde{D}(k)$ by its cumulant expansion as given in (9.16) we find that taking the Nth power of the distribution function simply multiplies all cumulants c_m by N and we obtain

$$\tilde{\Phi}_N(k) = \exp\left[\sum_{m=0}^{\infty} \frac{Nc_m}{m!}(ik)^m\right] . \tag{9.18}$$

Inverse Fourier transforming then yields the distribution function $\Phi_N(x)$

$$\Phi_N(x) = \frac{1}{2\pi} \int_{-\infty}^{\infty} e^{-ikx} \tilde{\Phi}_N(k)dk . \tag{9.19}$$

We know that $\Phi_N(x)$, being the sum of N random numbers, has spread out by a factor $\sqrt{N}$ compared to the underlying distribution $D(x)$, where the width was given by the variance, which is equal to the second cumulant c_2. If, on the other hand, we rescale $\Phi_N(x)$ by $x = \sqrt{N}y$, the width or cumulant of the rescaled distribution function is again equal to c_2. The Fourier-Transform then becomes

$$\sqrt{N} \int_{-\infty}^{\infty} e^{iky} \Phi_N(\sqrt{N}y)dy = \tilde{\Phi}_N\left(\frac{k}{\sqrt{N}}\right) , \tag{9.20}$$

where the factor $\sqrt{N}$ on the left-hand side preserves normalization. Inspection shows that rescaling in real space by a factor $\sqrt{N}$ rescales the variable k in Fourier-space by the inverse factor $1/\sqrt{N}$.

Applying this procedure to $\tilde{\Phi}_N(k)$ in (9.18), namely replacing k by $k/\sqrt{N}$ we arrive at the rescaled function $\tilde{\Phi}_{Nr}(k)$ where we added r to the subscript to indicate rescaling

$$\tilde{\Phi}_{Nr}(k) = \exp\left[\sum_{m=0}^{\infty} \frac{Nc_m}{m!}\left(\frac{ik}{\sqrt{N}}\right)^m\right] = \exp\left[\sum_{m=0}^{\infty} N^{1-m/2}\frac{c_m}{m!}(ik)^m\right] . \tag{9.21}$$

We observe that the Fourier-transform of the distribution resulting from summing N random numbers and rescaled to the original scale by $\sqrt{N}$ is given in terms of the cumulant expansion where the cumulants are rescaled by

$$c_m \rightarrow N^{1-m/2}c_m . \tag{9.22}$$

In the limiting case of large $N \rightarrow \infty$, only the first two cumulants survive, all those with $m > 2$ vanish. In the limit we obtain

$$\lim_{N\to\infty} \tilde{\Phi}_{Nr}(k) = e^{-c_2 k^2/2} , \tag{9.23}$$

where we tacitly assumed that the distributions are centered with $c_1 = 0$. For the limiting distribution in real space we find

$$\Phi_r(y) = \frac{1}{2\pi} \int\limits_{-\infty}^{\infty} e^{-iky} e^{-c_2 k^2/2} = \frac{1}{\sqrt{2\pi c_2}} e^{-y^2/2c_2} , \tag{9.24}$$

which is a Gaussian with standard deviation given by the second cumulant of the underlying distribution function $D(x)$. Note that there are very few prerequisites on the distribution function, only that it can be expressed as a cumulant expansion, which requires the existence of at least the first and second cumulant. In fact, this covers all distributions with existing second moment or equivalently second cumulant.

But this is the essence of the *central limit theorem*, which states that sums of random numbers drawn from a distribution that has a second moment will converge towards a Gaussian in case sufficiently many random numbers are drawn and the spreading of the distribution is counteracted by rescaling, which is just what we did. We can also describe this behavior by considering an arbitrary distribution function with existing second moment that is repeatedly convoluted with itself and rescaled. If that procedure is iterated sufficiently often we obtain a Gaussian distribution function. Once we arrive at a Gaussian, convoluting with itself and rescaling reproduces the same Gaussian. Basically we map distributions onto other distributions and eventually get stuck at some general distribution, which is a fixed point of the mapping rule. Note that this is a very pedestrian description of the *renormalization group*.

This discussion, however, does not cover power laws—they do not have a first or a second moment. In particular random numbers drawn from a Cauchy or Lorentz distribution in (9.4) will not converge to a Gaussian distribution. In the next section we will figure out what they do.

9.8 Lévy-Stable Distributions

It turns out that the process of repeatedly convoluting and rescaling the fat-tailed power law distributions discussed in Sect. 9.5 converge [13] to so-called Lévy-stable distributions. They obey the fixed-point condition

$$L_\mu(y; N)dy = L_\mu(x; 1)dx \quad \text{with} \quad y = a_N x + b_N , \tag{9.25}$$

where $L_\mu(x; N)$ is $L_\mu(x; 1)$ convoluted N times with itself but retains the same form as the original distribution. There are very few explicitly known representations of Lévy-stable distributions in real space known, but the Fourier-transforms of symmetric distributions with $L_\mu(x; 1) = L_\mu(-x; 1)$ have the form

$$\bar{L}_\mu(k; 1) = e^{-a|k|^\mu} \quad \text{with} \quad 0 < \mu < 2 \tag{9.26}$$

which resembles the cumulant expansion from (9.16), but uses fractional powers μ smaller than 2. In fact $\mu = 2$ describes a Gaussian with $a = \sigma^2/2$, which is easy to see when comparing to (9.13). The case $\mu = 1$ leads to the Cauchy distribution in (9.4). The general form for asymmetric distributions can be found in [13]. In the following we will, however, restrict ourselves to symmetric distributions, whose relevance to describe financial data was illustrated in [14].

First we show that the Lévy distributions, as described by (9.26), actually obey the scaling relation from (9.25). Since all distributions are i.i.d, we can calculate the Fourier transform of the N times convoluted form by

$$\tilde{L}_\mu(k; N) = \left[\tilde{L}_\mu(k; 1)\right]^N = e^{-aN|k|^\mu} \tag{9.27}$$

and obtain the real space version from the inverse Fourier-transform

$$
\begin{aligned}
L_\mu(y; N) &= \frac{1}{2\pi} \int_{-\infty}^{\infty} e^{-lky} e^{-aN|k|^\mu} dk \\
&= \frac{1}{N^{1/\mu}} \frac{1}{2\pi} \int_{-\infty}^{\infty} e^{-ik'x} e^{-a|k'|^\mu} dk' \\
&= \frac{1}{N^{1/\mu}} L_\mu(x; 1)
\end{aligned}
\tag{9.28}
$$

where we substituted $k' = kN^{1/\mu}$ and $x = yN^{-1/\mu}$ in the second equality, such that $k'x = ky$. The last equality is just the definition of the Fourier-transform of $L_\mu(x; 1)$. We thus see that the shape of the distribution is preserved by repeated convolution, just as required by (9.25).

Now it remains to be shown that the Lévy distributions actually lead to power law distributions. To show this we calculate their asymptotic behavior for large x and that is determined by the behavior of the Fourier-transform near $k = 0$. We therefore Fourier-transform only the first order term of the Taylor-expansion of (9.26)

$$\frac{1}{2\pi} \int_{-\infty}^{\infty} e^{-ikx} \left(1 - a|k|^\mu\right) dk \approx \delta(x) - \frac{a}{\pi} \int_{0}^{\infty} k^\mu e^{-ikx} dk \tag{9.29}$$

$$\approx \delta(x) - \frac{1}{x^{\mu+1}} \frac{a}{\pi} \int_{0}^{\infty} z^\mu e^{-iz} dz \,,$$

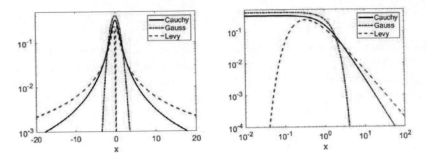

Fig. 9.9 The normalized Cauchy ($\mu = 1$, black) Gauss ($\mu = 2$, red) and symmetrized Lévy distribution ($\mu = 1/2$, blue). On the left the vertical axis is logarithmic and on the right plot both axes are logarithmic with only positive x−values. The strong power-law tails for Cauchy and Lévy-distributions with $1/x^{\mu+1}$ dependence are clearly visible

where we use the substitution $z = kx$ in the last step and see that the asymptotic x-dependence is indeed a power law tail with $1/x^{\mu+1}$. Comparing with (9.1) we see that the tail-order α used in Sect. 9.5 is related to μ by $\alpha = \mu + 1$.

It is instructive to plot the three known real-space examples of Lévy-stable distributions. In Fig. 9.9, we show normalized Cauchy (9.4), Gaussian (9.10), and the symmetrized Lévy-distribution

$$L_{1/2}(x; 1) = \frac{1}{\sqrt{\pi}} \frac{e^{-1/2|x|}}{(2|x|)^{3/2}} . \tag{9.30}$$

On the left-hand side the distributions are shown with a vertical logarithmic axis, and on the right-hand side with double-logarithmic axes and somewhat expanded horizontal range. We note in particular the power law dependence for Cauchy and Lévy-distributions. Note also that the Lévy distribution $L_{1/2}$ has a hole at the origin. Moreover, $\mu = 1/2$ implies that the Lévy distribution possesses neither first nor second moment.

Now we know how fat-tailed power laws are characterized and we use that information to investigate how the maximum value drawn from a distribution, fat-tailed or otherwise, grows with repeatedly sampling numbers from the distribution. This addresses the question of how long we expect to wait until we encounter an extreme event.

9.9 Extreme-Value Theory

Extreme-value theory addresses, among other things, the question how fast the maximum value of the samples, drawn from given distribution, grows as a function of repeated drawings. Conversely, how often do we have to sample until we can expect to exceed a certain large value? Furthermore, we might be curious about the distribution of extreme values. This is basically the histogram of values that exceed all previous samples. We will discuss both topics but start with the first question and follow [9] in doing so.

So, first we ask ourselves how fast the value of the maximum sample grows with the number of drawings n. We start from a normalized distribution function $D(x)$ and its cumulative distribution function

$$C(x) = \int_{\hat{x}}^{x} D(x')dx' , \tag{9.31}$$

where $\hat{x}$ is the minimum possible value, which could be $-\infty$, for example, for a Gaussian distribution. For a power law distribution it would be the minimum value, for example 1. Note that we could equally well have introduced the tail fraction $T(x) = 1 - C(x)$ which is the area under the distribution function from x to the maximum value, which often is infinity. The probability to find a sample between x and $x + dx$ after n drawings is given by

$$\Xi(x)dx = nC(x)^{n-1}D(x)dx \tag{9.32}$$

because $n - 1$ times we have to find a value smaller than x, which accounts for the factor $C(x)^{n-1}$ and once a value between x and $x + dx$, which accounts for $D(x)dx$. Moreover, there are n different ways to choose when the maximum is drawn.

The sought-after expected maximum value after n drawings can therefore be calculated as the expectation value of x with respect to the distribution function $\Xi(x)$

$$\langle x_n \rangle = \int_{\hat{x}}^{\infty} x \Xi(x)dx = n \int_{\hat{x}}^{\infty} xC(x)^{n-1}D(x)dx . \tag{9.33}$$

Let us now calculate $\langle x_n \rangle$ for different distributions $D(x)$, in order to explore their scaling for large n. First, we consider an exponential distribution, defined on the interval from 0 to infinity

$$D(x)dx = e^{-x}dx \quad \text{such that} \quad C(x) = 1 - e^{-x} . \tag{9.34}$$

The expectation value for the maximum after n drawings is then

$$\langle x_n \rangle_{\exp} = n \int_{0}^{\infty} xe^{-x}(1 - e^{-x})^{n-1}dx , \tag{9.35}$$

which we integrate numerically for $n = 10^m$ with $m = 1, \ldots, 10$ and show the result as the solid black line on Fig. 9.10. We observe that $\langle x_n \rangle_{\exp}$ grows linearly with $\log(n)$.

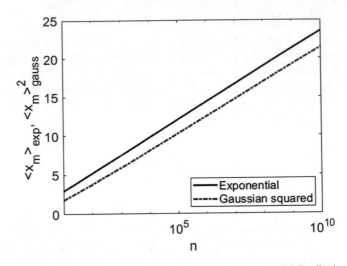

Fig. 9.10 The average maximum expected value $\langle x_n \rangle_{\exp}$ for an exponential distribution is shown as a solid black line. The red dot-dashed line shows the square of the average maximum expected value $\langle x_n \rangle^2_{\text{gauss}}$ for a Gaussian distribution. This implies for the exponential distribution $\langle x_n \rangle_{\exp} \propto \log(n)$ and $\langle x_n \rangle_{\text{gauss}} \propto \sqrt{\log(n)}$ for a Gaussian

In a second example, we use a simplified Gaussian, defined only for positive x to evaluate the scaling of the maximum value for large n. For the Gaussian we use

$$D(x) = \frac{2}{\sqrt{\pi}} e^{-x^2} \quad \text{such that} \quad C(x) = \text{erf}(x) \tag{9.36}$$

which, upon inserting into (9.33) leads to the expression

$$\langle x_n \rangle_{\text{gauss}} = \frac{2n}{\sqrt{\pi}} \int_0^\infty x e^{-x^2} \text{erf}(x)^{n-1} dx . \tag{9.37}$$

Numerically integrating this expression and *squaring* the result, we obtain the dot-dashed red line in Fig. 9.10, which shows a linear dependence. This indicates that for a Gaussian the expected maximum value after n drawings $\langle x_n \rangle_{\text{gauss}}$ scales as $\sqrt{\log(n)}$.

For the power laws, discussed in Sect. 9.5, we write the distribution function $D(x)$ as

$$D(x) = \frac{\mu \hat{x}^\mu}{x^{\mu+1}} , \tag{9.38}$$

where we use the parameter $\mu = \alpha - 1$ to characterize the tail order to be consistent with Sect. 9.8. For the cumulative tail distribution function $T(x)$ we then obtain

$$T(x) = 1 - C(x) = \int_x^\infty D(x')dx' = \left(\frac{x}{\hat{x}}\right)^{-\mu} , \tag{9.39}$$

such that the expression for the expectation $\langle x_n \rangle$ value becomes

$$\langle x_n \rangle = n\mu \int_{\hat{x}}^\infty \frac{\hat{x}^\mu}{x^\mu} \left[1 - \left(\frac{x}{\hat{x}}\right)^{-\mu}\right]^{n-1} dx . \tag{9.40}$$

Substituting $y = 1 - (x/\hat{x})^{-\mu}$ changes the integration boundaries and we arrive at

$$\langle x_n \rangle = n\hat{x} \int_0^1 (1 - y)^{1/\mu} y^{n-1} dy = n\hat{x} B\left(n, \frac{\mu - 1}{\mu}\right) , \tag{9.41}$$

where $B(z, w)$ is the beta function we already encountered in Sect. 7.7. Now we need to find the asymptotic dependence of the beta function for large arguments n. We therefore express the beta function through the Gamma function and use Stirling's formula (6.1.37 in [15]). To simplify the writing we set $z = (\mu - 1)/\mu$ and get

$$B(n, z) = \frac{\Gamma(n)\Gamma(z)}{\Gamma(n + z)} \approx \Gamma(z)\frac{e^{-n}n^n}{e^{-(n+z)}(n + z)^{n+z}} \approx \Gamma(z)\frac{n^n}{(n + z)^{n+z}} \approx \Gamma(z)n^{-z} , \tag{9.42}$$

where we used several times that $z \ll n$. For the asymptotic dependence of $\langle x_n \rangle$ we therefore find

$$\langle x_n \rangle \sim n^{1+(\mu-1)/\mu} \sim n^{1/\mu} . \tag{9.43}$$

The maximum value that we can expect to obtain after n samples thus grows as $n^{1/\mu}$. Note that this grows much more rapidly than the logarithmically growing expected values for the exponential and Gaussian distributions from the first two examples.

In Fig. 9.11 we show the the dependence of the expected maximum value for several distributions with $\mu = 0.5, 1, 1.5$ and a Gaussian for comparison. Note that, for Gaussian distributions, large values (high on the vertical axis) are rarely expected, despite waiting for many samples, whereas those distributions with fat tails, and this includes the Cauchy distribution with $\mu = 1$, have a much higher expectancy of large values, even after a moderate number of samples.

It is instructive to investigate what fraction of an asset x with distribution $D(x)$ is actually distributed in a certain part of the distribution. This is what V. Pareto did when he analyzed what fraction of the wealth of the Italian population is held by what fraction of the population. Imagine that someone with a wealth of 100 Billion US\$ holds about 0.6% of the US gross domestic product (GDP) in 2015. The latter is approximately 18 000 Billion US\$.

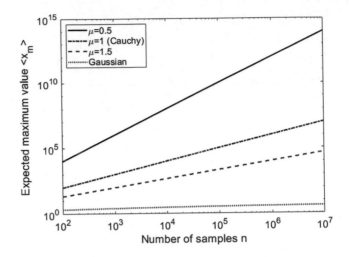

Fig. 9.11 The expected extreme value as a function of samples n for several distributions. Note that small μ cause $\langle x_n \rangle$ to rapidly grow, whereas the Gaussian changes weakly

The cumulative asset $A(x)$ in the tail of a power law distribution is given by

$$A(x) = \int_x^\infty x' D(x')dx' = \frac{\mu \hat{x}}{\mu - 1} \left(\frac{\hat{x}}{x} \right)^{\mu - 1} , \qquad (9.44)$$

where we inserted the power law from (9.38). The fraction of the asset (the "wealth fraction") in the tail is

$$W(x) = \frac{A(x)}{A(\hat{x})} = \left(\frac{\hat{x}}{x} \right)^{\mu - 1} . \qquad (9.45)$$

We already calculated the fraction in the tails in (9.39) and found it to be $T(x) = (x/\hat{x})^{-\mu}$. Expressing $W(x)$ through $T(x)$, we find

$$W = T^{(\mu - 1)/\mu} , \qquad (9.46)$$

which we plot in Fig. 9.12 for $\mu = 1.1, 1.2, 1.5, 2.0,$ and 2.5. In [9] the author quotes that the power law exponent for the US-wealth distribution is $\alpha = \mu + 1 = 2.1$. Thus, from the curve for $\mu = 1.1$ we see that a fraction $T = 20\%$ of the population holds over 80% of the wealth. This observation is consistent with Pareto's original analysis in the 19th century and is coined *Pareto's 80/20 law.*

Finally, we follow [16] and calculate the probability that in n drawings the maximum value stays below x. But this distribution is given by

$$H(x) = C(x)^n , \qquad (9.47)$$

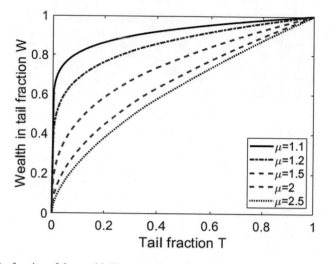

Fig. 9.12 The fraction of the wealth W contained in the tail-fraction T for power law distributions with $\mu = 1.1, 1.2, 1.5, 2.0, 2.5$.

where $C(x)$ is the cumulative distribution function, given by the integral over $D(x)$. Remarkably, in the limit of $n \to \infty$ the distribution $H(x)$ approaches a constant distribution. There are three different cases. First, if the underlying distribution has tails that drop faster than a power law, such as a Gaussian, it can be shown [13] to approach the *Gumbel-distribution*

$$G(x) = \exp\left(-e^{-(x-m)/a}\right) , \qquad (9.48)$$

where m describes the position, and a is a scale parameter. If the underlying distribution $D(x)$ follows a power law with $D(x) \propto 1/x^{\mu+1}$, the limiting distribution can be shown [13] to be the *Fréchet distribution* defined by

$$F(x) = \exp\left(-\frac{1}{\max(0, 1 + (x - m)/(\mu a))^{\mu}}\right) . \qquad (9.49)$$

If the distribution $D(x)$ has a finite maximum value $x_f = m + a/|\xi|$ the limiting distribution can be shown [13] to be the *Weibull distribution*, given by

$$W(x) = \exp\left(-\max\left(0, \frac{m + (a/|\xi|) - x}{a}\right)^{1/|\xi|}\right) , \qquad (9.50)$$

where ξ is negative and x approaches the maximum value x_f as $(x_f - x)^{1/|\xi|}$.

In Fig. 9.13 we show the dependence of the Gumbel, Fréchet and Weibull distributions for $a = 1$ and $m = 0$. We chose $\xi = -1/2$ for the Weibull distribution, and $\mu = 1.1$ for the Fréchet distribution. We point out that the probability of finding a

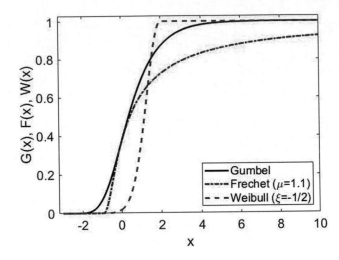

Fig. 9.13 Examples of the Gumbel, Fréchet and Weibull distributions

value larger than the abscissa x is given by $1 - G(x)$, or $1 - F(x)$, or $1 - W(x)$, respectively. The Fréchet distribution, having an underlying power-law distribution from which to draw the samples, rises very slowly at large values of x and is far from unity, even at $x = 10$. This implies that there is a significant probability of finding even more extreme values. In contrast, the Weibull distribution has an endpoint at $x = 2$ and the Gumbel distribution approaches unity already for moderate values of x.

After having discussed the distributions of potential day-to-day variations and analyzed their behavior to some extent, we will now turn back to consider speculative bubbles.

9.10   Finite-Time Divergence and Log-Periodic Oscillations

The time from the start of a speculative bubble to the time it grows without bounds and crashes is finite. This is in contrast to an exponential growth, which needs an infinitely long time to "reach infinity." Thus bubbles and crashes are characterized by a finite-time divergence and a larger-than-linear growth rate $\nu > 1$. We investigate the behavior by considering the following simple system for a dynamical variable s

$$\frac{ds}{dt} = \frac{1}{\tau} \frac{s^{\nu}}{\hat{s}^{\nu-1}} \, , \tag{9.51}$$

where τ is a time constant and $\hat{s}$ a scale factor. Rewriting the equation as $ds/s^{\nu} = dt/(\hat{s}^{\nu-1}\tau)$ we can integrate it with the result

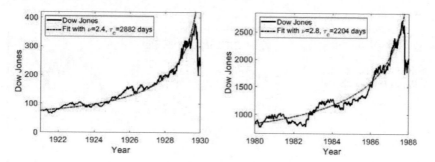

Fig. 9.14 On the left-hand side we show the bubble leading to the crash of 1929 and an "eye-balled" fit to (9.52) with $\nu = 2.4$ and $\tau_c = 2862$ days. The plot on the right-hand side shows the bubble leading to the crash of 1987 with a fit with $\nu = 2.8$ and $\tau_c = 2204$ days. In both cases the singularity is located 180 days after the end of the plotted data

$$s(t) = \frac{s_0}{(1 - t/\tau_c)^{1/(\nu-1)}} \quad \text{with} \quad \tau_c = \frac{\tau}{\nu - 1}\left(\frac{\hat{s}}{s_0}\right)^{\nu-1}, \tag{9.52}$$

where s_0 is the value of s at $t = 0$. We observe that s has a singularity at the finite time $t = \tau_c$. We also observe that τ_c depends on the initial value s_0 in the way that larger s_0 cause τ_c to be smaller, such that the singularity is reached earlier. The power of the non-linearity ν describes the rate at which the singularity is approached. In particular $\nu = 2$ leads to a hyperbolic singularity with $s(t) \sim 1/(t - \tau_c)$.

On the left-hand plot in Fig. 9.14 we show the Dow Jones Industrial Index during the "roaring twenties" leading to the crash of 1929. The red line is a fit done by "eye-balling" and manually adjusting the parameters ν and τ_c until the fit "looks good." Likewise, the right-hand plot shows the bubble leading to the crash of 1987. On both plots the values are specified in the legend. The time of the singularity τ_c was in this way determined to be 180 days after the end of the displayed data. The interpretation in that the market in 1929 was super-heated and in this way became unstable and any small perturbation could (and did) lead to the crash.

We note that despite quite reasonably looking fits the predictive power is limited, because today ("ex-post") we already know that there was a crash at the respective times and we could select the range for the fit suitably. Had we observed the raw data while they were produced in the 1980s ("ex-ante") we were likely unable to extrapolate the noisy raw data towards the future and predict, when the singularity will happen. One helping observation, pointed out by Sornette [17], are systematic oscillations on top of the fitted curves that appear to have an increasingly shorter period as the singularity is approached. These oscillations are nicely visible in Fig. 9.14 and even more pronounced preceding the soft-crash in 1962, shown in Fig. 9.15, that was caused by investors, enthusiastic about the first wave of consumer electronics, after the invention of the transistor.

As it turns out, the origin of the oscillations can be traced to a scale invariance defined by $p(\lambda x) = q(\lambda)p(x)$ that we already discussed in Sect. 9.5, but with a twist.

Fig. 9.15 The bubble from the early 1960s that fizzled out 1962. The oscillations, however, are very clearly expressed in this case and a fit to (9.58) works rather well

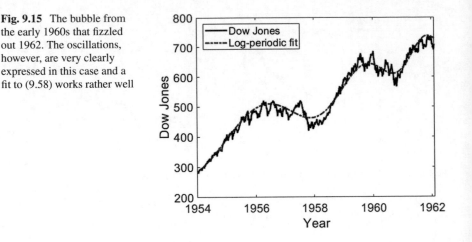

Here the scale changes in a discrete fashion by a factor λ, rather than continuously, where the scale-invariant distributions led us to power laws. In the case of a *discrete scale invariance* the most general solution to (9.5) is

$$p(x) = \hat{C} x^\gamma Q\left(\frac{\ln x}{\ln \lambda}\right) , \tag{9.53}$$

where the exponent γ is just the negative of α in (9.1) and $Q(x)$ is a periodic function with a period of unity $Q(z + 1) = Q(z)$. This is easy to verify by calculating

$$p(\lambda x) = \hat{C} \lambda^\gamma x^\gamma Q\left(\frac{\ln(\lambda x)}{\ln \lambda}\right) = \hat{C} \lambda^\gamma x^\gamma Q\left(\frac{\ln x}{\ln \lambda} + 1\right)$$

$$= \hat{C} \lambda^\gamma x^\gamma Q\left(\frac{\ln x}{\ln \lambda}\right) = \lambda^\gamma p(x) \tag{9.54}$$

where we identify $q(\lambda) = \lambda^\gamma$.

Since we know that $Q(z)$ is periodic with unit periodicity we can write it as a Fourier-series

$$\hat{C} Q(z) = \sum_{n=-\infty}^{\infty} c_n e^{2\pi i n z} , \tag{9.55}$$

where we absorbed the constant $\hat{C}$ in the Fourier coefficients c_n. Upon inserting the representation of $Q(z)$ in (9.53), we find

$$p(x) = x^\gamma \sum_{n=-\infty}^{\infty} c_n e^{2\pi i n \ln(x)/\ln(\lambda)} = \sum_{n=-\infty}^{\infty} c_n x^{\gamma + 2\pi i n \ln(x)/\ln(\lambda)} , \tag{9.56}$$

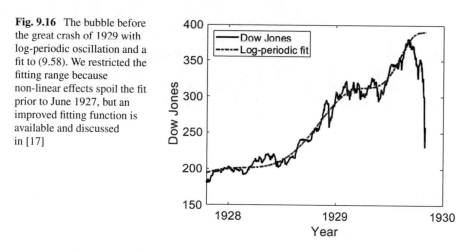

Fig. 9.16 The bubble before the great crash of 1929 with log-periodic oscillation and a fit to (9.58). We restricted the fitting range because non-linear effects spoil the fit prior to June 1927, but an improved fitting function is available and discussed in [17]

which shows that for discrete scale invariances the function $p(x)$ is not a simple power law, but an infinite series of power laws with *complex exponents*.

In Sect. 9.6 we showed how the evolution of stocks can be described in terms of a multi-fractal random walk which was explicitly constructed by repeatedly applying the same rule on each temporal subdivision. This is indicative of a discrete scale transformation and we follow [17] and hypothesize that the stocks can be described by (9.56). If we only use the first two modes with $n = 0$ and $n = 1$, we obtain

$$p(x) = Bx^\gamma + Cx^\gamma \cos\left(2\pi \frac{\ln x}{\ln \lambda} - \phi\right) . \tag{9.57}$$

We now realize that the discrete subdivisions of the time series of our stock happens in time and thus x corresponds to time. We chose to measure time as time-until-crash and thus have $x = t_c - t$. Moreover we identify $\omega = 2\pi/\ln \lambda$ with the log-periodicity of the oscillations. If we also include an additional ad-hoc fit-parameter A to account for an offset in the stock values, we arrive at the following description for the stocks before the crash

$$p(t) = A + B(t_c - t)^\gamma + C(t_c - t)^\gamma \cos(\omega \ln(t_c - t) - \phi) , \tag{9.58}$$

which resembles our previous simplified model in (9.52), provided that $A = 0$ and $C = 0$. Moreover, the exponent γ in (9.58) is related to v in (9.52) by $\gamma = -1/(v - 1)$.

In Fig. 9.15, the red line shows a fit to (9.58) with $A = 960$, $B = -120$, $C = -14.9$, $\gamma = 0.68$, $t_c = 164.83$, $\omega = 12.1$, and $\phi = 4.1$, where the parameters are taken from [17]. The last 30 month prior to the crash of 1929 can also be fitted with (9.58) with parameters taken from [17] ($A = 571$, $B = -267$, $C = 14.3$, $\gamma = 0.45$, $t_c = 1930.22$, $\omega = 7.9$, $\phi = 1$). The fit is shown as the red line in Fig. 9.16.

Fitting the entire period shown on the left-hand plot in Fig. 9.14 requires a non-linear theory (see [17] and references therein) that is, however, beyond our scope.

We have come a long way discussing speculative bubbles and crashes all the way from a historic presentation, via behavioral concepts to extreme value theory, fractals and eventually a theory that attempts to quantify and even predict the time of a crash. But after all the havoc created by the bubbles and crashes we now turn back to the quieter realm of option pricing, but use quantum mechanical tools and path integrals for the task.

Exercises

1. Calculate the generating function of the probability distribution function $R(x)$ that is given by $R(x) = 1/a$ for $-a/2 < x < a/2$ and $R(x) = 0$ otherwise.
2. Calculate the first four cumulants of $R(x)$ from Exercise 1.
3. Calculate the convolution of two Cauchy distributions.
4. Generate 10^5 random numbers, uniformly distributed between -1 and 1, split them into 10^4 groups of ten numbers, sum the ten numbers, and prepare a histogram of the sums. What do you observe? Can you reconcile the result with the second cumulant, calculated in Exercise 2?
5. What is the fractal dimension of a modified Cantor set, where you always remove the central quarter of each line segment?
6. Consider the distribution function $D_v(x) = A(v)e^{-x^v}$ for $1/2 < v < 2$ in the range $0 < x < \infty$ and calculate the expected maximum value of the random numbers after drawing n numbers and how it scales with n. Therefore,

    a. determine $A(v)$ such that $\int_0^\infty D_v(x')dx' = 1$;
    b. calculate the cumulative distribution function $C_v(x) = \int_0^x D_v(x')dx'$;
    c. numerically calculate $\langle x_n \rangle$ from (9.33);
    d. plot $\langle x_n \rangle^v$ versus $\log_{10}(n)$ and convince yourself that this approximately follows a straight line, just as the data in Fig. 9.10 does.

    Hint: in (a) and (b) try the substitution $t = x^v$.
7. The population N of a simplified biological system changes with time according to the rate equation $dN/dt = -\delta N + \beta N^2$ with the death rate δ and the birth rate β. The latter enters quadratically, because two members need to meet in order to procreate. The resemblance to (9.51) stimulates the question how δ affects the divergence. Discuss and analyse how the time τ_c until the singularity is affected.
8. Fat-tailed distributions: Where does the derivation of the Fokker-Planck equation in Sect. 4.5 break down if we use a fat-tailed distribution for what was called $\phi(\xi)$?
9. Research why the crash from 1929 led to a long depression, but the crash 1987 did not.

References

1. https://finance.yahoo.com/quote/^DJI/history. Accessed 19 May 19 2020
2. P. Charles, *Kindleberger, Manias, Panics, and Crashes, A History of Financial Crises*, revised edn. (Basic Books, New York, 1989)
3. J. Galbraith, *The Great Crash of 1929* (Houghton Mifflin, New York, 2009)
4. R. Shiller, *Irrational Exuberance*, 2nd edn. (Broadway Books, New York, 2005)
5. R. Thaler, C. Sunstein, *Nudge* (Penguin, 2009)
6. A. Tversky, D. Kahneman, Prospect theory: an analysis of decision under risk. Econometrica **47**, 263 (1979); A. Tversky, D. Kahneman, Advances in prospect theory: cumulative representation of uncertainty. J. Risk Uncertainty **5**, 297 (1992); D. Kahneman, *Thinking, Fast and Slow* (Farrar, Straus and Giroux, New York, 2011)
7. B. Mandelbrot, The variation of certain speculative prices. J. Bus. **36**, 394 (1963)
8. E. Fama, The behavior of stock-market prices. J. Bus. **38**, 34 (1965)
9. M. Newman, Power laws, Pareto distributions and Zipf's law. Contemporary Phys. **46**, 323 (2005)
10. Swedish Statisiska Centralbyrå. http://www.scb.se
11. F. Clementi, T. Di Matteo, M. Gallegati, The power-law tail exponent of income distributions. Physica A **370**, 49 (2006)
12. B. Mandelbrot, *Multifractals and 1/f Noise* (Springer, Heidelberg, 1999)
13. B. Gnedenko, A. Kolmogorov, *Limit Distributions for Sums of Independent Random Variables* (Addison Wesley, 1968)
14. R. Mantegna, H. Stanley, Scaling behavior in the dynamics of an economic index. Nature **376**, 46 (1995)
15. M. Abramowitz, I. Stegun, *Handbook of Mathematical Functions* (Dover, New York, 1972)
16. D. Sornette, *Critical Phenomena in Natural Sciences*, 2nd edn. (Springer, Heidelberg, 2006)
17. D. Sornette, *Why Stock Markets Crash* (Princeton University Press, Princeton, 2004); D. Sornette, Critical market crashes. Phys. Rep. **378**, 1 (2003)

Chapter 10
Quantum Finance and Path Integrals

Abstract Stimulated by the resemblance of the Black-Scholes equation and the Schrödinger equation, this chapter uses quantum-mechanical methods to determine the pricing kernel, which turns out to be equivalent to the Green's function found in earlier chapters. Using the quantum mechanical methods the down-and-out barrier option is treated in some detail. After covering Feynman's description of quantum mechanics in terms of path integrals, they are used to re-derive the pricing kernel, found earlier. Since path integrals are very amenable to numerical evaluations, we introduce Monte-Carlo methods, including the Metropolis-Hasting algorithm, to evaluate the multi-dimensional integrals.

At first sight it appears strange to relate finance and option pricing to quantum mechanical concepts, but we will see that there are several common points. First note that near the end of Sect. 5.1 we derived the Black-Scholes equation in (5.6). Now we observe that the substitution $S/S_0 = e^x$ transforms it into the form of a quasi-Schrödinger equation

$$\frac{\partial c}{\partial t} = -\frac{\sigma^2}{2}\frac{\partial^2 c}{\partial x^2} + \left(\frac{\sigma^2}{2} - r_f\right)\frac{\partial c}{\partial x} + r_f c = H_{BS}c \qquad (10.1)$$

with the Black-Scholes Hamiltonian H_{BS}

$$H_{BS} = -\frac{\sigma^2}{2}\frac{\partial^2}{\partial x^2} + \left(\frac{\sigma^2}{2} - r_f\right)\frac{\partial}{\partial x} + r_f . \qquad (10.2)$$

Thus, it appears that there is at least a formal resemblance between the Schrödinger equation and the Black-Scholes equation, which we earlier found to be related to the diffusion equation in Sect. 5.2. This is no surprise, because the stochastic meandering of stock prices—the inability to predict them at a later time—and the quantum

© The Author(s), under exclusive license to Springer Nature Switzerland AG 2021
V. Ziemann, *Physics and Finance*, Undergraduate Lecture Notes in Physics,
https://doi.org/10.1007/978-3-030-63643-2_10

mechanical inability to predict the definite state of a system at a later time suggest the possibility to find a common mathematical framework to describe both. Note that in this framework $x = \log(S/S_0)$ assumes the role of the position coordinate in a quantum mechanical system.

Let us start with a short refresher of the basics of quantum mechanics, but focus on the points pertinent to the later discussion. We will then point out the similarities, but also the differences, of how these concepts are used in finance. For a complete treatment of quantum mechanics, the reader is referred to, for example [1].

10.1 Quantum Mechanics

We will extensively use the notation, first introduced by Dirac, to describe quantum-mechanical states as bra $\langle x|$ and ket $|\psi\rangle$ vectors, where $\langle x|$ and $|\psi\rangle$ are elements of dual vector spaces. Furthermore, the scalar product between vectors in these dual vector spaces is denoted by $\langle x|\psi\rangle = \psi(x)$, where $\psi(x)$ is the *wave function* of state $|\psi\rangle$. If we exchange the order of entries, we obtain the complex conjugate $\langle \psi|x\rangle = \langle x|\psi\rangle^* = \psi(x)^*$ of the scalar product, where the asterisk denotes the complex conjugate. The wave function $\psi(x)$ has the interpretation that the probability of finding a particle in the range between x and $x + dx$ is given by the $\psi(x)\psi(x)^*dx$. Moreover, the states $|x\rangle$ form a complete basis for the vector space, which is a consequence of

$$1 = \int dx |x\rangle\langle x| = \int dx \hat{P}(x) , \qquad (10.3)$$

where $\hat{P}(x) = |x\rangle\langle x|$ is the projection operator onto state $|x\rangle$. The integral simply states that the basis comprising all position vectors $|x\rangle$ is complete. The scalar product of two basis vectors $\langle x'|$ and $|x\rangle$ is given by $\langle x'|x\rangle = \delta(x' - x)$, where $\delta(y)$ is Dirac's delta-function. Using the representation of the delta-function as an integral over a complex exponential, we can write

$$\langle x'|x\rangle = \delta(x' - x) = \frac{1}{2\pi} \int\limits_{-\infty}^{\infty} e^{ik(x'-x)}dk = \frac{1}{2\pi} \int\limits_{-\infty}^{\infty} \langle x'|k\rangle\langle k|x\rangle dk \qquad (10.4)$$

with $\langle k|x\rangle = e^{-ikx}$. Note that k is related to the momentum p of a particle by $k = p/\hbar$ with $\hbar = h/2\pi$, where h is Planck's constant. The last equation indicates that also the vectors $|k\rangle$ form a complete basis and we have

$$1 = \frac{1}{2\pi} \int dk |k\rangle\langle k| . \qquad (10.5)$$

Here $|k\rangle\langle k|$ is a projector onto a state with momentum $p = \hbar k$.

In quantum mechanics all physical properties of states are described by hermitian operators, denoted by $\hat{O}$, and labeled with a caret. Measuring the property $\hat{O}$ of a state $|f\rangle$ involves probing the result of $\hat{O}|f\rangle$ with a second state $\langle g|$, which involves calculating the matrix-element $\langle g|\hat{O}|f\rangle$. This procedure describes calculating the overlap of $\hat{O}$'s action on $|f\rangle$ with $\langle g|$. The probability of finding $\hat{O}|f\rangle$ in $\langle g|$ is then given by the squared absolute value of the matrix element $|\langle g|\hat{O}|f\rangle|^2$. We can explicitly calculate these matrix elements by inserting the identity from (10.3) between the states and the operator

$$\langle g|\hat{O}|f\rangle = \int dx' \int dx \langle g|x'\rangle\langle x'|\hat{O}|x\rangle\langle x|f\rangle$$
$$= \int dx' \int dx \, g(x')^* O(x', x) f(x) \tag{10.6}$$

and $O(x', x) = \langle x'|\hat{O}|x\rangle$ is the operator $\hat{O}$ expressed in the basis of position vectors. We could have equally well used the momentum base vectors from (10.5) and had obtained the states $|f\rangle$ and $\langle g|$ and the operator $\hat{O}$ expressed in "momentum space."

All operators representing physically observable quantities are described by hermitian operators. Their action on a vector or its dual vector gives the same result. To see this, we first define the hermitian conjugate operator to operator $\hat{O}$ by a dagger $\hat{O}^\dagger$ and that it operates on the dual bra-state on its left-hand side via

$$\langle f|\hat{O}^\dagger|g\rangle = \langle g|\hat{O}|f\rangle^* , \tag{10.7}$$

where, as before, the asterisk denotes the complex conjugate.

The momentum operator $\hat{p}$ has the form $\hat{p} = -i\hbar\partial/\partial x$ and it is easy to see that

$$\langle x|\hat{p}|k\rangle = -i\hbar\frac{\partial}{\partial x}\langle x|k\rangle = -i\hbar\frac{\partial}{\partial x}e^{ikx} = \hbar k e^{ikx} = p\langle x|k\rangle . \tag{10.8}$$

Applying $\hat{p}$ to a state with $k = p/\hbar$ thus produces a real number $p = \hbar k$ and represents a measurement of the momentum of the state $|k\rangle$.

In the Schrödinger formulation of quantum mechanics, particles are described by waves $\Psi(x, t) \propto e^{ikx-i\omega t}$, characterized by a wave vector $k = p/\hbar$ and a frequency ω, which is related to the energy E of the wave by Planck's relation $E = \hbar\omega$. We see that we can determine the energy E by a partial derivative with respect to the time t, such that $i\hbar\partial\Psi(x, t)/\partial t = E\Psi(x, t)$. Furthermore, we know from classical mechanics that the (kinetic) energy of a freely moving particle is related to the momentum p by $E = p^2/2m$ with the mass m of the particle. Following Schrödinger's daring step to interpret this relation as an operator equation, we write $E = \hat{p}^2/2m = -(\hbar^2/2m)(\partial/\partial x)^2$. Inserting in the equation with the time-derivative we obtain the *Schrödinger equation* for a free particle

$$i\hbar\frac{\partial}{\partial t}\Psi(x, t) = -\frac{\hbar^2}{2m}\frac{\partial^2}{\partial x^2}\Psi(x, t) = \hat{H}\Psi(x, t) . \tag{10.9}$$

Here we added the last equality by introducing the energy operator $\hat{H}$, the Hamiltonian. As a matter of fact, we can even use Hamiltonians including potential energies. They are derived from their classical counterparts $H(q, p) = p^2/2m + V(q)$ by replacing q and p by their quantum-mechanical operator equivalents.

An important property of the Hamiltonian is that it moves the wave function Ψ forward in time. This can be seen by discretizing the time derivative on the left-hand side of the Schrödinger equation $\Psi(x, t + dt) - \Psi(x, t) = -i\hat{H}dt\Psi(x, t)/\hbar$, which leads to

$$\Psi(x, t + dt) = \Psi(x, t) - \frac{i\hat{H}}{\hbar}\Psi(x, t)dt = \left(1 - \frac{i\hat{H}dt}{\hbar}\right)\Psi(x, t). \qquad (10.10)$$

This is only valid for a small time step dt. For larger time steps $\Delta t = ndt$, subdivided into n infinitesimal time steps, we have to repeatedly apply the small time step n times

$$\Psi(x, t + \Delta t) = \left(1 - \frac{i\hat{H}\Delta t}{n\hbar}\right)^n \Psi(x, t) = e^{-i\hat{H}\Delta t/\hbar}\Psi(x, t) \qquad (10.11)$$

where we used $\lim_{n\to\infty}(1 - x/n)^n = e^{-x}$ and ruthlessly ignored questions of time-ordering and convergence. Note, however, that the Hamiltonian $\hat{H}$ generates the motion in time. It pushes the wave functions towards the future and this is the property that we will exploit when using quantum mechanical methods to describe finance.

10.2 Black-Scholes Hamiltonian

There are two particular points where the application of quantum concepts in finance differs from the treatment in physics. First, the wave functions in quantum mechanics are complex-valued and the physical relevant quantities—the probabilities—are the squared moduli of the wave function, whereas in finance the "wave functions" are real-valued and describe, for example, option prices. This is also apparent from the missing complex unit i in the Black-Scholes "Schrödinger equation" in (10.1).

Second, the operators in the Black-Scholes Hamiltonian are not necessarily hermitian, which we understand by considering the operator $\partial/\partial x$, which appears in (10.2) both as first and second power. We calculate

$$\langle f | \left(\frac{\partial}{\partial x}\right)^\dagger |g\rangle = \langle g | \frac{\partial}{\partial x} |f\rangle^* = \left(\int dx \langle g|x\rangle \langle x|\frac{\partial}{\partial x}|f\rangle\right)^* \qquad (10.12)$$

$$= \left(\int dx g^*(x)\frac{\partial f}{\partial x}\right)^* = -\int dx \frac{\partial g}{\partial x} f^*(x) = -\langle f|\frac{\partial}{\partial x}|g\rangle,$$

where we use partial integration to shift the derivative from f to g and assume that f and g vanish at the integral boundaries. Equation 10.12 implies that the derivative operator is *anti-hermitian*

$$\left(\frac{\partial}{\partial x}\right)^{\dagger} = -\frac{\partial}{\partial x} . \tag{10.13}$$

In a similar fashion, we can show that the position operator x is hermitian. This mixture of hermitian and anti-hermitian operators requires us to pay extra attention in the calculations.

Third, we found in Sect. 5.2 and, in particular, in (5.13) that the value of an option at a time τ before maturity can be written as the convolution of the payoff function and the Green's function. This concept resembles that of an operator—a propagator—pushing a quantum mechanical state forward in time, as illustrated in (10.11). Having established this correspondence, we now develop methods to pursue this analogy of the quantum-mechanical description to the stochastic description further. We will loosely base the discussion on [2].

Let us use the Hamiltonian H_{BS} to find the temporal evolution of the option c by integrating (10.1), which reads $\partial c/\partial t = H_{BS}c$. Using arguments, similar to those that led to (10.11), we write

$$c(x, t) = e^{tH_{BS}}c(x, 0) \quad \text{or} \quad |c(t)\rangle = e^{tH_{BS}}|c(0)\rangle , \tag{10.14}$$

where we recover the equation on the left-hand side by multiplying with $\langle x|$ from the left. At time $\tau = T - t$ before maturity, where the value of the option is the payoff function $g(x)$, we can write $|c(T)\rangle = e^{TH_{BS}}|g\rangle$, such that

$$|c(t)\rangle = e^{-(T-t)H_{BS}}|g\rangle = e^{-\tau H_{BS}}|g\rangle . \tag{10.15}$$

This equation has a rather intuitive interpretation of the Black-Scholes Hamiltonian H_{BS} mapping the payoff function g backwards in time to the present time $\tau = T - t$ prior to maturity. Different options are characterized by their individual payoff functions $g(x)$, but the dynamics of mapping the value back to the present time is common to all options and is determined by the Hamiltonian H_{BS}.

10.3   Pricing Kernel

In order to evaluate (10.15), we need to choose a basis and therefore multiply with $\langle x|$ from the left-hand side and also insert the identity from (10.3) between the operator $e^{-\tau H_{BS}}$ and $|g\rangle$. This results in

$$c(x,t) = \langle x|c(t)\rangle = \int dx' \, \langle x|e^{-\tau H_{BS}}|x'\rangle\langle x'|g\rangle$$

$$= \int dx' \, p_{BS}(x,\tau;x')g(x') \,, \tag{10.16}$$

where $\langle x'|g\rangle = g(x')$ is the payoff function and $p_{BS}(x,\tau;x') = \langle x|e^{-\tau H_{BS}}|x'\rangle$ is called the *pricing kernel* for the Black-Scholes Hamiltonian. We see that the value of an option c is given by the convolution of the pricing kernel p_{BS} with the payoff function $g(x)$, which is similar to the procedure described in Sect. 5.2. Just compare (10.16) to (5.13). This illustrates the functionality of the pricing kernel p_{BS} as the propagator between state $|x'\rangle$ and state $\langle x|$ some time τ earlier. The kernel p_{BS} is thus defined as the matrix element of an effective "interaction Hamiltonian" $e^{-\tau H_{BS}}$ sandwiched between the two states $\langle x|$ and $|x'\rangle$.

But we still need to find the functional dependence of p_{BS} on its arguments, which means that we need to evaluate the matrix element $\langle x|e^{-\tau H_{BS}}|x'\rangle$. We do that by inserting the identity in momentum space from (10.5). By using the symbol p instead of k in the integral, we obtain

$$p_{BS}(x,\tau;x') = \int \frac{dp}{2\pi} \langle x|e^{-\tau H_{BS}}|p\rangle\langle p|x'\rangle \,. \tag{10.17}$$

We note that we can write the matrix element as

$$\langle x|e^{-\tau H_{BS}}|p\rangle = e^{-\tau H_{BS}}\langle x|p\rangle = e^{-\tau H_{BS}}e^{ipx} \,. \tag{10.18}$$

In order to evaluate the exponential, we first calculate

$$H_{BS}e^{ipx} = \left[\frac{\sigma^2 p^2}{2} + i\left(\frac{\sigma^2}{2} - r_f\right)p + r_f\right]e^{ipx} \,, \tag{10.19}$$

where we use the Hamiltonian from (10.2) and note that every derivative with respect to x produces a factor ip in the same way Fourier-transforms do. Any function of H_{BS} in the momentum basis can therefore be written as the function of the Hamiltonian in the previous equation

$$\langle x|e^{-\tau H_{BS}}|p\rangle = e^{-\tau(\sigma^2 p^2/2 + i(\sigma^2/2 - r_f)p + r_f)}e^{ipx} \,. \tag{10.20}$$

Inserting into the (10.17), the pricing kernel becomes

$$p_{BS}(x,\tau;x') = \int \frac{dp}{2\pi} e^{-\tau(\sigma^2 p^2/2 + ip(\sigma^2/2 - r_f) + r_f)}e^{ip(x-x')} \,, \tag{10.21}$$

where the remaining integral is Gaussian and can be calculated by completing the square in the exponent. The final result for the pricing kernel then becomes

$$p_{BS}(x, \tau; x') = \frac{e^{-r_f \tau}}{\sqrt{2\pi \sigma^2 \tau}} \exp\left[-\frac{(x - x' + (r_f - \sigma^2/2)\tau)^2}{2\sigma^2 \tau}\right]. \qquad (10.22)$$

It can be shown that the pricing kernel actually equals the Green's function from (5.11), where we have to keep in mind that some of the variables are named differently.

For clarity we repeat the previously-made statement that the pricing kernel acts like a propagator in quantum mechanics, which gives the transition probability between initial and final states. This gives a rather intuitive picture of how options are priced. Since the kernel embeds all the dynamics of the stock market fluctuations, it enables us to calculate the value of any option with any given payoff function $g(x)$ by simply convoluting with the pricing kernel.

Instead of repeating the calculations from Sect. 5.2 to determine the pricing formulae for previously calculated options, we will use the new formalism to calculate the pricing of barrier options, which is difficult using the methods from Sect. 5.2.

10.4 Barrier Options

Here we consider the *down-and-out* barrier option, which becomes void, once the stock price passes a lower limit $S_{DO}/S_0 = e^{-B}$. We incorporate this property in the Hamiltonian by introducing a potential $V(x)$ that is infinite for $x \leq B$ and thus forces the wave function to become zero in that region. The lower limit thus acts like barrier for the wave function, hence the name. The Hamiltonian is then given by

$$H_{DO} = H_{BS} + V(x) = -\frac{\sigma^2}{2}\frac{\partial^2}{\partial x^2} + \left(\frac{\sigma^2}{2} - r_f\right)\frac{\partial}{\partial x} + r_f + V(x), \qquad (10.23)$$

where $V(x) = \infty$ for $x \leq B$ and $V(x) = 0$ for $x > B$.

In order to calculate the pricing kernel $\langle x|e^{-\tau H_{DO}}|x'\rangle$, we need to find the eigenfunctions and eigenvalues for H_{DO}. We already know that they are zero in the region $x \leq B$. Conversely, in the region $x > B$, where the potential is zero, we can assume that the eigenfunctions resemble those of the unperturbed Black-Scholes Hamiltonian H_{BS}. Moreover, they must vanish at the boundary $x = B$. We therefore write the eigenfunctions $\psi_E(x)$ with eigenvalue E in the following form

$$\langle x|E\rangle = \psi_E(x) = e^{(\alpha+ip)(x-B)} - e^{(\alpha-ip)(x-B)}$$
$$= 2ie^{\alpha(x-B)} \sin(p(x - B)) \qquad (10.24)$$

with unknown parameters α and p. By construction, $\psi_E(x)$ vanishes at $x - B$ and we assume that it is only defined for $x > B$.

We now use $\psi_E(x)$ as an Ansatz for the eigenfunction of H_{DO} and therefore need to calculate the derivatives of $\psi(x)$ with respect to x. Differentiating twice, we find

$$\psi'_E(x) = 2i e^{\alpha(x-B)} \left[\alpha \sin(p(x-B)) + p \cos(p(x-B)) \right] \qquad (10.25)$$

$$\psi''_E(x) = 2i e^{\alpha(x-B)} \left[(\alpha^2 - p^2) \sin(p(x-B)) + 2\alpha p \cos(p(x-B)) \right] ,$$

which allows us to calculate $H_{DO}\psi(x)$. After some algebra, this leads to

$$H_{DO}\psi_E(x) = 2i e^{\alpha(x-B)}$$

$$\left\{ \left[-\frac{\sigma^2}{2}(\alpha^2 - p^2) + \alpha \left(\frac{\sigma^2}{2} - r_f \right) + r_f \right] \sin(p(x-B)) \right.$$

$$\left. + \left[-\sigma^2 \alpha p + \left(\frac{\sigma^2}{2} - r_f \right) p \right] \cos(p(x-B)) \right\} \qquad (10.26)$$

$$= E\psi_E(x) .$$

In order for $\psi_E(x)$ to be an eigenfunction of H_{DO} the square bracket before $\cos(p(x-B))$ must vanish. After canceling the common factor p and solving for α, this results in

$$\alpha = \frac{\sigma^2/2 - r_f}{\sigma^2} , \qquad (10.27)$$

which determines α, one of the initially unknown parameters. The other parameter p follows from the requirement that the square bracket before $\sin(p(x-B))$ must be equal to the energy eigenvalue E, which gives us

$$E = -\frac{\sigma^2}{2}(\alpha^2 - p^2) + \alpha \left(\frac{\sigma^2}{2} - r_f \right) + r_f = \frac{\sigma^2}{2} (p^2 + \gamma^2) \qquad (10.28)$$

with $\gamma = (\sigma^2/2 + r_f)/\sigma^2$. Note that γ differs from α by the sign before r_f. Solving (10.28) for p results in $p = \sqrt{2E/\sigma^2 - \gamma^2}$.

We remember that the Hamiltonian is not hermitian, which entails that the eigenfunctions of the Hamiltonian H_{DO} and its adjoint $H^\dagger_{DO}$ are not the same. Performing a similar calculation as above, we find the adjoint eigenfunctions

$$\langle E|x \rangle = e^{-(\alpha+ip)(x-B)} - e^{-(\alpha-ip)(x-B)}$$

$$= -2i e^{-\alpha(x-B)} \sin(p(x-B)) , \qquad (10.29)$$

where α and p depend on $\tilde{E}$ and the eigenvalues fulfill $\langle E|H_{DO} = E\langle E|$. Normalization of the eigenvalues follows from

$$\langle E|E'\rangle = \int_B^\infty dx \langle E|x\rangle\langle x|E'\rangle = 4\int_B^\infty dx \sin(p(x-B))\sin(p'(x-B))$$

$$= 4\int_0^\infty dx \sin(px)\sin(p'x) = 2\pi\delta(p-p') = 2\pi\frac{\delta(E-E')}{|dp/dE|} \quad (10.30)$$

$$= 2\pi\sigma^2\sqrt{2E/\sigma^2 - \gamma^2}\,\delta(E-E'),$$

where we used (10.28) to write $dp/dE = \left(\sigma^2\sqrt{2E/\sigma^2 - \gamma^2}\right)^{-1}$. The normalization permits us to calculate the unit operator, expressed through the energy eigenstates

$$1 = \int_{\sigma^2\gamma^2/2}^\infty \frac{dE}{2\pi\sigma^2\sqrt{2E/\sigma^2 - \gamma^2}}|E\rangle\langle E|, \quad (10.31)$$

which we can prove by choosing the basis $|x\rangle$ to calculate

$$Q(x,x') = \int_{\sigma^2\gamma^2/2}^\infty \frac{\langle x|E\rangle\langle E|x'\rangle dE}{2\pi\sigma^2\sqrt{2E/\sigma^2 - \gamma^2}}. \quad (10.32)$$

Inserting the functions $\langle x|E\rangle = \psi_E(x)$ and $\langle E|x'\rangle$ from (10.24) and (10.29), respectively, and changing the integration variable from E to $p = \sqrt{2E/\sigma^2 - \gamma^2}$, we obtain

$$Q(x,x') = e^{\alpha(x-x')}\int_0^\infty \frac{dp}{2\pi}\left[e^{ip(x-B)} - e^{-ip(x-B)}\right]\left[e^{-ip(x'-B)} - e^{ip(x'-B)}\right]$$

$$= e^{\alpha(x-x')}\int_0^\infty \frac{dp}{2\pi}\left[e^{ip(x-x')} + e^{-ip(x-x')} - e^{ip(x+x'-2B)} - e^{-ip(x+x'-2B)}\right]$$

$$- e^{\alpha(x-x')}\int_{-\infty}^\infty \frac{dp}{2\pi}\left[e^{ip(x-x')} - e^{ip(x+x'-2B)}\right] \quad (10.33)$$

$$= e^{\alpha(x-x')}\delta(x-x') - e^{\alpha(x-x')}\delta(x+x'-2B)$$

$$= \delta(x-x').$$

In the third equality the exponentials with $-ip$ cover the negative values of p such that we can omit them at the same time as extending the lower integration boundary to $-\infty$. The second delta function is zero, because $x > B$ and $x' > B$, such that $x + x' - 2B > 0$ and the argument of the delta function never becomes zero, such that its function value is zero. In summary, we have shown that $Q(x,x') = \delta(x -$

x'), which proves that the basis of energy eigenstates as used in (10.31) is indeed complete.

We immediately exploit this completeness relation to calculate the pricing kernel $p_{DO}(x, \tau; x') = \langle x|e^{-\tau H_{DO}}|x'\rangle$

$$p_{DO}(x, \tau; x') = \int_{\sigma^2\gamma^2/2}^{\infty} \frac{dE}{2\pi\sigma^2\sqrt{2E/\sigma^2 - \gamma^2}} \langle x|e^{-\tau H_{DO}}|E\rangle\langle E|x'\rangle$$

$$= \int_0^{\infty} \frac{dp}{2\pi} e^{-\tau E} \langle x|E\rangle\langle E|x'\rangle , \qquad (10.34)$$

where we replaced the integration over E by an integration over p with $p = \sqrt{2E/\sigma^2 - \gamma^2}$, as before. Furthermore, we used $e^{-\tau H_{DO}}|E\rangle = e^{-\tau E}|E\rangle$, because $|E\rangle$ is an eigenstate of the Hamiltonian. We proceed by inserting the expressions for the eigenfunctions $\langle x|E\rangle$ and $\langle E|x'\rangle$ and express the energy in the exponent through p by $E = (p^2 + \gamma^2)\sigma^2/2$, such that after some further algebra, we get

$$p_{DO}(x, \tau; x') = e^{-\gamma^2\tau\sigma^2/2 + \alpha(x-x')} \int_0^{\infty} \frac{dp}{2\pi} e^{-p^2\tau\sigma^2/2} \qquad (10.35)$$

$$\times \left[e^{ip(x-x')} + e^{-ip(x-x')} - e^{ip(x+x'-2B)} - e^{-ip(x+x'-2B)} \right]$$

$$= e^{-\gamma^2\tau\sigma^2/2 + \alpha(x-x')} \int_{-\infty}^{\infty} \frac{dp}{2\pi} e^{-p^2\tau\sigma^2/2} \left[e^{ip(x-x')} - e^{ip(x+x'-2B)} \right] .$$

After evaluating the Gaussian integrals over p, we finally obtain

$$p_{DO}(x, \tau; x') = \frac{1}{\sqrt{2\pi\sigma^2\tau}} \exp\left[-\frac{(\sigma^2/2 + r_f)^2\tau}{2\sigma^2} + \alpha(x - x') \right] \qquad (10.36)$$

$$\times \left\{ \exp\left[-\frac{(x - x')^2}{2\sigma^2\tau} \right] - \exp\left[-\frac{(x + x' - 2B)^2}{2\sigma^2\tau} \right] \right\} .$$

The first term in the curly braces, combined with the exponential forefactors, yields the following result

$$\frac{(\sigma^2/2 + r_f)^2\tau}{2\sigma^2} - \frac{\sigma^2/2 - r_f}{\sigma^2}(x - x') + \frac{(x - x')^2}{2\sigma^2\tau} \qquad (10.37)$$

$$= \frac{(x - x' + (r_f - \sigma^2/2)\tau)^2}{2\sigma^2\tau} + r_f\tau ,$$

where we completed the square to absorb the term proportional to $x - x'$ into the quadratic term. The second term can be handled in much the same way, but we need to add and subtract terms proportional to $2B - 2x$ in order to be able to express the quadratic term in the form discussed below. For the sum of the terms in the exponent, we have

$$\frac{(\sigma^2/2 + r_f)^2 \tau}{2\sigma^2} - \frac{\sigma^2/2 - r_f}{\sigma^2}(x - x') + \frac{(2B - x - x')^2}{2\sigma^2\tau} \tag{10.38}$$

$$= \frac{(2B - x - x' + (r_f - \sigma^2/2)\tau)^2}{2\sigma^2\tau} + r_f\tau + 2\frac{r_f - \sigma^2/2}{\sigma^2}(x - B) .$$

Collecting the two terms and inserting in (10.36) the pricing kernel $p_{DO}(x, \tau; x')$ now reads

$$p_{DO}(x, \tau; x') = \frac{1}{\sqrt{2\pi\sigma^2\tau}} \exp\left[-\frac{(x - x' + \tau(r_f - \sigma^2/2))^2}{2\sigma^2\tau} - r_f\tau\right]$$

$$- \frac{1}{\sqrt{2\pi\sigma^2\tau}} \exp\left[-\frac{(2B - x - x' + \tau(r_f - \sigma^2/2))^2}{2\sigma^2\tau} - r_f\tau\right]$$

$$\times \exp\left[-2\frac{r_f - \sigma^2/2}{\sigma^2}(x - B)\right] . \tag{10.39}$$

Comparison with (10.22) reveals that the first term is equal to the plain Black-Scholes pricing kernel $p_{BS}(x, \tau; x')$ and the second term is given by $p_{BS}(2B - x, \tau; x')$ with an additional forefactor

$$\exp\left[-2\frac{r_f - \sigma^2/2}{\sigma^2}(x - B)\right] = \left(\frac{e^x}{e^B}\right)^{-\frac{2(r_f - \sigma^2/2)}{\sigma^2}} \tag{10.40}$$

Thus the pricing kernel for our barrier option can finally be written as

$$p_{DO}(x, \tau; x') = p_{BS}(x, \tau; x') - \left(\frac{e^x}{e^B}\right)^{-\frac{2(r_f - \sigma^2/2)}{\sigma^2}} p_{BS}(2B - x, \tau; x') \tag{10.41}$$

with $p_{BS}(x, \tau; x')$ given by (10.22). All prices for dropout options with any given payoff function $g(x)$ can now be calculated by convoluting the payoff with the pricing kernel from (10.41). All information about the dynamics of the process, such as r_f and σ, but also what happens along the way—the absorbing boundary—is ecapsulated in $p_{DO}(x, \tau, x')$. That's why different dropout options use the same kernel, but depend on the specific payoff function. Before moving on to path integrals, let us briefly reflect on the way that the dropout kernel p_{DO} is assembled from two Black-Scholes kernels p_{BS}.

The difference of the two Black-Scholes kernels in (10.41) is constructed in such a way that $p_{DO}(x, \tau, x') = 0$ at $x = B$. This follows the same spirit as placing image charges in electrostatic problems in order to satisfy the boundary conditions on, for example, a conducting plane, as shown on the left-hand side in Fig. 10.1.

Fig. 10.1 The left-hand image illustrates the use of image charges in order to satisfy the boundary conditions on a conducting surface and the right-hand side shows the use of image sources for diffusive problems

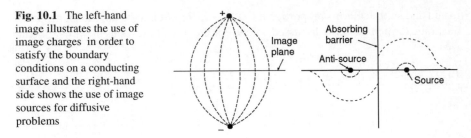

Image sources can always be used, if the underlying partial differential equations are linear, which is also the case for the Black-Scholes equation with its underlying diffusive process as defined by the diffusion equation (4.15). Placing an anti-source, equally strong as the original source, on the other side of an absorbing boundary forces the solution to be zero on the boundary. This is illustrated in the sketch on the right-hand side in Fig. 10.1. The two terms in (10.41) thus correspond to a pair of diffusion processes that ensure that the boundary condition $p_{DO}(x, \tau, x') = 0$ on the absorbing boundary at $x = B$ is satisfied. The additional forefactor from (10.40) is needed, because the Black-Scholes equation includes an additional drift term and uses the log-normal form of the distribution function, rather than a plain Gaussian.

After this digression on image charges, let us now have a look at another advanced concept—path integrals. First we review their use in quantum mechanics before addressing applications in finance.

10.5 Path Integrals in Quantum Mechanics

Path integrals provide yet another, and quite remarkable, way to characterize the dynamics of quantum systems by Green's functions $K(x_b, t_b; x_a, t_a)$, which describe how wave functions $\psi(x, t)$ evolves from time t_a to time t_b

$$\psi(x_b, t_b) = \int\limits_{-\infty}^{\infty} K(x_b, t_b; x_a, t_a)\psi(x_a, t_a)dx_a . \tag{10.42}$$

Due to Feynman's tremendous intuition, we now know that we can write the Green's function K, also referred to as propagator, as the sum over all paths [3] that start at time t_a at location x_a and end at time t_b at location x_b. But we need to put a weight to each possible path and Feynman conjectured that this weight depends on the classical action

$$S(b, a) = \int\limits_{t_a}^{t_b} L(x, \dot{x})dt \quad \text{with} \quad x(t_a) = x_a \quad \text{and} \quad x(t_b) = x_b , \tag{10.43}$$

where $L(x, \dot{x}) = m\dot{x}^2/2 - V(x)$ is the Lagrange function that characterizes the dynamics of the system. Here we only consider Lagrangians without explicit time-dependence. The weight assigned to each path is $e^{iS(b,a)/\hbar}$, such that we express the propagator $K(x_b, t_b; x_a, t_a)$ as

$$K(x_b, t_b; x_a, t_a) \propto \sum_{all\ paths} e^{iS(b,a)/\hbar} , \tag{10.44}$$

which means that each path from a to b is weighted by a phase factor that depends on the action $S(b, a)$ *on that path*. Since we are dealing with quantum mechanical systems, we divide the action $S(b, a)$ by $\hbar$, which has the same units (Joule-seconds) as the action and therefore makes the exponent dimensionless.

Owing to the smallness of $\hbar$ the value of $S/\hbar$ varies a lot and contributions of different paths interfere destructively, unless many paths in the vicinity have similar values of the action $S(b, a)$. But this happens near a path $\bar{x}(t)$ for which the action $S(b, a)$ is stationary when varying that path a little bit, or $\delta S = 0$. But this requirement for stationarity is just Hamilton's principle, which we know leads to the Euler Lagrange equations for the equations of motion that determine the trajectory $\bar{x}$ of the equivalent classical system. This means that the paths near the solution $\bar{x}$ of the equations of motion for the classical system contribute most to the propagator K, all the others paths will interfere destructively and average out.

We still have to work out how to actually calculate the "sum over all paths." This can be, however, accomplished by discretizing time into n small steps $\varepsilon = t_{k+1} - t_k$, which allows us to write

$$S(b, a) = \varepsilon \sum_{k=1}^{n} L(\dot{x}_k, x_k) \quad \text{or} \quad e^{iS(b,a)/\hbar} = \prod_{k=1}^{n} e^{i\varepsilon L(\dot{x}_k, x_k)/\hbar} , \tag{10.45}$$

where the index k labels the time slices. We also identify $x_0 = x_a$ and $x_n = x_b$. For the propagator K, we find

$$K(x_b, t_b; x_a, t_a) = \lim_{\varepsilon \to 0} \frac{1}{A} \int \cdots \int \prod_{k=1}^{n} e^{i\varepsilon L(\dot{x}_k, x_k)/\hbar} \frac{dx_1}{A} \cdots \frac{dx_{n-1}}{A} \tag{10.46}$$

with $n\varepsilon = t_b - t_a$ and $A = \sqrt{2\pi i \hbar \varepsilon / m}$ is a normalization factor that we later show to have this form. Note that the integrals extend over all intermediate points $x_1 \ldots x_{n-1}$ and $e^{i\varepsilon L(\dot{x}_k, x_k)/\hbar}$ is the weighting factor to go from slice k to slice $k + 1$. The Lagrangian depends on both the positions x_k and the velocities $\dot{x}_k$, but the latter can be expressed in terms of positions in adjacent slices by $\dot{x}_k = (x_{k+1} - x_k)/\varepsilon$, which causes the integral to depend on the positions only. Here we see that the weight to go from widely separated positions causes $\dot{x}_k$ to be large and thus increases the phase factor in the exponent by a large amount. But summing over all combinations of intermediate points in the slices parameterizes all paths and gives each one the proper weight and phase factor.

It is instructive to calculate the propagator $\hat{K}$ for a free particle (potential $V(x) = 0$), which turns the propagator into

$$\hat{K}(x_b, t_b; x_a, t_a) = \lim_{\varepsilon \to 0} \frac{1}{A^n} \int \cdots \int \exp\left[\frac{im}{2\hbar\varepsilon} \sum_{k=1}^{n} (x_k - x_{k-1})^2\right] dx_1 \ldots dx_{n-1} ,$$

(10.47)

where we see that the integrals are convolutions of Gaussian leading to a new Gaussian. In this way we obtain

$$\hat{K}(x_b, t_b; x_a, t_a) = \left(\frac{m}{2\pi i\hbar(t_b - t_a)}\right)^{1/2} \exp\left[\frac{im(x_b - x_a)^2}{2\hbar(t_b - t_a)}\right] ,$$

(10.48)

provided we somewhat ruthlessly convolute Gaussians with complex argument and observe that the width of each individual Gaussian is proportional to $\sqrt{\varepsilon}$. Recall from (9.14) that the sum of the squared widths determines the widths of the convolution, which is $N\varepsilon = t_b - t_a$. We refer to Chap. 3.1 in [4] for the details of the calculation.

It remains to be shown that the formulation of quantum mechanics using path integrals is actually equivalent to the formulation using the Schrödinger equation. We follow Sect. 4.1 of [4] and write (10.42) for an infinitesimal temporal increment ε

$$\psi(x, t + \varepsilon) = \frac{1}{A} \int_{-\infty}^{\infty} \exp\left[\frac{i\varepsilon}{\hbar} L\left(\frac{x - y}{\varepsilon}, x\right)\right] \psi(y, t) dy$$

(10.49)

$$= \frac{1}{A} \int_{-\infty}^{\infty} \exp\left[\frac{im(x - y)^2}{\hbar\varepsilon}\right] \exp\left[-\frac{i\varepsilon V(x)}{\hbar}\right] \psi(y, t) dy .$$

As discussed above, if the original point y and the new point x are vastly different, the velocity $(x - y)/\varepsilon$ makes the action large and the phase varies wildly. We therefore assume that y and x only differ by a small amount $\xi = y - x$. Inserting ξ in the previous equation and changing the integration variable from y to ξ, we arrive at

$$\psi(x, t + \varepsilon) = \frac{1}{A} \int_{-\infty}^{\infty} \exp\left[\frac{im\xi^2}{\hbar\varepsilon}\right] \exp\left[-\frac{i\varepsilon V(x)}{\hbar}\right] \psi(x + \xi, t) d\xi .$$

(10.50)

To proceed, we now utilize that the time step ε is small and expand the previous expression to first order in ε with the result

$$\psi(x,t) + \varepsilon\frac{\partial\psi}{\partial t} = \frac{1}{A}\int_{-\infty}^{\infty}\exp\left[\frac{im\xi^2}{\hbar\varepsilon}\right]\left(1 - \frac{i\varepsilon V(x)}{\hbar}\right)\psi(x+\xi,t)d\xi \tag{10.51}$$

$$= \frac{1}{A}\int_{-\infty}^{\infty}\exp\left[\frac{im\xi^2}{\hbar\varepsilon}\right]\left(1 - \frac{i\varepsilon V(x)}{\hbar}\right)\left[\psi(x,t) + \xi\frac{\partial\psi}{\partial x} + \frac{\xi^2}{2}\frac{\partial^2\psi}{\partial x^2}\right]d\xi\,,$$

where we expand to second order in ξ because the first order averages to zero and would lead to a triviality. We see that we are left with Gaussian integrals over powers of ξ which are easily calculated, as we do below. But first we consider the zeroth order in ε, from which we deduce

$$\psi(x,t) = \frac{1}{A}\int_{-\infty}^{\infty}\exp\left[\frac{im\xi^2}{\hbar\varepsilon}\right]\psi(x,t)d\xi = \frac{1}{A}\left(\frac{2\pi i\hbar\varepsilon}{m}\right)^{1/2}\psi(x,t)\,, \tag{10.52}$$

which implies the already stated result for the normalization function A is

$$A = \left(\frac{2\pi i\hbar\varepsilon}{m}\right)^{1/2}. \tag{10.53}$$

Note that we ruthlessly calculated the Gaussian integral over ξ ignoring that the exponent has an imaginary argument due to the factor i. Here, and in much of the presentation, we ignored many mathematical subtleties that are discussed and properly treated in the more rigorous literature about path integrals. But here we keep the heuristic attitude that is actually advocated in [4]. Returning to the derivation, we now compare terms in (10.51) that are linear in ε and recover the Schrödinger equation

$$\varepsilon\frac{\partial\psi}{\partial t} = -\frac{i\varepsilon V(x)}{\hbar} + \frac{1}{A}\int_{-\infty}^{\infty}\exp\left[\frac{im\xi^2}{\hbar\varepsilon}\right]\frac{\xi^2}{2}\frac{\partial^2\psi}{\partial x^2}d\xi \tag{10.54}$$

$$= -\frac{i\varepsilon V(x)}{\hbar} + \frac{1}{2}\frac{\partial^2\psi}{\partial x^2}\frac{i\hbar\varepsilon}{m}$$

or, after canceling ε and some reordering of terms

$$\frac{\partial\psi}{\partial t} = -\frac{i}{\hbar}\left[-\frac{\hbar^2}{2m}\frac{\partial^2\psi}{\partial x^2} + V(x)\psi\right] = -\frac{i}{\hbar}H\psi\,, \tag{10.55}$$

which is the well-known Schrödinger equation with the time derivative on the left hand side and the Hamiltonian H on the right hand side.

This very terse presentation in this section should serve as an introduction to the historic origin of the path integrals and to the concept of summing over paths. Furthermore, we emphasize the special relation of the action functional with the

Lagrangian to the Hamiltonian in the Schrödinger equation. Note also that the calculations we performed in the last part of this section resembled the steps to derived the Fokker-Planck equation from the Master equation in Sect. 4.5. This observation should justify the daring step to apply path integrals to finance.

10.6   Path Integrals in Finance

In this section we will re-derive the pricing kernel p_{BS} from (10.22) for the Black-Scholes Hamiltonian with the help of path integrals, first used in the context of finance by Dash [5], to illustrate the methodology. In the previous section, we found that the path integral depends on the action in the exponent, where the action is the integral over the Lagrangian from fixed starting and end points at fixed initial and final times. So how do we find the Lagrangian that corresponds to the Black-Scholes Hamiltonian from (10.2)? We start by considering an infinitesimally short period of time ε and write the transition probability in the form with the Lagrangian in the exponent

$$p_{BS}(x_i, \varepsilon; x_{i-1}) = \langle x_i | e^{-\varepsilon H_{BS}} | x_{i-1} \rangle = \mathcal{N}(\varepsilon) e^{\varepsilon L_{BS}(x_i, x_{i-1}, \varepsilon)} \qquad (10.56)$$

with some normalization constant $\mathcal{N}_i(\varepsilon)$ that depends on the short time interval ε. Comparing with the expression for the pricing kernel in (10.22) for $\tau = \varepsilon$ we find the following relations for $L_{BS}(x_i, x_{i-1}, \varepsilon)$ and $\mathcal{N}(\varepsilon)$

$$L_{BS}(x_i, x_{i-1}, \varepsilon) = -\frac{1}{2\sigma^2} \left[\frac{x_i - x_{i-1}}{\varepsilon} + r_f - \sigma^2/2 \right]^2 - r_f \qquad (10.57)$$

and $\mathcal{N}(\varepsilon) = 1/\sqrt{2\pi\sigma^2\varepsilon}$.

It is instructive to compare this Lagrangian with the one obtained from converting the Black-Scholes Hamiltonian directly into the equivalent Lagrangian with a Legendre transformation, well-known from classical mechanics [6]. It converts between Lagrangian $L(x, \dot{x})$, which depends on the position x and velocity $\dot{x}$, and Hamiltonian $H(x, p) = \dot{x}p - L(x, \dot{x})$, which depends on the position x and the momentum $p = \partial L/\partial \dot{x}$. The quantum-mechanical Hamiltonian from (10.2) is converted into the equivalent classical Hamiltonian by substituting $p = \partial/\partial x$

$$H_{BS}(x, p) = -\frac{\sigma^2}{2}p^2 - \left(r_f - \frac{\sigma^2}{2}\right)p + r_f . \qquad (10.58)$$

Applying one of Hamilton's equations leads us to the equation of motion

$$\dot{x} = \frac{\partial H_{BS}}{\partial p} = -\sigma^2 p - r_f + \frac{\sigma^2}{2} \quad \text{or} \quad p = -\frac{\dot{x} + r_f - \sigma^2/2}{\sigma^2} . \qquad (10.59)$$

The other equation of motion yields $\partial H_{BS}/\partial x = 0$ because the Black-Scholes Hamiltonian H_{BS} does not explicitly depend on x.

Now we can find the Lagrangian via the Legendre transformation

$$
\begin{aligned}
L_{BS} &= \dot{x}p - H_{BS} \\
&= \left(-\sigma^2 p - r_f + \frac{\sigma^2}{2}\right)p - \left[-\frac{\sigma^2}{2}p^2 - \left(r_f - \frac{\sigma^2}{2}\right)p + r_f\right] \\
&= -\frac{\sigma^2}{2}p^2 - r_f = -\frac{1}{2\sigma^2}\left(\dot{x} + r_f - \frac{\sigma^2}{2}\right)^2 - r_f,
\end{aligned}
\tag{10.60}
$$

where we substituted $p = (\dot{x} + r - \sigma^2/2)/\sigma^2$ in the last equation. We note that this Lagrangian is the same we found in (10.57) by using the pricing kernel p_{BS} for an infinitesimal time step and after substituting $\dot{x} = (x_i - x_{i-1})/\varepsilon$.

Having determined the Lagrangian L_{BS}, we are now ready to calculate the pricing kernel p_{BS} by subdividing the time interval $\tau = N\varepsilon$ into N short time-slices and using the infinitesimal Black-Scholes propagator to step from x_{i-1} to x_i. In order to sum over all possible paths we therefore need to integrate over all $N - 1$ intermediate coordinates

$$
\begin{aligned}
\langle x_N|e^{-\tau H_{BS}}|x_0\rangle &= \int dx_{N-1}\langle x_N|e^{-\varepsilon H_{BS}}|x_{N-1}\rangle \cdots \\
&\qquad \cdots \int dx_1\langle x_2|e^{-\varepsilon H_{BS}}|x_1\rangle\langle x_1|e^{-\varepsilon H_{BS}}|x_0\rangle \\
&= \frac{1}{\sqrt{2\pi\sigma^2\varepsilon}}e^{\varepsilon L_{BS}(N)} \\
&\qquad \times \int_{-\infty}^{\infty}\frac{dx_{N-1}}{\sqrt{2\pi\sigma^2\varepsilon}}e^{\varepsilon L_{BS}(N-1)} \cdots \int_{-\infty}^{\infty}\frac{dx_1}{\sqrt{2\pi\sigma^2\varepsilon}}e^{\varepsilon L_{BS}(1)} \\
&= \left(\frac{1}{\sqrt{2\pi\sigma^2\varepsilon}}\right)^N \int_{-\infty}^{\infty}dx_{N-1}\cdots\int_{-\infty}^{\infty}dx_1 e^{\varepsilon\sum_{i=1}^{N}L_{BS}(i)} \\
&= \left(\frac{1}{\sqrt{2\pi\sigma^2\varepsilon}}\right)^N \left(\prod_{i=1}^{N-1}\int_{-\infty}^{\infty}dx_i\right)e^{S_{BS}},
\end{aligned}
\tag{10.61}
$$

where we introduced the abbreviation $L_{BS}(i) = L_{BS}(x_i, x_{i-1}, \varepsilon)$ for the Lagarangian from (10.57). Moreover, we define the Black-Scholes *action* S_{BS} as the time-integral over the Lagrangian with end points x_0 and x_N kept fixed. For the small time steps ε, we can write the the integral as a sum

$$S_{BS} = \varepsilon \sum_{i=1}^{N} L_{BS}(i) = \varepsilon \sum_{i=1}^{N} L_{BS}(x_i, x_{i-1}, \varepsilon) \tag{10.62}$$

$$= -\frac{1}{2\sigma^2\varepsilon} \sum_{i=1}^{N} \left[x_i - x_{i-1} + \varepsilon \left(r_f - \frac{\sigma^2}{2} \right) \right]^2 - \varepsilon r_f N \ .$$

The last line of (10.61) suggests to introduce a path-integral measure DX through the following expression

$$\int DX = \left(\frac{1}{\sqrt{2\pi\sigma^2\varepsilon}} \right)^N \left(\prod_{i=1}^{N-1} \int_{-\infty}^{\infty} dx_i \right) . \tag{10.63}$$

Note that there are $N - 1$ integrations over the intermediate points, but the forefactor $1/\sqrt{2\pi\sigma^2\varepsilon}$ is raised to the power N because there is one factor for each time slice and there are N slices. In other words, there is one propagator for each slice, but only $N - 1$ intermediate points to integrate over. In particular, we do not need to integrate over x_0 or x_N because they represent the boundary conditions, and are therefore fixed. Finally, we see that we can formally write the pricing kernel as the path integral

$$p_{BS}(x_N, \tau; x_0) = \int DX e^{S_{BS}} , \tag{10.64}$$

where the Black-Scholes action S_{BS} is defined in (10.62) and the measure $\int DX$ must be interpreted by the limit of large N in (10.63).

It is instructive to verify that the path integral formulation actually recovers the pricing kernel from (10.22). To achieve this, we follow [2] and start from the last line of (10.61). We express the action S_{BS} through the last line of (10.62)

$$\int DX e^{S_{BS}} = \left(\frac{1}{\sqrt{2\pi\sigma^2\varepsilon}} \right)^N \int_{-\infty}^{\infty} dx_{N-1} \cdots \int_{-\infty}^{\infty} dx_1 \tag{10.65}$$

$$\times \exp\left[-\frac{1}{2\sigma^2\varepsilon} \sum_{i=1}^{N} \left(x_i - x_{i-1} + \varepsilon(r_f - \sigma^2/2) \right)^2 - r_f \varepsilon N \right]$$

$$= \left(\frac{1}{\sqrt{2\pi\sigma^2\varepsilon}} \right)^N \int_{-\infty}^{\infty} dy_{N-1} \cdots \int_{-\infty}^{\infty} dy_1$$

$$\times \exp\left[-r_f\tau - \frac{1}{2\sigma^2\varepsilon} \sum_{i=1}^{N} y_i^2 \right] ,$$

where we introduced new variables $y_i = x_i - x_{i-1} + \varepsilon(r_f - \sigma^2/2)$ and note that the Jacobian for this variable transformation is unity because of $dy_i = dx_i$. We have

N terms y_i^2 in the exponent, but only $N - 1$ integrals to solve. Moreover, we have to fulfill the constraints to satisfy the boundary conditions, namely that we need to reach x_N at the end. We accommodate this constraint by observing that the sum of the y_i must be

$$
\kappa = \sum_{i=1}^{N} y_i = x_N - x_0 + N\varepsilon(r_f - \sigma^2/2) = x_N - x_0 + \tau(r_f - \sigma^2/2) \quad (10.66)
$$

because in the sum the factors y_i for $i = 1, \ldots, N - 1$ appear once with positive and once with negative sign. Only x_N and x_0 are unpaired. This constraint we accommodate in (10.65) by adding a delta function with the constraint

$$
\delta\left(\kappa - \sum_{i=1}^{N} y_i\right) = \int_{-\infty}^{\infty} \frac{dp}{2\pi} e^{ip\left(\kappa - \sum_{i=1}^{N} y_i\right)} \quad (10.67)
$$

and one additional integration over y_N leads us to

$$
\int DX e^{S_{BS}} = e^{-r_f \tau} \left(\frac{1}{\sqrt{2\pi\sigma^2\varepsilon}}\right)^{N} \int_{-\infty}^{\infty} dy_N \int_{-\infty}^{\infty} dy_{N-1} \ldots \int_{-\infty}^{\infty} dy_1
$$

$$
\times \exp\left[-\frac{1}{2\sigma^2\varepsilon} \sum_{i=1}^{N} y_i^2\right] \int_{-\infty}^{\infty} \frac{dp}{2\pi} e^{-ip(\kappa - \sum_{i=1}^{N} y_i)} \quad (10.68)
$$

$$
= e^{-r_f \tau} \int_{-\infty}^{\infty} \frac{dp}{2\pi} e^{-ip\kappa} \left(\frac{1}{\sqrt{2\pi\sigma^2\varepsilon}} \int_{-\infty}^{\infty} dy_i \exp\left[-\frac{y_i^2}{2\sigma^2\varepsilon} + ipy_i\right]\right)^{N},
$$

where we combined the terms with y_i and observe that they are all equal, except for the name of the index $i = 1, \ldots, N$. The integral over y_i is Gaussian and evaluated in a straightforward way. After taking the Nth power, we obtain

$$
\int DX e^{S_{BS}} = e^{-r_f \tau} \int_{-\infty}^{\infty} \frac{dp}{2\pi} e^{-ip\kappa} e^{-\sigma^2\tau p^2/2} = \frac{e^{-r_f t}}{\sqrt{2\pi\sigma^2\tau}} e^{-\kappa^2/2\sigma^2\tau}, \quad (10.69)
$$

where the remaining integral over p is also of standard Gaussian type. Finally, inserting κ from (10.66) recovers the expression for the Black-Scholes pricing kernel from (10.22).

At this point, we have completed the circle and shown that path integrals solve the Black-Scholes problem of finding the pricing kernel that allows us to calculate the option pricing formulas by integration over the payoff-function. We only showed the equivalence of the path integral method with conventional theory, but nevertheless, we found a new tool—path integrals—at our disposal. They allowed us to solve a simple problem analytically. But for more complex problems, path integrals turn out to be very amenable to numerical methods based on Monte-Carlo methods, the topic of the next section.

10.7   Monte-Carlo Integration

In previous sections, the evaluation of path integrals relies on splitting a long time interval into many short intervals, introducing spatial coordinates x_i at all intermediate times, then using the transition probability from one time slice to the next, and finally integrating over all intermediate coordinates. This is the rational behind the first line of (10.65), which also shows that the path integral is essentially a multi-dimensional integral over the intermediate coordinates with the action S_{BS} in the exponent being the integrand. Monte-Carlo integration is a powerful method to evaluate such multi-dimensional integrals.

We start, however, by exploring Monte-Carlo methods for a one-dimensional example, where we evaluate the numerical value of a Riemann integral of a function $f(x)$ over an interval $a < x < b$. The conventional method is based on splitting the interval into a large number n of sub-divisions dx between equidistant points x_i, summing up the area of the "bars" with height $f(x_i)$ and width dx and in the end taking the limit $n \rightarrow \infty$. Instead, we can also approximate the integral by simply generating a large number N of random numbers x_j in the interval $[a, b]$, and summing up all values $f(x_j)$. We then have to multiply the sum by an average interval width, which we estimate to be $(b - a)/N$. For the Monte-Carlo estimate of the integral, we therefore find

$$\int_a^b f(x)dx \approx \frac{b-a}{N} \sum_{j=1}^N f(x_j) . \tag{10.70}$$

If we "roll the dice" a sufficient number of times, thus for large N, we can expect the approximation to approach the real value of the integral. In order to get an impression of the rate of convergence as a function of N, we resort to an example: we calculate the integral of $f(x) = 15(x^2 - x^4)/4$ in the interval $[-1, 1]$ by using random numbers and compare it with the correct value of $\int_{-1}^1 f(x)dx = 1$. The result of the evaluation is shown in Fig. 10.2. After a few thousand random numbers we approximate the correct value within the percent level.

The advantage of this method is the simplicity of coding it. Only two lines of MATLAB suffice

```
x=-1+2*rand(1,N);      % random numbers between -1 and 1
Imc=(2.0/N)*sum(f(x));
```

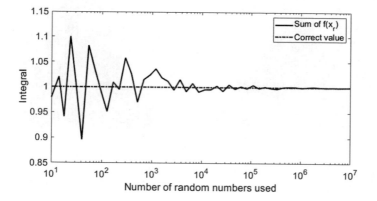

Fig. 10.2 Monte-Carlo approximation to the integral a function of number of random numbers used

The first line generates N random numbers between -1 and 1 and second line sums up all the function evaluations. The disadvantage is the slow convergence, especially when integrating functions that are small over large sub-regions of the integration interval. Integrals of Gaussians over infinite intervals are an example that is problematic. We therefore need to find a better method; one that pays more attention to the regions where the function is large, or even better, generate random numbers whose distribution mimics the integrand. In other words, we try to build a random number generator that produces random numbers, whose histogram reproduces the function $f(x)$. This procedure of building a tailor-made random number generator is called *importance sampling*.

One way of generating random numbers that mimic the integrand is the *acceptance-rejection* method which is based on generating two random numbers x_j and y_j, where y_j must lie between the minimum and maximum of the function $f(x)$. We then select random numbers by taking those x_j for which $y_j < f(x_j)$. A visualization of this method is based on throwing darts onto a target plane with x and y and the line $f(x)$ drawn on the target. If we hit below the line we pick the value of x, if it is above the line, we ignore it. If we repeat this procedure sufficiently often, the random numbers x_j are clustered around values where $f(x)$ is large. The MATLAB code to produce such a distribution is also rather compact

```
x=-1+2*rand(1,N);   % random numbers between -1 and 1
y=rand(1,N);        % random numbers between  0 and 1
x=x(y<f(x));        % select only values under f(x)
hist(x,30);         % display histogram
```

where the last line only serves to display the distribution of random numbers. Figure 10.3 shows the output of running the code for $N = 10^4$ iterations.

But this way of generating the distribution still suffers from a large rejection rate of random numbers and is still very inefficient in regions where $f(x)$ is small. We

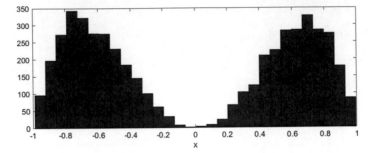

Fig. 10.3 The histogram of random numbers generated by the acceptance-rejection method for $f(x) = 15(x^2 - x^4)/15$. Only about half of the initially $N = 10^4$ random numbers are accepted for the histogram

thus need an algorithm that generates new random numbers, but has a preference to stay near values where $f(x)$ is large.

The *Metropolis-Hastings algorithm* is such an algorithm. It is based on generating a new sample x_{j+1} from the previous sample x_j by adding a random increment of magnitude $\beta \hat{P}$, where $\hat{P}$ is a sample drawn from any random number generator that produces numbers symmetrically distributed around zero, for example, a generator producing uniformly distributed random numbers between -1 and 1. The parameter β determines the magnitude of the step. It should be large enough to avoid getting stuck in local maxima and give the algorithm a chance to explore all possible values. The algorithm then works as follows: first a new candidate is generated by $y = x_k + \beta \hat{P}$ and we calculate $\alpha = f(y)/f(x_k)$. If $\alpha > 1$ the new value y gets us closer to the maximum and is accepted as the new sample $x_{k+1} = y$. If, on the other hand $\alpha \leq 1$, we give y a second chance by comparing α with a uniformly distributed random number u between zero and unity and only select y if $\alpha > u$. If the second chance fails, we simply re-use x_k. One can start the algorithm from any random value, but it is recommended to iterate for a number (100s to 1000s) of burn-in iterations before accepting values. This avoids getting stuck in a region due to an unfortunate choice of initial value. A MATLAB rendition of the algorithm is shown in Appendix B.5. It has the bonus that it does not waste a lot of effort creating random numbers where they are not needed, but lingers around the maxima of the distribution function.

Since the integrands of the path integrals are typically of the Gaussian type, we run the acceptance-rejection algorithm 10000 times while sampling $x-$ values between ± 5 times the rms value of the Gaussian. About 8000 values are rejected and we can use only 2000 values whose histogram we show in the upper plot of Fig. 10.4. We then use the Metropolis algorithm and can use all generated 10000 random numbers after an initial burn-in period of 100 iterations and show the resulting histogram in the lower plot of Fig. 10.4. Obviously, the Metropolis algorithm is more efficient, because it spends less time in regions where the function value is small.

In the following section we will use these methods to evaluate path integrals.

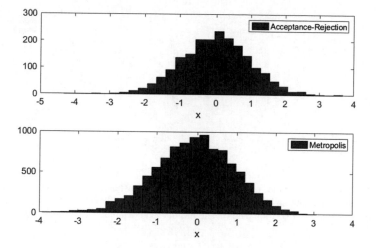

Fig. 10.4 Histograms of random numbers from a Gaussian distribution, generated by the acceptance-rejection method (top) and by the Metropolis-Hastings algorithm (bottom)

10.8   Numerical Evaluation of Path Integrals

We now turn to evaluating the pricing kernel $p_{BS}(x)$ from (10.22) using the representation as path integral from (10.64) with the definition of the actions S_{BS} from (10.62). After evaluating the path integral using uniformly distributed paths, we then use paths generated with the Metropolis-Hastings algorithm and finally, with paths, mimicking the dynamics of the system.

a large number of uniformly distributed random paths in the range of $\pm 4\,\sigma$. The following code snippet illustrates the generation of Npath paths, which are stored in the matrix x. The first index labels the nslice points and the second index labels the different paths. Next, the contribution of each slice and for each path to S_{BS} is calculated and stored in the variable term1. The contribution of the first slice, starting at $x_0 = 0$, is calculated separately and stored in term2.

```
Npath=20000000;
x=-4*sigma+8*sigma*rand(nslice,Npath);
term1=sum(((x(2:end,:)-x(1:end-1,:)+dt*rfhat).^2,1);
term2=(x(1,:)+dt*rfhat).^2;
eSBS=exp(-(term1+term2)/(2*dt*sigma^2));
```

Both terms enter in the calculation of $e^{S_{BS}}$, in the script denoted by eSBS, with S_{BS} given in (10.62). Note that we have not included the discount factor $\varepsilon r_f N$ and postpone the normalization of the path integral until later.

We complete the calculation of the path integral in the following code segment, where we first cast the end positions of the paths onto a grid xx with spacing $\sigma/5$,

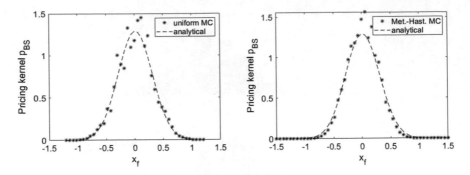

Fig. 10.5 The asterisks show the pricing kernel, evaluated by using 2×10^7 paths constructed from uniformly distributed random numbers (left), and evaluated using 10^5 paths that are generated by the Metropolis-Hastings algorithm (right). The red dashed lines indicate the analytical result from (10.22)

allocate space for the `path_integral`, and loop over all paths. Inside the loop we add the contribution of each path to the appropriate grid point.

```
ix=round(x(end,:)/(sigma/5));  ixmin=min(ix);  ixmax=max(ix);
xx=(ixmin:ixmax)*sigma/5;
path_integral=zeros(1,ixmax-ixmin+1);
for k=1:Npath
    ipos=ix(k)-ixmin+1;
    path_integral(ipos)=path_integral(ipos)+eSBS(k);
end
i0=(xx(2)-xx(1))*sum(path_integral);  % normalize
path_integral=exp(-rf*t)*path_integral/i0;
```

Finally, we calculate the normalized path integral using the previously determined normalization constant `i0` and include the discount factor $e^{-r_f t}$. The left-hand side in Fig. 10.5 shows the numerically evaluated pricing kernel p_{BS} as asterisks, whereas the red dashed line shows the analytic result from (10.22) or (10.69). For this plot, we used 2×10^7 sample path to evaluate the path integral. Smaller numbers lead to significant deviations of the numerical result from the analytical results. Despite the large number of path is the running time only a few seconds on a desktop computer. The MATLAB script used to prepare the plot is reproduced in Appendix B.6. The reader is encouraged to vary the parameters and explore how the plot changes.

   In a second example we employ the Metropolis-Hastings algorithm to generate the paths. All we have to do is to replace the path-generation algorithm by the following code, which first defines the desired number of paths and then specifies the distribution of random numbers to follow a Gaussian distribution with a rms of 3σ. Next, the starting value $x0$ and β for the `metropolis3()` function, discussed in the previous section and reproduced in Appendix B.5, are defined.

```
Npath=100000;                                    % sample paths
h=@(x)exp(-x.^2/(2*(2*sigma)^2));   % Metropolis-Hastings
x0=0.01; beta=3*sqrt(sigma^2*dt);
y=metropolis3(h,beta,1000,x0);          % burn-in
x=metropolis3(h,beta,Npath*nslice,y(1000));
x=reshape(x,[nslice,Npath]);
```

After 1000 burn-in iterations, `metropolis3()` returns `Npath*nslice` random numbers that are generated consecutively and therefore describe paths that are more correlated than the uniformly generated random numbers from the previous example. Most of the paths in his example therefore produce larger contributions to the path integral. The `reshape()` command is used to recast the one-dimensional array `x`, returned from `metropolis3()`, into the form where each column contains the positions for one path.

After the preparation of the paths, we can re-use the calculation of the path integral, discussed for the previous example, to calculate the path integral that describes the pricing kernel. The right-hand side in Fig. 10.5 shows the numerically evaluated kernel as asterisks and the analytical values by a dashed line. Note that using only 10^5 different paths in this example is sufficient to obtain a comparable accuracy to the previous example where we use 2×10^7 uniformly distributed random numbers.

In a third example we prepare the sample paths by directly simulating the random walk, where the step size is given by $\sigma/\sqrt{N}$ with the number of slices N. We artificially double the step size to give the process a chance to explore the tails of the distribution; should the additional paths meander too far astray their large actions suppresses their contribution to the path integral. Then we use the built-in `cumsum()` function to cumulatively add the steps for each path.

```
Npath=10000;            % sample paths
x=2*randn(nslice,Npath)*sigma/sqrt(nslice);
x=cumsum(x,1);
```

The rest of the simulation stays the same. Exploring different number of paths, we find that even as few as 10^4 paths are sufficient to approximate p_{BS} reasonably well.

In practice, it is not possible to calculate the price for many options with analytical methods and one has to resort to Monto-Carlo methods, such as those discussed in this chapter. To explore this wide field further is, however, beyond our scope. Instead we turn to the control of dynamical systems, either physical or macro-economic systems. As opposed to the stock values on their random walk, which we cannot control, let us explore whether we can do something about the economy as a whole? Can we control that instead? Let us find out in the next chapter.

Exercises

1. Is the operator $x\frac{\partial}{\partial x}$ hermitian or anti-hermitian? Prove your answer!
2. (a) Verify that the expectation value of the Hamilton operator $H = p^2/2m + m\omega^2x^2/2$ of the harmonic oscillator, where $p = i\hbar\partial/\partial x$ and for the wave function $\psi(x) = \left(\beta^2/\pi\right)^{1/4} e^{-\beta^2x^2/2}$ with $\beta^2 = m\omega/\hbar$ is $\langle\psi|H|\psi\rangle = \hbar\omega/2$. Calculate the expectation values of (b) the position $\langle\psi|x|\psi\rangle$, (c) the momentum $\langle\psi|p|\psi\rangle$, (c) the kinetic energy $\langle\psi|(p^2/2m)|\psi\rangle$, and (d) the potential energy $\langle\psi|(m\omega^2x^2/2)|\psi\rangle$.
3. Show that the pricing kernel from (10.22) and the Green's function from (5.11) describe the same quantity.
4. Calculate the integral I using Monte-Carlo methods with

$$I = \int\limits_1^2 \ln(x)e^{-x^{3/4}}dx \ . \tag{10.71}$$

How many random numbers do you need until the result stabilizes within $\Delta I/I \approx 10^{-3}$? Check your result with another numerical integration tool of your choice. Document, how well your Monte-Carlo compares to it.
5. Use the Metropolis-Hastings algorithm to build a random number generator that generates random numbers according to the Cauchy distribution. Make a histogram of the numbers and verify that the numbers are distributed according the Cauchy distribution. Explore different values of β.

References

1. C. Cohen-Tannoudji, B. Diu, F. Laloe, *Quantum Mechanics*, vol. 2 (Wiley, New York, 1977)
2. B. Baaquie, *Quantum Finance* (Cambridge University Press, Cambridge, 2004)
3. R. Feynman, Space-time approach to non-relativistic quantum mechanics. Rev. Mod. Phys. **20**, 367 (1948)
4. R. Feynman, A. Hibbs, *Quantum Mechanics and Path Integrals*, emended edn. (Dover, New York, 2005)
5. J. Dash, *Path Integrals and Options, Part I*, CNRS Preprint CPT-88, PE.2206, see also J Dash, *Quantitative Finance and Risk Management* (World Scientific, Singapore, 1988), p. 2004
6. H. Goldstein, J. Safko, C. Poole, *Classical Mechanics* (Pearson, Harlow, 2014)

Chapter 11
Optimal Control Theory

Abstract After introducing the Solow and Robinson-Crusoe models as examples of real business cycle models—analogous to the equations of motion in physics—we derive the Bellman equation to find a control law for a parameter that optimizes a performance measure. A simple mechanical model, based on a donkey pulling a mass across rough ground, illustrates the basics of the state-space formalism in optimal control theory. Following a review of the relation between Lagrange and Hamilton functions in classical mechanics, this chapter derives Hamilton's equations for general dynamical systems that minimize a performance measure. Here similarities to Hamilton's principle of minimizing the action are pertinent. At this point we can use the newly-developed methods to optimize the donkey's progress, before using the same methods to derive the Riccati equation and analyze linear quadratic regulators that are subsequently used to control a Robinson-Crusoe economy close to its equilibrium.

In this chapter we will discuss methods to control parameters that affect a dynamical system in order to achieve some desirable objective. Examples from engineering are control valves that regulate the flow rate of liquids needed to maintain a chemical reaction; or centrifugal governors, regulators that automatically control the speed of steam engines, the ancestor of the cruise controller found in modern cars. Another group of systems that can be controlled are macroeconomic systems. Central banks control the interest rate at which commercial banks can borrow money. This directly affects the availability of loans to companies, which they use to expand, for example, by opening an additional factory. As a consequence, the unemployment rate drops, because the new jobs in the factories must be filled. The overall effect of the actions of the central bank is (hopefully) an increase of the gross domestic product and the wealth of the population as a whole. At least that is the theory, and a prerequisite to analyze this theory is a mathematical model for the dynamics of the system. In physics, the models are described by equations of motion. In economics, rate equations that describe the dynamics are at the heart of so-called *real business cycle* models [1].

© The Author(s), under exclusive license to Springer Nature Switzerland AG 2021
V. Ziemann, *Physics and Finance*, Undergraduate Lecture Notes in Physics,
https://doi.org/10.1007/978-3-030-63643-2_11

In the next section we will discuss some of these models, and how they are cast into a mathematical form [2] that is amenable to the optimization methods we will discuss further down.

11.1 Macroeconomic Models

First, we consider a simplified version of the *Solow model*, [1] which considers the output y of a company with capital base k, and the question what fraction σ of the output to re-invest in the capital. A larger capital base will enable the company to produce a higher output in the future, but will leave little profit to distribute to share holders today. We analyze this model with suitably chosen discrete time steps, labeled by t. For a company, this could be a three-month period that coincides with common fiscal reporting practice. We therefore label all quantities with a subscript t to denote the time step t. The output $y_t = \lambda f(k_t)$ is then some function $f(k_t)$ of the available capital k_t, multiplied by a parameter λ, which describes the technological level of the company. It is large for a company producing high-tech products and small for a sweat shop in the third world. A commonly used model for the function f is the Cobb-Douglas model $f(k_t) = k_t^{\Theta}$ with the *output elasticity* Θ in the range $0 < \Theta < 1$. Since Θ is always smaller than unity, it describes the effect of *diminishing returns*. A 100 000 Euro increase in capital will make a huge difference for a small computer shop but is hardly noticeable in a large company, such as Apple. The output y_t is thus given in terms of the capital base k_t at time step t by

$$y_t = \lambda f(k_t) = \lambda k_t^{\Theta} . \tag{11.1}$$

But how does the capital k_t change from period to period? This is governed by two effects. First, the capital k_t deprecates at a rate δ, for example, due to equipment that wears down and eventually breaks. Machines used to produce goods lose their value when used and need to be replaced after some time. This is commonly referred to as *writing off* part of the purchasing price for the machines. We describe it by the deprecation parameter δ. The second effect that changes the capital base are investments i_t during period t, such that the capital base k_{t+1} for the subsequent period is given by

$$k_{t+1} = (1 - \delta)k_t + i_t . \tag{11.2}$$

And here the rate of reinvestment σ comes into play. In this model the investments i_t are given by a fraction σ of the output y_t

$$i_t = \sigma y_t = \sigma \lambda f(k_t) . \tag{11.3}$$

After inserting (11.3) into (11.2), we obtain

$$k_{t+1} = (1 - \delta)k_t + \sigma \lambda f(k_t) , \tag{11.4}$$

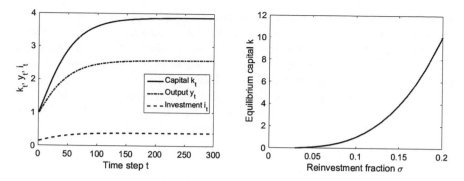

Fig. 11.1 Left: the time evolution of the Solow model from (11.2) with parameters $\lambda = 1, \delta = 0.1, \sigma = 0.15$, and $\Theta = 0.7$. Right: the equilibrium value of the capital $\bar{k}$ as a function of the re-investment rate σ

which describes a dynamical system that maps the capital k_t from one period to the next. It is straightforward to code this equation in MATLAB and follow the dynamical variable k_t, and also y_t and i_t, as a function of the period t. On the left-hand side of Fig. 11.1, we show the result of the simulation that was generated by the code from Appendix B.7 and uses the parameters $\lambda = 1, \delta = 0.1, \sigma = 0.15$, and $\Theta = 0.7$. Starting from initial value $k_0 = 1$, we observe that the capital k_t increases as a consequence of re-investing and approaches an equilibrium after about 150 periods. Note that the capital at the end of the simulation period is about four times larger than it was initially.

Because we are interested in a high output y_t on the long run, let us find out how the equilibrium level depends on the system parameters. We therefore introduce equilibrium values $\bar{k} = k_{t+1} = k_t$ and $\bar{i} = i_{t+1} = i_t$. They follow from requiring that these values do not change from one period to the next. Inserting the equilibrium values into (11.4), we find

$$\bar{k} = (1 - \delta)\bar{k} + \sigma\lambda f(\bar{k}) \quad \text{or} \quad \bar{k} = \frac{\sigma\lambda}{\delta} f(\bar{k}) = \frac{\sigma\lambda}{\delta}\bar{k}^{\Theta} . \tag{11.5}$$

This is an equation for the capital at equilibrium $\bar{k}$, which depends on the system parameters σ, λ, and δ. For the Cobb-Douglas function $f(\bar{k}) = \bar{k}^{\Theta}$ we can solve the second equation with the result $\bar{k} = (\sigma\lambda/\delta)^{1/(1-\Theta)}$. The right-hand side in Fig. 11.1 shows $\bar{k}$ as a function of σ where observe that large values of σ increase the equilibrium values $\bar{k}$ significantly. So, why not increase the reinvestment even further? The answer is, of course, that we want to use the not-reinvested part of the output, the profit $p_t = y_t - i_t$, for something else.

Let us therefore extend the model to account for the profit p_t and the fact that we value it. Since this economic model only has a single agent, it is called a *Robinson Crusoe economy* [1] (you'll guess why). The equations that describe the dynamics of this economy are very similar to those of the Solow model

$$k_{t+1} = (1 - \delta)k_t + i_t \quad \text{and} \quad p_t = \lambda f(k_t) - i_t . \tag{11.6}$$

What we actually want to do is to maximize is the "joy" we derive from the profits
during the period t, which is quantified by the *utility* function $u(p_t)$. An often-used
form is the logarithmic utility function $u(p_t) = \log(p_t)$. But we do not only maximize
the utility for a single period, but into the foreseeable future. We do, however, care
more about the near future than about the distant future. We take this effect into
account by introducing the parameter β, which "discounts" the future utility $u(p_t)$.
This is encapsulated in the *objective functional* $V_t[p]$, which depends on the time
series of all future profits, rather than a single value. We indicate this by using square
brackets. $V_t[p]$ is then defined by

$$V_t[p] = \sum_{j=0}^{\infty} \beta^j u(p_{t+j}) = \sum_{j=0}^{\infty} \beta^j \log(p_{t+j}) , \qquad (11.7)$$

where we replaced the utility u by the logarithm in the second equality. (11.6)
describes the dynamics of the system and (11.7) is the objective functional that
we seek to maximize. It depends on the time series of profits, which in turn, depend
on the control parameter, the investment i_t. The question to answer is now: which
sequence of investments i_t maximizes $V_t[p]$? Thus, what is the best *policy* or control
law to split the output $y_t = p_t + i_t$ into investment i_t and profit p_t? If we pull out
too much profit to enjoy, we will have less capital in the future, such that there will
be less profit to enjoy in the future. Conversely, re-investing too much, leaves too
little profit to enjoy, despite having a huge capital base. Obviously there should be
an optimum investment policy.

Let us now try to find this optimum policy by splitting the objective functional
into the most recent term and the rest. Using $p_t = \lambda f(k_t) + (1 - \delta)k_t - k_{t+1}$, we
can write

$$V_t[p] = \sum_{j=0}^{\infty} \beta^j u(p_{t+j}) = u(p_t) + \sum_{j=1}^{\infty} \beta^j u(p_{t+j}) = u(p_t) + \beta \sum_{i=0}^{\infty} \beta^i u(p_{t+1+i})$$
$$= u(p_t) + \beta V_{t+1}[p] , \qquad (11.8)$$

where we first split the sum into terms with $j = 0$ and $j \geq 1$ and then introduce the
new variable $i = j - 1$, which allows us to rewrite the second term as $\beta V_{t+1}[p]$.
This equation is called *Bellman equation*. It recursively expresses the objective
functional $V_t[p]$ at time t through the utility that is closest in time $u(p_t)$ and the
objective functional $V_{t+1}[p]$ that encodes our minimization objective for the future,
starting at $t + 1$.

This recursive description of the objective functional $V_t[p]$ gives us a handle to
find the optimum policy. If we assume that we already had found the optimum policy
for $V_{t+1}[p]$, let us call it $V_{t+1}^*[p]$, all we have to do in order to find the optimum
for $V_t[p]$, is to minimize the additional term $u(p_t)$. This algorithm, which forms
the basis of *dynamic programming* [3], requires, however, to start from the distant
future and work ourselves back towards the point closest in time. The difficulties
of the infinite time horizon (the sum extends to infinity) notwithstanding, we will

investigate the optimum by calculating the derivative of $V_t[p]$ with respect to k_{t+1}, where we have to express all profits p_t through the values of the capital k_t via $p_t = \lambda f(k_t) + (1 - \delta)k_t - k_{t+1}$. Using k_{t+1} instead of p_t is irrelevant, because both contain the information about how much of the output $y_t = \lambda f(k_t)$ is carried over into the next time period. We find conditions for the p_t or equivalently for the k_{t+1} by setting the derivative to zero

$$
\begin{aligned}
0 &= \frac{\partial u \left(\lambda f(k_t) + (1 - \delta)k_t - k_{t+1} \right)}{\partial k_{t+1}} + \beta \frac{\partial V_{t+1}[p]}{\partial k_{t+1}} \\
&= -u' \left(\lambda f(k_t) + (1 - \delta)k_t - k_{t+1} \right) \\
&\quad + \beta u' \left(\lambda f(k_{t+1}) + (1 - \delta)k_{t+1} - k_{t+2} \right) \left[\lambda f'(k_{k+1}) + 1 - \delta \right] \\
&= -u'(p_t) + \beta u'(p_{t+1}) \left[\lambda f'(k_{k+1}) + 1 - \delta \right]
\end{aligned}
\tag{11.9}
$$

where $u'(p)$ denotes the derivative of the utility function with respect to its argument. Note that $V_{t+1}[p]$ contains the k_t shifted by one period, which accounts for the inner derivative in the square bracket in the second equality. The last equality only re-expresses the arguments of the utility functions in terms of the profits, rather than the capital in order to make the equation easier to read.

Using the recursive description of the objective functional, we could now progress to find an equation for the next period and by repeating this procedure we arrive at an infinite sequence of equations like (11.9). Even taking the initial capital k_0 into account, it is still impossible to solve this infinite sequence of equations. We can, however, close the equations by assuming to take out all capital as profit at a far-away time. This terminates the infinite regression and leads to a finite-dimensional set of non-linear and coupled equations, which is difficult to solve. Therefore, instead of analyzing the transient system, we continue with a discussion of the steady-state solution that the system approaches asymptotically.

In the steady state, we will have $k_t = k_{t+1} = \bar{k}$ and therefore also $u(p_t) = u(p_{t+1})$, which allows us to simplify the last line in (11.9) to

$$
0 = -1 + \beta \left[\lambda f'(\bar{k}) + 1 - \delta \right] \quad \text{or} \quad \frac{1}{\beta} - 1 + \delta = \lambda f'(\bar{k}) .
\tag{11.10}
$$

If we assume a Cobb Douglas production function $f(k) = k^\Theta$, we can solve for the equilibrium capital base $\bar{k}$, the profit $\bar{p}$ and the investment $\bar{i}$ rate.

$$
\bar{k} = \left[\frac{1}{\lambda \Theta} \left(\frac{1}{\beta} - 1 + \delta \right) \right]^{\frac{1}{\Theta - 1}} , \quad \bar{p} = \lambda \bar{k}^\Theta - \delta \bar{k} , \quad \text{and} \quad \bar{i} = \delta \bar{k}
\tag{11.11}
$$

These equations allow us to evaluate, or example, the effect of the deprecation δ on the steady-state profits. This provides us with information on how the failure rate of the equipment affects profits. We therefore plot $\bar{k}$, $\bar{p}$, and $\bar{i}$ as a function of δ on the left-hand side in Fig. 11.2. Here we use $\beta = 0.8$, $\Theta = 0.7$, and $\lambda = 1$ in the simulation. We observe that the sustainable capital $\bar{k}$, shown as the solid black

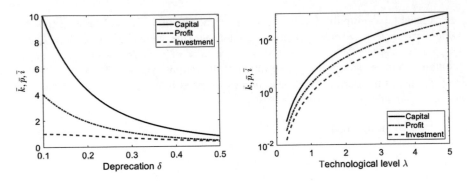

Fig. 11.2 The steady-state values of the capital $\bar{k}$, the profit $\bar{p}$, and the investment $\bar{i}$ from (11.11) for $\beta = 0.8$, $\Theta = 0.7$, $\lambda = 1$ as a function of the deprecation δ (left) and as a function of the technological level λ (right), where δ was set to $\delta = 0.2$

line, decreases from about 10 to a little over 2 when δ increases from 0.1 to 0.3. Simultaneously the profits decrease by approximately the same factor. Apparently it pays off to buy high-quality equipment and maintain it well. The plot on the right-hand side in Fig. 11.2 shows the capital, profit, and investment for $\delta = 0.2$ as a function of the technological level λ, which allows us to assess the difference in profits when comparing a sweat shop with a small value of $\lambda \approx 0.3$ to that of a much more advanced company with $\lambda = 3$. The latter sustains a more than 2000 times higher capital base (solid black line) and correspondingly higher profits, shown as the dash-dotted red line.

Before moving on to optimize the models, let us briefly discuss a few extensions of the basic model. One extension comprises of adding random effects, where we make the technological level λ of our process a random variable that meanders around an average value. This simulates, for example, equipment breaking down at random moments in time. The random process for λ then obeys the following dynamics

$$\lambda_{t+1} = \gamma \lambda_t + (1 - \gamma) + a\varepsilon_t \quad \text{with} \quad \langle \varepsilon \rangle = 0 \text{ and } \langle \varepsilon^2 \rangle = 1 . \tag{11.12}$$

Here ε_t is a sequence of random shocks with the statistical properties specified. Moreover, γ specifies the time scale over which the process has "memory" and the term $1 - \gamma$ ensures that the process has a mean of unity. Note that this process is a variant of the AR(1) model that we discussed in Chap. 8. The dynamics of the Robinson Crusoe economy with random shocks is then given by the equations

$$
\begin{aligned}
k_{t+1} &= (1 - \delta)k_t + i_t \\
\lambda_{t+1} &= \gamma \lambda_t + (1 - \gamma) + a\varepsilon_t \\
p_t &= \lambda_t f(k_t) - i_t \\
V[p] &= \sum_{j=0}^{\infty} \beta^j \langle \log(p_{t+j}) \rangle ,
\end{aligned}
\tag{11.13}
$$

where the only difference to the previous model is the additional random process for λ and the need to use the expectation value of the utility, expressed by the angle brackets, in the definition of the objective functional.

The second extension we consider adds the effect of labor h_t that is required to produce the output y_t. This extension describes the *student's dilemma*, which is also known as *Hansen's model*. For the output y_t at time step t we use a Cobb-Douglas production function, given by

$$y_t = \lambda f(k_t, h_t) = \lambda k_t^\Theta h_t^{1-\Theta} . \tag{11.14}$$

The dilemma of the student comes from the fact that she has to split her time between work h_t and leisure $l_t = 1 - h_t$ to goof off. The leisure l_t is valued in the utility function, but the the work h_t must be expended to achieve an objective, which now values both profits p_t and leisure l_t and has the form $V[p, l] = \sum_{j=0}^{\infty} \beta^j u(p_{t+j}, l_{t+j})$. The complete dynamical system, including the objective functional is then given by

$$
\begin{aligned}
k_{t+1} &= (1 - \delta)k_t + i_t \\
p_t &= \lambda f(k_t, h_t) - i_t \\
l_t &= 1 - h_t
\end{aligned}
\tag{11.15}
$$

$$V[p, l] = \sum_{j=0}^{\infty} \beta^j u(p_{t+j}, l_{t+j}) ,$$

where $u(p_t, l_t)$ is often assumed to have a logarithmic dependence

$$u(p_t, l_t) = \log(p_t) + w \log(l_t) \tag{11.16}$$

with some weight w to assign a relative weight to the two contributions. Some people value money higher than their spare time; they use a small value of w. Others do not care much about money, but prefer to goof off instead; they use a large value of w.

In the third extension we illustrate how to make the evolution equations, which so far use discrete time steps, time-continuous. This involves making the duration of a time step Δt infinitely small, such that we can treat differences from period to period as derivatives. Let us therefore denote the profit per unit time by $p' = p_t/\Delta t$ and the other quantities λ', δ' and i' are likewise the corresponding quantities, divided by Δt. As before, β describes a factor to discount future utilities and p_0' is a reference profit to make the argument of the logarithm unitless. The continuous-time version of (11.6) and (11.7) is

$$
\begin{aligned}
\dot{k} &= -\delta'k + i' \\
p' &= \lambda' f(k) - i'
\end{aligned}
\tag{11.17}
$$

$$V[p'] = \int_0^\infty \beta^t \log(p'(t)/p_0')dt ,$$

where the derivative of the capital $\dot{k}$ in the limit of $\Delta t \to 0$ is given by $\dot{k} = (k_{t+1} - k_t)/\Delta t$.

Let us now briefly introduce a generic notation that is widely used in the literature. Discrete-time optimization problems are often defined using a framework, where the dynamics of the system is described by difference equations $\mathbf{x}_{t+1} = F(\mathbf{x}_t) + G(\mathbf{u}_t)$ for the *state variables* $\mathbf{x}$ and the *controllers* $\mathbf{u}$. Based on these equations of motion that step the state variables forward in time, we then try to affect *observables* $\mathbf{y}_t = H(\mathbf{x}_t)$, in such a way that an objective functional $V[\mathbf{y}, \mathbf{u}]$ is minimized. Here $V[\mathbf{y}, \mathbf{u}]$ depends on the times series of both observables $\mathbf{y}$ and controllers $\mathbf{u}$. For linear systems the functions F, G, and H can be represented by matrices.

If the system is continuous in time, the difference equations are replaced by differential equations of the form $\dot{\mathbf{x}} = F(\mathbf{x}) + G(\mathbf{u})$ for the state vectors $\mathbf{x}$ and the controllers $\mathbf{u}$. Likewise, observables $\mathbf{y} = H(\mathbf{x})$ that depend on the state vectors $\mathbf{x}$ are introduced and the objective of the optimization is to minimize the objective functional $V[\mathbf{y}, \mathbf{u}]$, which is typically an integral over a function of $\mathbf{y}$ and $\mathbf{u}$ for some time interval. Representing control problems by state variables, controllers and observables is often referred to as *state-space formalism*. Such systems do not only appear in economics, but also in physics and engineering. Let us therefore consider a simple example, based on a donkey pulling a mass across rough ground.

11.2   Control and Feedback

The donkey pulling a mass m is shown in Fig. 11.3. The dynamics of the somewhat simplified system is determined by the equations of motion for the mass

$$m\ddot{x} + \alpha\dot{x} = F ,\qquad\qquad(11.18)$$

where α describes the friction between the mass and the rough ground. F is the force—the controller—with which the donkey pulls the mass. To bring this equation into the form mentioned at the end of the previous section, we convert it to state-space form by introducing the state variables x_1 and x_2, defined by

$$x = x_1 \quad \text{and} \quad \dot{x} = \dot{x}_1 = x_2 .\qquad\qquad(11.19)$$

This allows us to write the equations of motion as a set of first-order differential equations

$$\begin{aligned} \dot{x}_1 &= x_2 \\ \dot{x}_2 &= -\frac{\alpha}{m}x_2 + \frac{F}{m} , \end{aligned}\qquad\qquad(11.20)$$

which has the form given at the end of the previous section, provided we identify the force as the controller $u = F$. Since this system of differential equations is linear, it can also be written in matrix form

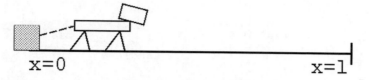

Fig. 11.3 The donkey pulls a mass across rough ground, which is modeled by a velocity-dependent friction force

$$\begin{pmatrix} \dot{x}_1 \\ \dot{x}_2 \end{pmatrix} = \begin{pmatrix} 0 & 1 \\ 0 & -\frac{\alpha}{m} \end{pmatrix} \begin{pmatrix} x_1 \\ x_2 \end{pmatrix} + \begin{pmatrix} 0 \\ \frac{1}{m} \end{pmatrix} u \, , \tag{11.21}$$

which makes it easy to adapt for a numerical treatment. We will return to the donkey in Sect. 11.5, but we first need to discuss different types of objective functionals and other constraints that we may have to satisfy in the optimization.

Besides the equations of motion that relate the state variables x_1 and x_2 to the controller u, these parameters might be constrained by limits. Examples of such limits are

$$0 \le x = x_1 \le l \qquad \text{stay inside limits}$$
$$0 \le \dot{x} = x_2 \le v_{max} \qquad \text{speed limit}$$
$$-F_{max} \le F = u \le F_{max} \qquad \text{limited force.}$$

Another constraint might be to require the mass to be delivered on time. This makes the final time t_f a constraint, which has a major influence on the objective functional $J[\mathbf{x}, \mathbf{u}]$. The specific objective depends on the particular case, and further examples come to mind

$$\text{Minimum time}: J[x, u] = t_f - t_0 = \int_0^l \frac{dx}{\dot{x}}$$
$$\text{Minimum fuel}: J[x, u] = \int_{t_0}^{t_f} |u(t)| dt$$
$$\text{Minimum power}: J[x, u] = \int_{t_0}^{t_f} u(t)^2 dt \tag{11.22}$$
$$\text{Reach the end}: x(t_f) = l$$
$$\text{...at speed zero}: \dot{x}(t_f) = 0$$
$$\text{Use optimum speed } v_d: J[x, u] = \int_{t_0}^{t_f} \left[(\dot{x}(t) - v_d)^2 + wu(t)^2 \right] dt$$

Let us illustrate the different objective functionals with an example. Consider air travel from one point $x = 0$ to another $x_f = l$, traveling at height h_0, which is illustrated in Fig. 11.4. The state variables are thus the distance $x_1 = x$ and the height $x_2 = h$ and the controllers are the thrust $u_1 = T$ of the jet engine and the elevator angle $u_2 = \phi$ of a control surface of the airplane. There are multiple constraints to satisfy: one of them is to reach the destination on time, which requires t_f to be

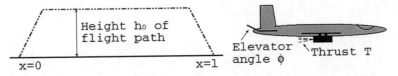

Fig. 11.4 Optimizing air traffic involves a number of constraints, such as covering a specified distance $0 \le x \le l$, arriving at a specific time final time t_f, flying at a predetermined height h_0, or minimizing the expended fuel, among others. The controllers to achieve these, sometimes conflicting requirements, are, for example, the elevator angle and the thrust of the jet engines

fixed. Moreover, we want to reach the destination, which implies $x_f = l$ and we want to reach it with speed zero. This implies $\dot{x}_f = 0$ and, more importantly, $\dot{h}_f = 0$. For the objective functional we have to fly at h_0 as much as possible, which results in the objective functional

$$J[\mathbf{x}, \mathbf{u}] = \int_{t_0}^{t_f} (x_2(t) - h_0)^2 dt \ . \tag{11.23}$$

If we want to fly as economical as possible, we might want to minimize fuel consumption by minimizing the integral over the thrust $u_1 = T$

$$J[\mathbf{x}, \mathbf{u}] = \int_{t_0}^{t_f} |u_1| dt \ . \tag{11.24}$$

In reality a weighted sum of several objectives would probably be used.

Equipped with a good understanding of the constraints, we turn to the generic formulation of an *optimal control problem*, which is based on dynamic equations

$$\dot{\mathbf{x}} = \mathbf{a}(\mathbf{x}, \mathbf{u}, t) \tag{11.25}$$

to express the equations of motion. It is slightly more general than the one we introduced at the end of the previous section. Here we assume that the observables $\mathbf{y}$ are identical to one or several of the state variables. The dynamical system, described by (11.25), has to be controlled in order to minimize the objective functional $J[\mathbf{x}, \mathbf{u}]$ that has the form

$$J[\mathbf{x}, \mathbf{u}] = h(\mathbf{x}(t_f), t_f) + \int_{t_0}^{t_f} g(\mathbf{x}(t), \mathbf{u}(t), t) dt \ . \tag{11.26}$$

It contains one term that integrates over a performance measure $g(\mathbf{x}, \mathbf{u})$; in the macroeconomic examples this was the utility function. Here $g(\mathbf{x}, \mathbf{u}, t)$ depends both on state variables $\mathbf{x}$ and controllers $\mathbf{u}$. It is an *integral constraint*. The second contribution is characterized by the function h that depends only on the state variables $\mathbf{x}_f$ at the final time t_f and is often called *end-point constraint*.

In many applications the equations of motion and their dependence on the controllers is given by a linear set of equations

$$\dot{\mathbf{x}} = A\mathbf{x} + B\mathbf{u} \tag{11.27}$$

where the matrices A and B can depend on time, but often are constant. Moreover, in many applications, the objective functional can be phrased as an integral over a quadratic form, given by

$$J[\mathbf{x}, \mathbf{u}] = \int_{t_0}^{t_f} \left(\mathbf{x}^t Q \mathbf{x} + \mathbf{u}^t R \mathbf{u} \right) dt , \tag{11.28}$$

where Q and R are positive semi-definite, possibly time-dependent, matrices. Quadratic forms are computationally attractive, because they lead to solutions that can be found by analytic means, as we shall see in a little while.

Solving an optimal control problem involves determining the controller $\mathbf{u}(t)$ given as a function of time. We thus seek to minimize a functional $J[\mathbf{x}, \mathbf{u}]$ in order to determine a function $\mathbf{u}(t)$, which is vaguely similar to finding the equations of motion from the action integral $S[q] = \int_{t_0}^{t_f} L(q, \dot{q}) dt$ that depends on a Lagrangian $L(q, \dot{q})$. In Chap. 3 we used methods from variational calculus to determine the Euler-Lagrange equations. They are the equations of motion, whose solution gives the trajectory $q(t)$. Here we have a similar problem. The sought controller $\mathbf{u}$ takes the role of the trajectory, the objective $J[\mathbf{x}, \mathbf{u}]$ takes the role of the action, and $g(\mathbf{x}, \mathbf{u}, t)$ that of the Lagrangian. The only difficulty is that here we have (11.25) changing the state variables simultaneously. It turns out that this difficulty can be overcome by using a Hamiltonian—instead of a Lagrangian—framework. We therefore briefly recapitulate the basics of Hamiltonian mechanics [4, 5].

11.3 Hamiltonian Mechanics

In Sect. 3.1 we found that minimizing the action integral $S[q]$ that is defined in terms of the Lagrangian $L(q, \dot{q})$

$$\delta S[q] = \delta \left[\int_{t_0}^{t_f} L(q, \dot{q}) dt \right] = 0 \quad \text{leads to} \quad 0 = \frac{d}{dt} \frac{\partial L}{\partial \dot{q}} - \frac{\partial L}{\partial q} , \tag{11.29}$$

which are the Euler-Lagrange equations to describe the equations of motion for the trajectory $q(t)$. Whereas the Lagrangian $L(q, \dot{q})$, here assumed to be time invariant, depends on the state variable q and its derivative $\dot{q}$, we want the Hamiltonian to depend on q and the canonical momentum $p = \partial L/\partial \dot{q}$ instead [4, 5]. Furthermore, we require the partial derivatives of $H(q, p)$ to return the time-derivatives of q and p. As a prerequisite, we note that the Euler-Lagrange equations allow us to write

$$0 = \frac{d}{dt}\frac{\partial L}{\partial \dot{q}} - \frac{\partial L}{\partial q} = \frac{d}{dt}p - \frac{\partial L}{\partial q} \quad \text{or} \quad \dot{p} = \frac{\partial L}{\partial q} . \tag{11.30}$$

Using these relations we can express the total differential of the Lagrangian in the following way

$$dL = \frac{\partial L}{\partial q}dq + \frac{\partial L}{\partial \dot{q}}d\dot{q} = \dot{p}dq + pd\dot{q} = \dot{p}dq + [d(p\dot{q}) - \dot{q}dp] , \tag{11.31}$$

where we replace the partial derivatives of the Lagrangian by p and $\dot{p}$ with the help of the definition of the canonical momentum p and (11.30). We remove the term proportional to $d\dot{q}$ with the help of $pd\dot{q} = d(p\dot{q}) - \dot{q}dp$. After collecting the terms proportional to dq and dp on the left-hand side we arrive at

$$\dot{q}dp - \dot{p}dq = d(p\dot{q} - L) = dH = \frac{\partial H}{\partial p}dp + \frac{\partial H}{\partial q}dq , \tag{11.32}$$

where we define the terms that are left over on the right-hand side as the Hamiltonian $H = p\dot{q} - L$. Moreover, by writing $dH(q, p)$ through its partial derivatives and comparing coefficients with the left-hand side, we recover Hamilton's equations, summarized in the following equations

$$H = p\dot{q} - L, \quad \dot{q} = \frac{\partial H}{\partial p}, \quad \dot{p} = -\frac{\partial H}{\partial q} . \tag{11.33}$$

Note that the requirement to use the canonical momentum p instead of $\dot{q}$ leads to the introduction of the Hamiltonian $H(q, p)$ and replaces the Euler-Lagrange equations, which are of second order, by pairs of first-order equations—Hamilton's equations. This transformation from Lagrangian to Hamiltonian is usually referred to as a *Legendre transformation*.

Let us now explore how this formalism helps us to solve the optimal-control problem.

11.4   Hamiltonians for Optimal Control

In this section, we omit the end-point constraints in order to make the problem more manageable, which can thus be summarized by minimizing

$$J[\mathbf{x}, \mathbf{u}] = \int_{t_0}^{t_f} g(\mathbf{x}, \mathbf{u})dt \quad \text{subject to} \quad \dot{\mathbf{x}} = \mathbf{a}(\mathbf{x}, \mathbf{u}) . \tag{11.34}$$

We now treat the equations of motion as a constraint and introduce Lagrange multipliers $\mathbf{p}$, which are usually called *costate variables*. We will later see that we can interpret the costates as the "momentum" corresponding to the state variables $\mathbf{x}$. The objective functional, now including the equations of motion, can be written as

$$J[\mathbf{x}, \mathbf{u}, \mathbf{p}] = \int_{t_0}^{t_f} \left\{ g(\mathbf{x}, \mathbf{u}) + \mathbf{p}^t \left[\mathbf{a}(\mathbf{x}, \mathbf{u}) - \dot{\mathbf{x}} \right] \right\} dt \ . \tag{11.35}$$

This objective functional depends on $\mathbf{x}$, $\mathbf{u}$, and the costates $\mathbf{p}$, which can all vary independently. Let us therefore calculate the variation δJ and and collect terms proportional to $\delta \mathbf{x}$, $\delta \mathbf{u}$, and $\delta \mathbf{p}$ independently. We then arrive at

$$\delta J[\mathbf{x}, \mathbf{u}, \mathbf{p}] = \int_{t_0}^{t_f} \left\{ \frac{\partial g}{\partial \mathbf{x}} \delta \mathbf{x} + \frac{\partial g}{\partial \mathbf{u}} \delta \mathbf{u} + \delta \mathbf{p}^t (\mathbf{a} - \dot{\mathbf{x}}) \right. $$

$$\left. + \mathbf{p}^t \left(\frac{\partial \mathbf{a}}{\partial \mathbf{x}} \delta \mathbf{x} + \frac{\partial \mathbf{a}}{\partial \mathbf{u}} \delta \mathbf{u} - \delta \dot{\mathbf{x}} \right) \right\} dt \tag{11.36}$$

$$= \int_{t_0}^{t_f} \left\{ \left[\frac{\partial g}{\partial \mathbf{x}} + \mathbf{p}^t \frac{\partial \mathbf{a}}{\partial \mathbf{x}} \right] \delta \mathbf{x} - \mathbf{p}^t \delta \dot{\mathbf{x}} + \delta \mathbf{p}^t (\mathbf{a} - \dot{\mathbf{x}}) \right. $$

$$\left. + \left[\frac{\partial g}{\partial \mathbf{u}} + \mathbf{p}^t \frac{\partial \mathbf{a}}{\partial \mathbf{u}} \right] \delta \mathbf{u} \right\} dt \ . $$

where $\partial/\partial \mathbf{x}$ denotes the gradient with respect to the components of $\mathbf{x}$. Careful inspection shows that there is also a term proportional to $\delta \dot{\mathbf{x}}$, which we can rewrite with the help of a partial integration

$$\int_{t_0}^{t_f} \left(\mathbf{p}^t \delta \dot{\mathbf{x}} \right) dt = \int_{t_0}^{t_f} \left[\frac{d}{dt} \left(\mathbf{p}^t \delta \mathbf{x} \right) - \dot{\mathbf{p}}^t \delta \mathbf{x} \right] dt = - \int_{t_0}^{t_f} \left(\dot{\mathbf{p}}^t \delta \mathbf{x} \right) dt \ , \tag{11.37}$$

where the total derivative vanishes, because we assume that t_0 and t_f are fixed. Finally we can write the total variation of the objective $\delta J[\mathbf{x}, \mathbf{u}, \mathbf{p}]$ as

$$\delta J[\mathbf{x}, \mathbf{u}, \mathbf{p}] = \int_{t_0}^{t_f} \left\{ \left[\frac{\partial g}{\partial \mathbf{x}} + \mathbf{p}^t \frac{\partial \mathbf{a}}{\partial \mathbf{x}} + \dot{\mathbf{p}}^t \right] \delta \mathbf{x} \right. \tag{11.38}$$

$$\left. + \delta \mathbf{p}^t \left[\mathbf{a} - \dot{\mathbf{x}} \right] + \left[\frac{\partial g}{\partial \mathbf{u}} + \mathbf{p}^t \frac{\partial \mathbf{a}}{\partial \mathbf{u}} \right] \delta \mathbf{u} \right\} dt \ . $$

Since the variations $\delta\mathbf{x}$, $\delta\mathbf{u}$, and $\delta\mathbf{p}$ are independent, the terms in the square brackets must vanish individually. Collecting these equations, we find the equivalent of Hamilton's equations, which are the "equations of motion"

$$\dot{\mathbf{p}} = -\frac{\partial g}{\partial \mathbf{x}} - \mathbf{p}^t \frac{\partial \mathbf{a}}{\partial \mathbf{x}}$$
$$\dot{\mathbf{x}} = \mathbf{a}(\mathbf{x}, \mathbf{u}) \tag{11.39}$$
$$0 = \frac{\partial g}{\partial \mathbf{u}} + \mathbf{p}^t \frac{\partial \mathbf{a}}{\partial \mathbf{u}}$$

for a Hamiltonian $\mathcal{H}(\mathbf{x}, \mathbf{u}, \mathbf{p})$, given by

$$\mathcal{H}(\mathbf{x}, \mathbf{u}, \mathbf{p}) = g(\mathbf{x}, \mathbf{u}) + \mathbf{p}^t \mathbf{a} \ . \tag{11.40}$$

With this definition of the Hamiltonian, we can finally write the equations of motion from (11.39) as

$$\dot{\mathbf{p}} = -\frac{\partial \mathcal{H}}{\partial \mathbf{x}} \qquad \dot{\mathbf{x}} = \frac{\partial \mathcal{H}}{\partial \mathbf{p}} \qquad 0 = \frac{\partial \mathcal{H}}{\partial \mathbf{u}} \ . \tag{11.41}$$

Now we see that the costates $\mathbf{p}$, originally introduced as Lagrange multipliers, assume the role of the momentum corresponding to the states $\mathbf{x}$. The Hamiltonian $\mathcal{H}$, defined in (11.40), reminds of a Legendre transform of g. Moreover, minimizing the objective functional is now written as the requirement of the partial derivative of $\mathcal{H}(\mathbf{x}, \mathbf{u}, \mathbf{p})$ with respect to $\mathbf{u}$ (the gradient) being zero.

So, what have we gained by rephrasing the problem in a Hamiltonian framework? First, we found a consistent way to handle the optimization when the state vector $\mathbf{x}$ moves around while we change the controls $\mathbf{u}$. Second, the minimization now looks like a normal minimization; we only have to calculate the gradient of the Hamiltonian with respect to the controls $\mathbf{u}$ and set the result equal to zero.

In order to illustrate this new formalism, we will have a second look at the donkey from Sect. 11.2.

11.5 Donkey Revisited

To make the problem as simple as possible (but not simpler than that), we require the donkey to arrive at $x = l$ at time $t = t_f$ and minimize the integral of the power expended, which is proportional to $g(\mathbf{x}, u) = u^2/2$ integrated from $t = 0$ to $t = t_f$. Moreover, we assume that the variables are suitably scaled, such that $m = 1$ and $F = u$. The equations of motion from (11.20) then assume the form

$$\dot{x}_1 = x_2 \tag{11.42}$$
$$\dot{x}_2 = -\alpha x_2 + u$$

and we can directly use (11.40) to write the Hamiltonian as

$$\mathcal{H}(x_1, x_2, u, p_1, p_2) = \frac{1}{2}u^2 + p_1 x_2 + p_2(-\alpha x_2 + u) . \tag{11.43}$$

Note that this Hamiltonian depends on two state variables x_1 and x_2, one controller u, and two costate variables p_1 and p_2. Applying Hamilton's equations (11.41) to this Hamiltonian, we obtain the following set of five equations

$$\dot{p}_1 = -\frac{\partial \mathcal{H}}{\partial x_1} = 0$$

$$\dot{p}_2 = -\frac{\partial \mathcal{H}}{\partial x_2} = -p_1 + \alpha p_2$$

$$\dot{x}_1 = \frac{\partial \mathcal{H}}{\partial p_1} = x_2 \tag{11.44}$$

$$\dot{x}_2 = \frac{\partial \mathcal{H}}{\partial p_2} = -\alpha x_2 + u$$

$$0 = \frac{\partial \mathcal{H}}{\partial u} = u + p_2$$

The first four equations relate the state and costate vectors, but the fifth equation, which contains the minimization constraint $0 = \partial \mathcal{H}/\partial u$ allows us to express the control u through the costate p_2. Replacing $u = -p_2$ in the first four equations results in

$$\dot{p}_1 = 0$$
$$\dot{p}_2 = -p_1 + \alpha p_2$$
$$\dot{x}_1 = x_2 \tag{11.45}$$
$$\dot{x}_2 = -\alpha x_2 - p_2 ,$$

which is a coupled system of four ordinary differential equations for the four variables $x_1, x_2, p_1,$ and p_2, subject to the four boundary conditions $x_0 = 0$ and $\dot{x}_0 = 0$ at $t = 0$ and $x_f = L$ and $\dot{x}_f = 0$ at $t = t_f = T$. Note that there are conditions for both initial and final values, which in general makes it very difficult to satisfy them simultaneously. The donkey, however, is lucky, because the original equations of motion for the two state variables x_1 and x_2 alone are linear and g is quadratic. This makes the equations in (11.45) linear and straightforward to solve, and then to determine the integration constants from satisfying the boundary conditions.

The first equation immediately leads to $p_1 = c_1$, where c_1 is an integration constant. After inserting into the second equation and separating the variables, the integral is elementary and leads to $p_2 = c_1/\alpha + (c_2 - c_1/\alpha)e^{\alpha t}$ with a second integration constant c_2. Inserting this into the last equation produces an inhomogeneous linear differential equation that has the homogeneous solution $z_h = c_3 e^{-\alpha t}$ with a third integration constant c_3. A particular solution z_p to the inhomogeneous equation can be found by the Ansatz $z_p(t) = g(t)e^{-\alpha t}$ which leads to $g(t) = -(c_1/\alpha^2)e^{\alpha t} - (1/2\alpha)(c_2 - c_1/\alpha)e^{2\alpha t}$, such that we find for $x_2(t) = z_h(t) + z_p(t)$, or

$$x_2(t) = c_3 e^{-\alpha t} - \frac{c_1}{\alpha^2} - \frac{1}{2\alpha}\left(c_2 - \frac{c_1}{\alpha}\right)e^{\alpha t} . \tag{11.46}$$

Using this expression allows us to integrate the third equation and determine x_1

$$x_1(t) = c_4 - \frac{c_3}{\alpha}e^{-\alpha t} - \frac{c_1}{\alpha^2}t - \frac{1}{2\alpha^2}\left(c_2 - \frac{c_1}{\alpha}\right)e^{\alpha t} \tag{11.47}$$

with a fourth integration constant c_4. The boundary conditions for $t = 0$, inserted into the equations for $x_1(t)$ and $x_2(t)$, lead to

$$0 = c_4 - \frac{c_3}{\alpha} - \frac{1}{2\alpha}\left(c_2 - \frac{c_1}{\alpha}\right) \quad \text{and} \quad 0 = c_3 - \frac{c_1}{\alpha^2} - \frac{1}{2\alpha}\left(c_2 - \frac{c_1}{\alpha}\right), \tag{11.48}$$

which allows us to solve for $c_1 = 2\alpha^2 c_3 - \alpha^3 c_4$ through c_3 and c_4 and likewise for $c_2 = \alpha^2 c_4$. With these simplifications the boundary conditions for $t = T$ can be written as

$$L = c_4\left[1 + \alpha T - e^{\alpha T}\right] + c_3\left[\frac{e^{\alpha T} - e^{-\alpha T} - 2\alpha T}{\alpha}\right]$$
$$0 = c_4\left[\alpha - \alpha e^{\alpha T}\right] + c_3\left[e^{\alpha T} + e^{-\alpha T} - 2\right] . \tag{11.49}$$

After solving this linear system of equations for c_3 and c_4 with the MATLAB code from Appendix B.8, we are ready to insert the four integration constants into (11.46) and (11.47) and then plot the trajectory $x_1(t)$ and the speed of the donkey $x_2(t)$ in Fig. 11.5.

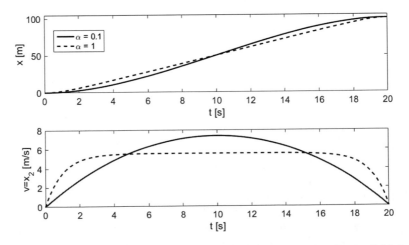

Fig. 11.5 The upper plot shows position of the mass during the traveling time for $\alpha = 0.1/s$ (solid) and $\alpha = 1/s$ (dashed). The lower plot shows the speed for the corresponding cases

In the simulation, we require the donkey to travel to $L = 100\,\text{m}$ in $T = 20\,\text{s}$ and plot the solutions for weak friction $\alpha = 0.1$ as a solid line and for ten times stronger friction as a dashed line. We see that for strong friction it is advantageous to quickly reach a lower and more constant speed, compared to the situation with weak friction, which is characterized by an almost constant acceleration until the mid-point and an equally constant deceleration towards the destination.

With the donkey safely back home in the stable, we now turn to general linear systems, mentioned at the end of Sect. 11.1. Objective functionals that are based on integrals over quadratic forms are rather common and we therefore devote the next section to *linear quadratic regulators*.

11.6 Linear Quadratic Regulators

The dynamics of state vectors $\mathbf{x}$ for linear systems is determined by (11.27) and (11.28)

$$\dot{\mathbf{x}} = A(t)\mathbf{x} + B(t)\mathbf{u} \quad \text{and} \quad J[\mathbf{x}, \mathbf{u}] = \frac{1}{2} \int_{t_0}^{t_f} \left(\mathbf{x}^t Q(t)\mathbf{x} + \mathbf{u}^t R(t)\mathbf{u} \right) dt$$

with matrices $A(t)$ and $B(t)$. The matrices $Q(t)$ and $R(t)$ that appear in the objective functional $J[\mathbf{x}, \mathbf{u}]$ are positive definite. Based on these definitions, we use (11.40) to construct the Hamiltonian

$$\mathcal{H}(\mathbf{x}, \mathbf{u}, \mathbf{p}) = \frac{1}{2}\mathbf{x}^t Q\mathbf{x} + \frac{1}{2}\mathbf{u}^t R\mathbf{u} + \mathbf{p}^t (A\mathbf{x} + B\mathbf{u})$$

$$= \frac{1}{2}\mathbf{x}^t Q\mathbf{x} + \frac{1}{2}\mathbf{u}^t R\mathbf{u} + \mathbf{x}^t A^t \mathbf{p} + \mathbf{u}^t B^t \mathbf{p} . \tag{11.50}$$

Applying (11.41) gives us the equations of motion for the state and costate variables, as well as the constraint condition.

$$\dot{\mathbf{p}} = -\frac{\partial \mathcal{H}}{\partial \mathbf{x}} = -Q\mathbf{x} - A^t \mathbf{p}$$

$$\dot{\mathbf{x}} = \frac{\partial \mathcal{H}}{\partial \mathbf{p}} = A\mathbf{x} + B\mathbf{u} \tag{11.51}$$

$$0 = \frac{\partial \mathcal{H}}{\partial \mathbf{u}} = R\mathbf{u} + B^t \mathbf{p} .$$

Solving the last equation for the controller $\mathbf{u}$ allows us to express it through the costates $\mathbf{p}$ via

$$\mathbf{u} = -R^{-1} B^t \mathbf{p} , \tag{11.52}$$

provided that the matrix R is invertible, which rules out zero eigenvalues of R and the matrix must be positive definite. Replacing the controller $\mathbf{u}$ in the equations for the states and costates, we arrive at

$$\begin{pmatrix} \dot{\mathbf{x}} \\ \dot{\mathbf{p}} \end{pmatrix} = \begin{pmatrix} A & -BR^{-1}B^t \\ -Q & -A^t \end{pmatrix} \begin{pmatrix} \mathbf{x} \\ \mathbf{p} \end{pmatrix} . \tag{11.53}$$

Note that if all matrices A, B, R, and Q are constant, this is a set of linear ordinary differential equations with constant coefficients, which is straightforward to solve.

Instead of attempting a direct numerical solution, we will seek to find a special solution in which the state and costate vectors are related through a matrix $K(t)$ with $\mathbf{p} = K(t)\mathbf{x}$, because jointly with (11.52) this allows us to find a *control law* of the form

$$\mathbf{u} = -R^{-1}B^t K \mathbf{x} , \tag{11.54}$$

which determines the required control value $\mathbf{u}$ directly derived from the current state $\mathbf{x}$. Thus it makes both $\mathbf{x}$ and $\mathbf{u}$ small, as required by the objective functional. Let us now find K. From the definition

$$\mathbf{p} = K(t)\mathbf{x} \quad \text{we obtain} \quad \dot{\mathbf{p}} = \dot{K}\mathbf{x} + K\dot{\mathbf{x}} . \tag{11.55}$$

Solving for $\dot{K}\mathbf{x}$, and repeatedly using (11.51) and (11.53), we arrive at

$$\begin{aligned} \dot{K}\mathbf{x} = \dot{\mathbf{p}} - K\dot{\mathbf{x}} &= \dot{\mathbf{p}} - K(A\mathbf{x} + B\mathbf{u}) \\ &= -Q\mathbf{x} - A^t\mathbf{p} - KA\mathbf{x} + KBR^{-1}B^t\mathbf{p} \\ &= -Q\mathbf{x} - A^t K\mathbf{x} - KA\mathbf{x} + KBR^{-1}B^t K\mathbf{x} . \end{aligned} \tag{11.56}$$

Since this equation must be satisfied for all states $\mathbf{x}$, we can omit the states $\mathbf{x}$ from the equality and obtain the *Riccati equation*

$$\dot{K} = -Q - A^t K - KA + KBR^{-1}B^t K , \tag{11.57}$$

which is a condition for the matrix $K(t)$ in order to satisfy (11.55). Solving this non-linear matrix-valued equation is in general non-trivial, but if we succeed, K determines the optimal-control law via (11.54). This is the controller that maintains small values of $\mathbf{x}$ with the least control effort in the sense that the controllers $\mathbf{u}$ are as small as possible, where the relative weight between the state $\mathbf{x}$ and controllers $\mathbf{u}$ is encoded in the matrices R and Q. For time-invariant systems with all matrices being constant, the steady-state solution with $\dot{K} = 0$, can be found from solving

$$0 = -Q - A^t K - KA + KBR^{-1}B^t K . \tag{11.58}$$

which is an algebraic equation. This makes it easier to solve for K than to determine K from the full Riccati equation (11.57), which is a non-linear differential equation.

In the following section we apply these methods to stabilize the Robinson Crusoe economy close to its equilibrium.

11.7 Controlling the Robinson-Crusoe Economy

Let us now use the optimal feedback controller from the previous section to stabilize the profits of the Robinson-Crusoe economy under stady-state conditions, should the system be perturbed. This is the task of a chief executive officer of a company in order to ensure a reliable and constant dividend paid to the share holders. But here we develop a strategy to do this automatically. Recall that the steady-state conditions of the Robinson-Crusoe economy are given in (11.11). Moreover, Fig. 11.2 shows their dependence on the deprecation δ and on the technological level λ. In order to adapt the model to the continuous-time framework, used in the previous section, we use the dynamics described by (11.17), reproduced here

$$\dot{k} = -\delta'k + i' \quad \text{and} \quad p' = \lambda'k^\Theta - i' ,$$

where the prime denotes the per-unit-time quantities, introduced just before (11.17). Eliminating the investment i' from the two equations leads to

$$\dot{k} = -\delta'k + \lambda'k^\Theta - p' . \tag{11.59}$$

In this equation we recognize the capital k as the state variable. Furthermore, we can control the capital by adjusting the profits p' suitably. Unfortunately, this system is non-linear, but we still attempt to stabilize it, once it has reached its steady state, which is defined by $\dot{\bar{k}} = 0$ and leads to the condition $\bar{p}' = \lambda'\bar{k}^\Theta - \delta'\bar{k}$. Expanding both $k = \bar{k} + x$ and $p' = \bar{p}' + u$ and inserting in (11.59), we obtain

$$\dot{x} = \left(\Theta\lambda'\bar{k}^{\Theta-1} - \delta'\right)x - u . \tag{11.60}$$

With the definitions $a = \Theta\lambda'\bar{k}^{\Theta-1} - \delta'$ and $b = -1$ this equation can be cast into the canonical form, defined by (11.27), of a control problem $\dot{x} = ax + bu$.

A suitable objective functional for this problem is given by the one-dimensional version of (11.28)

$$J[x, u] = \frac{1}{2} \int_{t_0}^{t_f} \left(qx^2 + ru^2\right) dt \tag{11.61}$$

with weights q and r for the state variable x and the controller u, respectively. If we are eager to return to the steady state as quickly as possible, we assign a greater weight to the first term $q \gg r$. Conversely, if we want to keep the share holders happy and maintain a constant level of profits, we use $q \ll r$, which, however, will lead to a longer time to reach the stady state.

Using the formalism from the previous section we solve the steady-state Riccati equation, given by (11.58), and find the factor K, which is done in Exercise 2. Subsequently inserting it in (11.52) leads to the control law

$$u = -\frac{a}{b}\left[1 \pm \sqrt{1 + \frac{qb^2}{ra^2}}\right]x \, . \tag{11.62}$$

Choosing the positive sign for the root and inserting into $\dot{x} = ax + bu$ leads to

$$\dot{x} = -a\sqrt{1 + \frac{qb^2}{ra^2}}x = -\frac{1}{\tau}x \tag{11.63}$$

which shows that the feedback causes perturbations to damp with an exponential time scale τ given by $1/\tau = a\sqrt{1 + qb^2/ra^2}$

Figure 11.6 shows two simulations of the perturbed steady-state of a Robinson Crusoe economy with parameters $\lambda = 1$, $\Theta = 0.7$, and $\delta = 0.2$. The system starts from a perturbed state with $x = k - \bar{k} = -0.1$ and then evolves while applying the controller $u = p' - \bar{p}'$, calculated by the control law specified in (11.62). The upper plot shows the x and u as a function of time t for $q = 5$ and $r = 1$ chosen as weights in the objective functional from (11.61). Here the controller tries to return to the steady state $x = 0$ and $u = 0$ very quickly, but penalizes large values of the controller u five times less. Indeed, we see that deviation from the stady-state profits u, shown as the

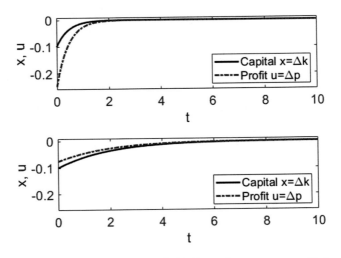

Fig. 11.6 Using a linear quadratic regulator on the Robinson Crusoe economy with $\lambda = 1$, $\Theta = 0.7$, and $\delta = 0.2$. In the upper plot a quick return to the stady state is favored by choosing $q = 5$ and $r = 1$ in (11.61). In the lower plot a small deviation of the profits is favored by choosing $q = 1$ and $r = 5$, which causes the recovery time τ to be much longer

dot-dashed red line assumes large negative values, which implies that a significant portion of the profits is used to replenish the capital, shown by the solid black line. The lower plot in Fig. 11.62 shows the situation when chosing the weights $q = 1$ and $r = 5$. Here the controller tries to use small excitations u, or equivalently, it tries to keep the profits reasonably constant. We observe that it now takes much longer to return to the steady state—the time constant τ is much larger—but the profits, shown as the dash-dotted red line, are much less affected, when compared to the situation depicted in the upper plot. We thus find that chosing suitable weights q and r allows us to flexibly tailor the system's behavior, depending on our preferences.

After having made a fortune with stocks and options and learnt how to control all that wealth, it's about time we talk about money. And, considering that we are living in the twentyfirst century, we will talk about *cryptocurrencies*.

Exercises

1. Simulate the Robinson Crusoe economy, defined by the equations

$$k_{t+1} = (1 - \delta)k_t + i_t , \quad y_t = k_t^\Theta , \quad i_t = \sigma y_t \quad \text{and} \quad p_t = y_t - i_t$$

with $\delta = 0.1, \Theta = 0.7, \sigma = 0.2$. Here we assume a Cobb-Douglas production function $y_t = k_t^\Theta$ and that a fraction σ of the output y_t is devoted to investments i_t, the rest we take out as profit to goof off. Today, you derive a certain amount of joy $V[p]$ from all future profits, which is characterized by $V[p] = \sum_{j=0}^{\infty} \beta^j \log(p_j)$, where we discount future joy by a "discount factor" $\beta = 0.9$. Write a simulation and vary the investment fraction σ to maximize V after 100 iterations.

2. Simulate the stochastic process

$$\lambda_{t+1} = \gamma \lambda_t + (2 - \gamma) + a\varepsilon_t$$

for $1 < t < 500$ generations. Here ε_t are random number drawn from a Gaussian distribution with zero mean and an rms of unity. Use the parameters $\gamma = 0.9$ and $a = 0.2$. Towards what value does λ tend on average? Find an analytic expression for this asymptotic value.

3. Determine the Hamiltonian $H(x, \theta, p_x, p_\theta)$ for a mechanical system with the Lagrange function $L(x, \theta, \dot{x}, \dot{\theta}) = 3m\dot{x}^2/2 + mr\dot{x}\dot{\theta} + mr^2\dot{\theta}^2/2 - kx^2 - mgr\theta^2/2$, where x and θ are the coordinates, $\dot{x}$, and $\dot{\theta}$ are the corresponding velocities, and p_x and p_θ are the corresponding momenta. Moreover, and m, r, k, and g are constants.

4. Consider the unstable one-dimensional process, defined by $\dot{x} = ax + bu$, where a and b are positive constants and the performance measure is given by (11.61). Show that the "gain" κ of the feedback, which defines the optimal controller through $u = \kappa x$ is given by (11.62).

5. Consider the the dynamical system, defined by $\dot{x} = \alpha x^\beta + u$ with $\alpha, \beta > 0$, and the performance measure given by (11.61). Derive the Hamiltonian from (11.40) and the equations of motion, given by (11.41), for this system. Assuming boundary conditions $x = x_0$ at $t = 0$ and $x = 0$ at $t = t_f$, explore whether you can use methods, similar to those from Sect. 11.5, to solve this problem.

References

1. G. McCandless, *The ABCs of RBCs* (Harvard University Press, Cambridge, Ma, 2008)
2. R. Dorfman, An economic interpretation of optimal control theory. Am. Econ. Rev. **59**, 817 (1969)
3. D. Kirk, *Optimal Control Theory* (Dover Publications, New York, 2004)
4. L. Landau, E. Lifschitz, *Lehrbuch der theoretischen Physik, Band I: Mechanik* (Akademie Verlag, Berlin, 1979)
5. H. Goldstein, J. Safko, C. Poole, *Classical Mechanics* (Pearson, Harlow, 2014)

Chapter 12
Cryptocurrencies

Abstract Based on the notion that today monetary exchanges are based on the transfer of information, this chapter provides a basic introduction into Shannon's theory of information, with its central concept—the entropy. After covering different ways to encode information the relation of Shannon's entropy to the entropy, known from thermodynamics, is illustrated with a physical system that has discrete energy levels. The chapter then discusses the transmission of information across binary symmetric channels and develops the necessary concepts to establish the maximum of the mutual information as the channel capacity. After moving information through continuous channels, limited by the signal-to-noise ratio, a section introduces the basics of cryptography, which is needed to provide the security when transmitting sensitive information. Public-key systems, Diffie-Helman and RSA, are discussed in some detail, before covering the basics of elliptic function cryptography, which provides the basis for both Bitcoin and Ethereum blockchains, the subject of later sections in the chapter. An Ethereum smart contract is discussed as an example of a distributed application—a DApp. The last section touches upon the basics of quantum computing and illustrates the key concepts of Shor's algorithm, which might one day pose a threat to cryptographic systems.

After initial developments from bartering goods, such as pigs and pultry, to coins and paper money, which we briefly discussed in Sect. 2.4, monetary transactions in today's world are predominantly handled electronically. Student loans and salaries are deposited in our accounts by bank transfer. Likewise, we pay our utility bills for electricity or gas electronically. Moreover, when eating out or on a shopping spree, we pay with credit cards or with our smartphones using Google pay, Apple pay, Venmo, or Swish. None of these methods to transfer funds involve the actual exchange of coins or paper money. Apparently money has evolved from the exchange of tangible assets—pigs, coins, or paper money—to the exchange of *information* about who has the right to buy things up to a certain value.

In the following we will therefore treat money as an entry in a *decentralized database of who has what purchasing power.* Even the ownership of a pig represents the purchasing power up to the equivalent value of other quantities, and a coin in your pocket associates you with the purchasing power equivalent to the face value of the coin. In these examples both pig and coin are part of some database of tradeable goods. Apparently, bank accounts are databases, maintained and verified by the banks that associate the purchasing power, represented by the account balance, with the owner of the account. Furthermore, the databases of different banks are linked via the internationally agreed IBAN system under the auspices of the Society for Worldwide Interbank Transactions (SWIFT).

In all cases the transactions are supervised by authorities to verify and guarantee that the transactions complete correctly. When trading a pig, the two partners of the transaction ensure that the pig is healthy and only hand over the pig if both partners are satisfied. The faithfulness of gold coins was historically (at least in Western movies) checked by biting into the coin. Paper money is printed on special paper that makes the notes difficult to counterfeit. Moreover, it bears the seal of the issuing authority, for example a king, and forging the seal carries a heavy penalty. Today, banks supervise the transactions. They verify that the originator of a transfer has sufficient funds available and that the account into which the funds will be deposited actually exists.

Here we see that a bank transaction actually transfers the ownership of purchasing power from one bank account to another. Since the accounts are part of a database, we have to transfer database entries, which represent the information about the ownership and the value of the transaction. And that information has to move from one database to another. Moreover, we definitely want that transfer to be faithful and error-free. In the following sections, we therefore have to find out what information actually is in an abstract sense and how it is affected by perturbations when transmitted from one database to the next.

To start, let us briefly summarize the key functions of money. Based on the concept of the distributed database from the previous paragraphs we associate a chunk of purchasing power with the information H, represented by an entry in some database. First, H represents a value, which is a pig, the face value of a coin, or the amount in a bank account. Second, H has an owner, for example the owner of a pig or a bank account. Third, we need to be able to faithfully transmit H from one database to another. Fourth, we must be able to verify the ownership of H at any time. Fifth, transactions must be secure and tamper-proof; we try to protect our coins from pick-pockets and trust the banks that they protect our accounts from illicit manipulations, such as changing the amount of a transfer or initiating a fraudulent transfer.

The first three points, value, ownership, and transmission, are intrinsic to the money, but the fourth and fifth point involve supervisory functions; we require that ownership and transactions can be validated. Conventional money, typified by coins, credit cards, or bank counts, have guaranteeing authorities that follow up and prosecute any violation of their "terms of acceptable use". We shall see in Sect. 12.5 that crypto-currencies have built-in validation mechanisms that make external supervisory systems obsolete.

From the discussion it should be obvious that the central concept underlying the modern use of monies is "information," which is closely related to the thermodynamic entropy. This will be the subject of the next section.

12.1 Information, Probabilities, and Codes

Information is commonly quantified as a sequence of yes-no decisions. This leads to the well-known binary representation with the number "1" corresponding to "yes" and "0" corresponding to "no." Here, each decision carries the information of one *binary digit,* or *bit.* We therefore measure information by the number of bits needed to uniquely describe a particular configuration among a number of equivalent configurations. Visualizing a single bit as an electrical switch, we see that the switch describes one of two possible configurations—or micro-states. Either the switch is on, or it is off. If both states appear equally often, with probability $p = 1/2$, we can assume that simply guessing the micro-state of the switch at any time, we guess correctly about half the time. More generally, note that the number n of possible configurations, *all assumed to have the same probability p is $n = 1/p$.* This even holds if n is larger than two. As long as all of the n micro-states are equally probable, we can ask ourselves: how many switches are needed to uniquely describe one of the n states. The number of switches then describes the information $\hat{H}$ needed to identify one configuration among the n possible micro-states, which is given by

$$\hat{H} = \log_2(n) = \log_2(1/p) . \tag{12.1}$$

Note that $\hat{H}$ is the information measured in bits. This measure of information was originally introduced by Hartley [1] in 1928, although it only became widely known after Shannon published his seminal report [2] in 1948.

In the literature the quantity $\bar{H}$ is often referred to by "information", "uncertainty," or "entropy." Before identifying a particular one among the n possible selections, $\hat{H}$ refers to the "uncertainty" of our knowledge. It is also referred to as "entropy," by analogy to the thermodynamic entropy, which also describes our inability to identify a particular configuration of a physical system. We will address this analogy further in Sect. 12.2. On the other hand, once we know which one among the n different choices is selected, we can refer to this knowledge as "information." Usually, we can deduce from the context which interpretation is the intended one.

The emphasis on "equally probable" in the previous paragraphs is justified by the existence of systems, where the states are not equally probable. Consider, for example, the ASCII [3] encoding for characters used on most computer systems. It uses a group of eight bits to encode the normal characters and special characters, such as comma, period, but also line feed. To identify the holders of bank accounts, however, we only need the 26 capital letters A to Z plus a space to separate first and family name. For this limited character set, we only need to distinguish 27 symbols or, using the next larger power of two: five bits to distinguish 32 symbols. Moreover, the different symbols do not occur with the same probability; the letter "E" is much more likely than the letter "Q." Consistent with the observation, supported by (12.1), that events with a small probability p carry more information, we find that despite omitting the much more likely vowels *y cn stll ndrstnd ths sntnc*.[1] Therefore, we need to work out how to characterize information if the symbols are not equally probable.

Let us consider a source of characters that sequentially ejects one character of written English per unit time and find out how much information we can expect to receive. Following Shannon [2], we assume that the probabilities p_i, with which the characters appear, are known. Here i labels the characters, such that $i = 1$ corresponds to "A" and so forth. To find the *average* entropy H per ejected character we calculate the expectation value of the entropy $\log_2(1/p_i)$, carried by each character, which occurs with probability p_i

$$H = \sum_i p_i \log_2(1/p_i) = - \sum_i p_i \log_2(p_i) \,, \qquad (12.2)$$

where the sum extends over all possible characters that are ejected by the source. Note that in the special case, where all probabilities p_i are equal to $p_i = 1/n$ for $i = 1, \ldots, n$, (12.2) reverts to (12.1). When calculating expressions such as $T(x) = x \log_2 x$, we follow the convention that $T(0) = 0$, which is justified by continuity in the limit $x \to 0$. Moreover, this assumption guarantees that configurations with probability $p = 0$, which never occur, do not contribute to the entropy H.

To gain a better understanding of (12.2) and its implications we download the file `pg10.txt` with the ASCII text of the King James bible [4] from Project Gutenberg [5]. Using the simple script from Appendix B.9 we prepare the following file containing the number of occurrences of each character

```
751152
275735 A
 48880 B
    :
```

The space occurs 751152 times and the letter "A" occurs 272735 times in the cleaned-up file which contains the text of the King James bible. From the character frequencies n_i, shown in the first column, it is straightforward to calculate the probabilities $p_i = n_i / \sum_i n_i$ and the average information per character H from (12.2). For the King James bible we find $H_{KJB} \approx 4.04$. Repeating the same exercise for Shakespeare's Hamlet [6], we find a similar value of $H_{Hamlet} \approx 3.94$. We conclude that, instead of using one byte with eight bits, we could use only four bits per character, or half of a byte, also called *nibble*, to encode English texts.

But how does one send Hamlet using only four bits per character if there are more than 16 different characters in use? This is, in fact, accomplished by using fewer bits to encode the more frequently appearing characters and use longer "codes" for the rarer characters. This task, which is called *source encoding*, is easily illustrated with a simple example, where only four characters, say A, B, C, and D, or, more generally, four *symbols*, are transmitted. Let us assume that these symbols appear

---

[1] ..you can still understand this sentence.

Table 12.1 Two different encodings for four symbols, shown in the first column, each appearing with the probabilities, shown in the second column

Symbol	Probability	Encoding 1	Encoding 2
A	1/2	00	0
B	1/4	01	10
C	1/8	10	110
D	1/8	11	111

with the probabilities $1/2, 1/4, 1/8$, and $1/8$, respectively. In Table 12.1 we show the symbols along with their associated probabilities and two ways to encode them. The first encoding uses binary codes and the second uses a zero to terminate the code and an increasing number of ones to differentiate the symbols.

Note that we assign the shortest code, a single "0" to the most frequent symbol "A," which has the highest probability to appear in a message. It is easy to convince ourselves that any sequence of zeros and ones can be uniquely converted back to the original sequence of symbols: count ones up to a maximum of three consecutive or until a zero is found. For example, "0" without preceeding "1" indicates the symbol "A".

Let us now calculate the average number of bits $\langle l \rangle = \sum_i p_i l_i$, where l_i is the number of bits—the length—of the corresponding code and i identifies the four symbols. For the first, binary, encoding every symbol is represented by two bits, so the average length is $\langle l \rangle_1 = 2$. In the second encoding, the average length for the symbols is $\langle l \rangle_2 = 7/4$, which is smaller than $\langle l \rangle_1$. The gain is not very big in this simple example, but illustrates the idea, namely to use shorter codes for the more frequently appearing symbols.

As a matter of fact, in his *source coding theorem*, Shannon proved [2] that long messages with very many, say n, symbols, can be encoded in a sequence of length L_n, given by $nH < L_n < nH + 1$, where the different symbols come from a set of symbols with probabilities p_i and are therefore characterized by the entropy $H = -\sum_i p_i \log_2 p_i$. Here n must be sufficiently large in order to exploit the different probabilities of the symbols. Shannon's proof, which is outside our scope, only proves the existence of such an optimal code, but does not describe how to construct it. Of course, finding efficient codes for a set of symbols with associated probabilities that minimizes the length of transmitted messages is of considerable interest. The problem was solved comprehensively in 1952 and led to the construction of *Huffman codes* [7].

In his seminal report [2], Shannon centered his analysis around the "entropy" as defined in (12.2). In the following section we will explore its relation to the concept of entropy that is well-known in thermodynamics and statistical mechanics.

12.2 Relation to the Thermodynamic Entropy

It is illuminating to discuss the relation of the definition of the entropy H from
Sect. 12.1 to the entropy $S = k_b \sigma$, known from thermodynamics and statistical
mechanics [8]. Let us therefore consider a system with discrete energy levels E_i.
We recall that the occupancy n_i of a level i in such a system can be derived by
finding the most probable occupancies n_i, subject to two constraints; both the sum
of particles $n = \sum_i n_i$ and the total energy $U = \sum_i n_i E_i$ must be constant. We note
that the total number of combinations C to distribute the n particles across the energy
levels is given by

$$C = \binom{n}{n_1}\binom{n-n_1}{n_2}\cdots = \frac{n!}{n_1!n_2!\cdots} . \tag{12.3}$$

In thermodynamic equilibrium the system assembles into a configuration with the
largest number of combination C, which we can find by maximizing C, or, more conveniently maximizing $\log C$ with respect to the occupation number n_i. Using Stirling's
approximation for the factorial $\log m! \approx m \log m - m$ for large m, we rewrite the
previous equation as $\log C = n \log n - \sum_i n_i \log n_i$. Including the two constraints
with Lagrange multipliers α and β, we find that the n_i have to satisfy

$$0 = \frac{\partial}{\partial n_i}\left[\log C + \alpha\left(n - \sum_j n_j\right) + \beta\left(U - \sum_j n_j E_j\right)\right] \tag{12.4}$$
$$= -\log n_i - 1 - \alpha - \beta E_i .$$

Solving for n_i we find the Boltzmann distribution $n_i = Ae^{-\beta E_i}$, where $A = e^{-1-\alpha}$
is some normalization constant. We also identify $\beta = 1/k_B T$ as the inverse absolute
temperature in the Kelvin scale and k_B is Boltzmann's constant. Henceforth we use
$\tau = k_B T = 1/\beta$ to simplify the notation.

Knowing that the occupancies n_i follow Boltzmann's law allows us to introduce the *partition function* $Z = \sum_i e^{-E_i/\tau}$ and the probability that a level i is
occupied is given by $p_i = e^{-E_i/\tau}/Z$, where we use Z to normalize the sum of
all probabilities $\sum_i p_i = 1$ to unity. Moreover, the total energy U is given by
$U = \sum_i p_i E_i = \tau^2 (\partial \log Z/\partial \tau)$. In order to express the entropy in this framework,
we first have to introduce the Helmholtz free energy $F(\tau, V)$ as the Legendre transform of the energy $U(\sigma, V)$. From $dU = \tau d\sigma - p dV = d(\tau \sigma) - \sigma d\tau - p dV$,
we find that $dF = d(U - \tau \sigma) = -\sigma d\tau - p dV$, which implies $F = U - \tau \sigma$ and
$\sigma = -(\partial F/\partial \tau)_V$. Inserting the latter into $F = U - \tau \sigma = U + \tau (\partial F/\partial \tau)_V$ leads
to $U = F - \tau (\partial F/\partial \tau)_V = -\tau^2 \partial(F/\tau)/\partial \tau$. Expressing U in terms of Z leads to

$$\frac{\partial(F/\tau)}{\partial \tau} = -\frac{\partial \log Z}{\partial \tau} \quad \text{or} \quad F = -\tau \log Z \tag{12.5}$$

apart from a term linear in τ, which can be shown to be zero [8]. For the entropy σ we then obtain

$$\sigma = -\frac{\partial F}{\partial \tau} = \log Z + \frac{1}{\tau} \sum_i p_i E_i = -\frac{F}{\tau} + \frac{U}{\tau} . \tag{12.6}$$

Note that this equation is equivalent to the definition of the the Helmholtz free energy $F = U - \tau\sigma$. Instead of calculating the entropy by differentiating the free energy with respect to the temperature τ, we can also use the probabilities $p_i = e^{-E_i/\tau}/Z$ to directly calculate H from (12.2)

$$H = -\sum_i p_i \log p_i = -\sum_i \frac{e^{-E_i/\tau}}{Z}\left[-\frac{E_i}{\tau} - \log Z\right] = \frac{U}{\tau} - \frac{F}{\tau} \tag{12.7}$$

where we used $U = \sum_i p_i E_i$ and $F = -\tau \log Z$. Note that we used the natural logarithm in (12.7) instead of the logarithm with base 2, which simply means that we use a different unit to measure the information; here we measure the information in *nats* instead of bits. We find, however, that the thermodynamic entropy, as specified by $S = k_B\sigma$ agrees with the concept of information H, as derived from the probabilities p_i in (12.2) or (12.7).

With this understanding of the concept "entropy" we are ready to move it from one place to the next, which allows banks to transfer salaries from their digital vaults to the recipients.

12.3 Moving Information Through Discrete Channels

The process of moving information from a source to a receiver is schematically displayed in Fig. 12.1. The information, a message, flows from the source to an *encoder*, shown in red. This could be a Huffman encoder, mentioned in Sect. 12.1, where the message is converted into a sequence of symbols x_i. For simplicity, we assume that the data are represented as binary data with two symbols "1" and "0". The probabilities of the two symbols does not have to be equal; encoding 2, discussed in Sect. 12.1, uses six "1" and only three "0". After the encoder, the data are passed through a transmission channel, shown in blue in Fig. 12.1. There the data may be subject to noise that corrupts the data stream, for example, by randomly flipping a bit from "0" to "1" or vice versa. Figure 12.2 illustrates such a *binary symmetric channel* in which ε is the probability that a bit is flipped and α parameterizes the probabilities that "0" or "1" appear, where $\alpha = 1/2$ describes equal probabilities for "0" and "1". On the receiving end, a *decoder*, shown in magenta, undoes the coding, recovers the message using the symbols used by the source, and finally passes it on to the receiver.

The key question we need to answer is how much we can learn about the transmitted symbols x_i by observing a received symbol y_j. Or, since we a priori already

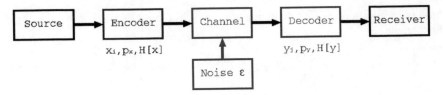

Fig. 12.1 A generic communication system with message source, encoder, transmission channel, decoder, and receiver. The encoder produces a data stream of symbols x_i with probabilities p_i. This stream is then corrupted by noise, parameterized by ε, on its way to the decoder

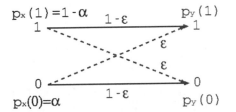

Fig. 12.2 Binary symmetric channel (bitflip channel) where the input symbols "1" and "0" (red) are perturbed by flipping the bits with probability ε before they are received by the decoder using the same symbols (magenta or almost red)

know that the transmitted stream originates from an encoder with uncertainty $H[x]$, by how much the uncertainty about x_i is reduced by observing y_j? And under what circumstances can we even restrict it to a single choice—the "real" value that was transmitted. It should be apparent that the answer is buried in the probabilities that one or the other alternative shows up in the decoder. Historically, all concepts were developed in [2], but we loosely follow the modern discussion in [9].

Therefore, let us analyze the channel described in Fig. 12.2 in parallel to the theoretical considerations. The encoder is characterized by the probabilities of the symbols "1" and "0," which can be read off from the figure as $p_x(1) = 1 - \alpha$ and $p_x(0) = \alpha$. Moreover, the channel is described by the *conditional probabilities* $p_{yx}(y_j|x_i)$, which describe the probabilities that a symbol y_j is received, provided that x_i was transmitted. Reading off from Fig. 12.2, we find

$$\begin{pmatrix} p_{yx}(1|1) & p_{yx}(1|0) \\ p_{yx}(0|1) & p_{yx}(0|0) \end{pmatrix} = \begin{pmatrix} 1 - \varepsilon & \varepsilon \\ \varepsilon & 1 - \varepsilon \end{pmatrix} . \tag{12.8}$$

Note that we need the subscript on p as a marker to distinguish the probability $p_{yx}(y_j|x_i)$ of y_j conditioned on x_i, from the probability $p_{xy}(x_i|y_j)$ of finding x_i, conditioned on y_j.

As a further refinement, we link the encoder, characterized by α, to the channel by introducing the *joint probability* $p_{xy}(x_i, y_j)$ of an event that that x_i and y_j occur simultaneously. It can be expressed as

$$p_{xy}(x_i, y_j) = p_{yx}(y_j|x_i)p_x(x_i) . \tag{12.9}$$

For the binary channel, for example, we find $p_{xy}(1, 1) = (1 - \varepsilon)(1 - \alpha)$. Summing (12.9) over all possible input symbols x_i gives us two ways to determine the probability $p_y(y_j)$ that the decoder finds a symbol y_j

$$p_y(y_j) = \sum_i p_{xy}(x_i, y_j) = \sum_i p_{yx}(y_j|x_i)p_x(x_i). \qquad (12.10)$$

In the binary channel, we have $p_y(1) = 1 - \gamma$ and $p_y(0) = \gamma$ with $\gamma = \varepsilon + \alpha - 2\varepsilon\alpha$. We observe that in (12.9) we can calculate the left-hand side $p_{xy}(x_i, y_j)$ in a second way, by multiplying the *backwards* conditional probability $p_{xy}(x_i|y_j)$ that the encoder sent x_i if the decoder found y_j by the probability $p_y(y_j)$ that symbol y_j arrives. We obtain

$$p_{xy}(x_i, y_j) = p_{xy}(x_i|y_j)p_y(y_j). \qquad (12.11)$$

Note that the left-hand sides of (12.9) and (12.11) are the same and equating the right-hand sides leads to *Bayes' theorem*

$$p_{xy}(x_i|y_j) = \frac{p_{yx}(y_j|x_i)p_x(x_i)}{p_y(y_j)}. \qquad (12.12)$$

which allows us to infer the probability that x_i was sent, provided we have received y_j. Note how the calculation takes probabilities with which x_i and y_j occur—the prior information—into account. For the binary channel, we find $p_{xy}(1|1) = (1 - \varepsilon)(1 - \alpha)/(1 - \varepsilon - \alpha + 2\varepsilon\alpha)$ and calculating the other probabilities is left as an exercise.

But we do not really want to make predictions about what symbols were sent. Instead, we want to assess the performance of the communication channel to transport information and therefore consider the entropy of the involved distributions. The entropy of the input $H[x]$ and output $H[y]$ are defined through their respective probabilities and are given by

$$H[x] = -\sum_i p_x(x_i) \log_2 p_x(x_i) \quad \text{and} \quad H[y] = -\sum_j p_y(y_j) \log_2 p_y(y_j),$$

$$(12.13)$$

where we use square brackets to denote that the entropy depends on the set of all x or y. In the binary channel, we have $H[x] = H_b(\alpha)$. Here $H_b(p) = -p \log_2 p - (1 - p) \log_2(1 - p)$ is the entropy of a binary system with probabilities p and $1 - p$. Using the probabilities $p_y(y_j)$ from above, we find $H[y] = H_b(\gamma)$ with $\gamma = \varepsilon + \alpha - 2\varepsilon\alpha$.

Analogously, the *joint entropy* $H[x, y]$ is defined in terms of the joint probabilities $p_{xy}(x_i, y_j)$ as

$$H[x, y] = -\sum_i \sum_j p_{xy}(x_i, y_j) \log_2 p_{xy}(x_i, y_j). \qquad (12.14)$$

Evaluating $H[x, y]$ for our binary channel by explicitly evaluating the sum yields $H[x, y] = H_b(\alpha) + H_b(\varepsilon)$. Note that $H[x, y]$ is given by the sum of the entropy of the input signal $H[x] = H_b(\alpha)$ and the entropy that is added by the random bit changes in the transmission channel $H_b(\varepsilon)$.

We can now replace $p_{xy}(x_i, y_j) = p_{yx}(y_j|x_i)p_x(x_i)$, known from (12.9), in the logarithm and write the product in the logarithm as a sum of two terms. We find the so-called *chain rule*

$$H[x, y] = H[x] + H[y|x] \quad \text{with} \quad H[y|x] = -\sum_i \sum_j p_{xy}(x_i, y_j) \log_2 p_{yx}(y_j|x_i),$$

$$(12.15)$$

where $H[y|x]$ is the *conditional entropy* of the set of y, provided that the set of x is already known. Equivalently we can use the backward conditional probabilities from (12.11) in the logarithm and find

$$H[x, y] = H[y] + H[x|y] \quad \text{with} \quad H[x|y] = -\sum_i \sum_j p_{xy}(x_i, y_j) \log_2 p_{xy}(x_i|y_j).$$

$$(12.16)$$

For our binary channel we find $H[x|y]$ by calculating $H[y|x] = H[x, y] - H[x] = H_b(\varepsilon)$ or by directly evaluating the sum in (12.15). Note that the additional entropy $H[y|x]$, needed to account for the difference of the joint entropy and the input entropy $H[x]$ is the entropy that is added due to the noisiness of the channel. And this noisiness is characterized by the entropy $H[y|x]$ which equals $H_b(\varepsilon)$ for the binary channel. Likewise, $H[x|y] = H[x, y] - H[y] = H_b(\alpha) + H_b(\varepsilon) - H_b(\gamma)$.

In order to find the amount of information we can transfer across a noisy communication channel we rephrase the problem by asking the question of how much we can actually learn about the source by observing the output of the decoder. To do so, we introduce the *mutual information $I[x; y]$*

$$I[x; y] = \sum_i \sum_j p_{xy}(x_i, y_j) \left[\log_2 \left(p_{xy}(x_i, y_j) \right) - \log_2 \left(p_x(x_i) p_y(y_j) \right) \right]$$

$$= \sum_i \sum_j p_{xy}(x_i, y_j) \log_2 \left(\frac{p_{xy}(x_i, y_j)}{p_x(x_i) p_y(y_j)} \right). \qquad (12.17)$$

As motivation for this definition consider the situation where the inputs x and outputs y are uncorrelated, which means we can not learn anything about the input from observing the output. But for uncorrelated variables, the distribution function factorizes and we would have $p_{xy}(x_i, y_j) = p_x(x_i) p_y(y_j)$, which would lead to $I[x; y] = 0$. If, on the other hand, the variables of input and output are correlated, $I[x; y]$ tells us how much information can be transferred.

It is straightforward to show that $I[x; y] = H[x] + H[y] - H[x, y]$. Moreover, by either inserting $H[x, y] = H[x] + H[y|x]$ from (12.15) or $H[x, y] = H[y] + H[x|y]$ from (12.16), we deduce

$$I[x; y] = H[y] - H[y|x] \quad \text{or} \quad I[x; y] = H[x] - H[x|y], \qquad (12.18)$$

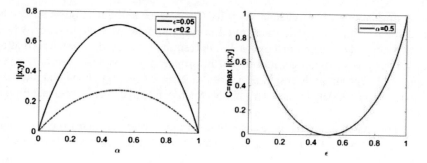

Fig. 12.3 The mutual information $I[x; y]$ for the binary symmetric channel as a function of α for $\varepsilon = 0.05$ and 0.2 (left) and the channel capacity C as a function of the error rate ε (right).

which Shannon used in [2] to characterize the communication channel. From our earlier discussion of identifying $H[y|x]$ with the amount of noise created in the channel, we interpret (12.18) as the amount of information $I[x; y]$ we can learn about the input x is given by whatever information $H[x]$ was sent minus the information $H[x|y]$, which was destroyed in the channel. For the binary channel, we find $I[x; y] = H_b(\gamma) - H_b(\varepsilon)$ with $\gamma = \varepsilon + \alpha - 2\varepsilon\alpha$.

Using the mutual information $I[x; y]$ we can now define the *channel capacity C* as the maximum achievable $I[x; y]$, where we maximize with respect to all possible input probability distributions $p_x(x_i)$

$$C = \max I[x; y] . \tag{12.19}$$

In general it is not possible to find this maximum explicitly, but for our binary channel all possible input distributions are parameterized by α. Therefore, we can either analytically or numerically determine the maximum of $I[x; y] = H_b(\alpha + \varepsilon - 2\varepsilon\alpha) - H_b(\varepsilon)$ as a function of α; or simply inspect the plot on the left-hand side in Fig. 12.3, where we display $I[x; y]$ as a function of α for two values of ε. Both curves assume their maximum at $\alpha = 1/2$. Using $\alpha = 1/2$ we then show the channel capacity C as a function of the bit-flip probability ε on the right-hand side in Fig. 12.3. We observe that for very small error rates the capacity approaches $C = 1$, which means that every bit that is sent is also received. As the bit-flip rate ε gets closer to $1/2$ the capacity approaches zero; if half the time the bits are randomly flipped, it is impossible to recover the input signal. As the bit-flip rate ε increases further and approaches unity, the capacity C again approaches $C = 1$ bit/bit; if each bit is flipped with certainty $\varepsilon = 1$, we can also recover the full information that was sent through the channel.

But what does a capacity $C = 0.5$ mean? It is instructive to write this as $C = 1$ bit/2 bits. In other words, we need to transmit two bits, in order to effectively transmit one bit reliably. That such error-correcting codes, in the limit of long messages, always exist for any non-zero capacity C is the essence of *Shannon's channel coding* theorem [2, 10]. He also showed that it is not possible to transmit more information than the channel capacity C across a communication channel. The noise in the channel destroys transmitted information and we have to detect which bits were flipped and then flip them back. This requires extra, redundant, information that we need to send along with the real information, the payload. We only name *Hamming codes* as an example of a class of error-correcting codes that stuff the payload into larger data frames. This allows the receiver to detect and correct flipped bits.

In the previous sections the transmitted symbols were selected from a discrete set. Next, we briefly discuss extending the theory to *continuous sets* in the following section.

12.4   Continuous Information Channels

An example where continuous signals appear is the transmission of the information—the music—as a continuous stream of voltages u, from an old-fashioned record player to the audio amplifier and on to the speakers. The voltages $x = u/u_0$, normalized to some reference voltage u_0, have a continuous probability distribution $p(x)$, which can be approximated by a histogram of the voltages measured in small time intervals, normalized by the total number of measurements. In Fig. 12.4 we show the amplitude-probability distributions for the first movement of Beethoven's fifth symphony on the left and for a sine on the right. The amplitude of the sine was chosen to have the same rms as that of the distribution on the left-hand side. The dashed red curves show Gaussians with the rms of the distributions and is the same on both plots. Having such probability distributions available, we can now generalize the concept of entropy, introduced in (12.2), to continuous variables x

$$H[x] = - \int p(x) \log \left(p(x) \right) dx , \qquad (12.20)$$

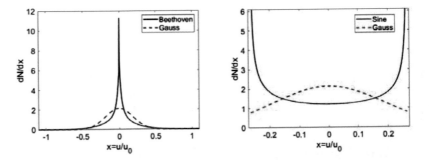

Fig. 12.4 The solid black lines show the amplitude-probability distribution of the first movement of Beethoven's fifth symphony (left) and that of a sine (right). The rms of the two Gaussians, shown as dashed red curves, are equal and also equal to the corresponding black curves

where the integral extends over the domain in which x is defined. Note that we use the natural logarithm with base e, as we did in Sect. 12.2, so the entropy has units of *nats*. The other concepts, introduced in Sect. 12.3, such as joint and conditional probabilities, joint and conditional entropies, as well as mutual information can be generalized likewise; we only have to replace the sums by integrals. In the following, we will liberally make use of this analogy and refer to the equations in the previous section that were proven for discrete variables. The entropies of the distributions, shown in Fig. 12.4 are $H = -0.460$ for the Beethoven symphony and a significantly smaller value of $H = -0.868$ for the sine. The Gaussian, shown as the red dashed curves, has the largest entropy $H = -0.259$.

Since we want to transfer the maximum information across a channel, we might ask ourselves: which probability distribution function $g(x)$ has the largest entropy? For physical reasons, we consider only voltages that dissipate a finite average power u^2/R in some resistance R. This implies that the distribution $g(x)$ must have a finite second moment, defined by $\sigma^2 = \int x^2 g(x)dx$. Moreover, since it is a probability distribution function, $g(x)$ must be non-negative and it must be normalized with $\int g(x)dx = 1$. Thus $g(x)$ can be found from minimizing the functional $J[g]$ with respect to $g(x)$

$$J[g] = \lambda_1 + \lambda_2\sigma^2 + \int \left[-g(x)\log(g(x)) - \lambda_1 g(x) - \lambda_2 x^2 g(x) \right] dx \quad (12.21)$$

with two Lagrange multipliers λ_1 and λ_2. The variation with respect to $\delta g(x)$ yields

$$0 = -\log(g(x)) - 1 - \lambda_1 - \lambda_2 x^2 \quad \text{or} \quad g(x) = e^{-\lambda_2 - 1}e^{-\lambda_2 x^2}. \quad (12.22)$$

Differentiating with respect to the Lagrange multipliers, we recover the constraints, which we use to determine λ_1 and λ_2. We thus find a Gaussian

$$g(x) = \frac{1}{\sqrt{2\pi}\sigma}e^{-x^2/2\sigma^2} \quad (12.23)$$

as the normalized probability distribution with σ^2 as second moment that maximizes the entropy given by (12.20). It is now straightforward to calculate the entropy of the Gaussian as

$$H_g = \frac{1}{2}\log\left(2\pi e\sigma^2\right). \quad (12.24)$$

The value calculated from (12.24) agrees with the value we determined numerically for the Gaussian in Fig. 12.4 and now we understand why it is the largest of the three entropies that we determined from the amplitude-probability distributions. On the coming pages we will use these Gaussians not only as models for the sources that produce the information-carrying signals, but also for the useless information, which is added by random voltages while the signals traverse a communication channel.

The distribution of voltages only defines the statistics of the voltages that may appear, but we also care about every possible voltage being able to immediately follow any other voltage. This means that we require that very sharp transitions from one voltage level to any other are faithfully transported. But sharp transitions are described by high Fourier harmonics. The highest useful frequency, called the bandwidth W, will therefore limit how many different voltages we can transmit across the channel per unit time T. According to Nyquist's theorem [11], we need at least two measurements per period to characterize a frequency. Therefore, a total of $n = 2TW$ distinct voltages can be transmitted during the time T through a channel with bandwidth W. The total entropy transported during time T is then given by $nH_g = 2TWH_g$ and the corresponding rate $\dot{H} = nH_g/T = 2WH_g$ is given by

$$\dot{H} = W \log \left(2\pi e \sigma^2\right) . \tag{12.25}$$

For a source, characterized by a rms signal level $\sigma^2 = P_S$, we find the rate $\dot{H}[x] = W \log \left(2\pi e P_S\right)$ and for the noise on the channel with $\sigma^2 = P_N$, it is $\dot{H}[N] = W \log \left(2\pi e P_N\right)$. Moreover, since the source and the noise are uncorrelated the total power P_t arriving at the decoder is given by $P_t = P_S + P_N$ and the corresponding entropy rate is $\dot{H}[y] = W \log \left(2\pi e (P_S + P_N)\right)$.

The capacity C of the continuous channel is defined analogously to (12.19) as the maximum of the mutual information rate $\dot{I}[x; y] = \dot{H}[x] + \dot{H}[y] - \dot{H}[x, y]$ over all possible inputs. By inserting $\dot{H}[x, y] = \dot{H}[x] + \dot{H}[y|x]$ from (12.16), we obtain $\dot{I}[x; y] = \dot{H}[y] - \dot{H}[y|x]$. Immediately following (12.16) we reasoned that $\dot{H}[y|x]$ is the entropy added to the signal by the noise, which is, as we know from the previous paragraph, equals $\dot{H}[N]$. Inserting all quantities into the expression for $C = \dot{I}[x; y] = \dot{H}[y] - \dot{H}[N]$, we find

$$C = W \log \left(2\pi e (P_S + P_N)\right) - W \log \left(2\pi e P_N\right) = W \log \left(1 + \frac{P_S}{P_N}\right) . \tag{12.26}$$

The entropy of both source and noise were assumed to be maximum-entropy Gaussian, such that we can identify the mutual information rate $\dot{I}[x; y]$ as the channel capacity C.

We observe that the channel capacity C is determined by the bandwidth W and the *signal-to-noise* ratio P_S/P_N. Higher bandwidth—the ability to transmit higher frequencies—will allow us to transmit more information. Moreover, a small noise power P_N allows us to distinguish smaller differences in the payload signal, which gives us more signal levels to encode information. As an example, consider the communication on a wireless WLAN network, which benefits from increasing the bandwidth by *channel bonding*, where two or more of the, by default, $W = 20\,\text{MHz}$ wide channels, are used simultaneously in order to increase the download speed. Moreover, the transmission amplifiers automatically increase the transmission power P_S, up to the legally allowed maximum, to maximize transmission capacity. Normally the power level is chosen to match the rate at which the information is created at the source—listening to an Internet radio station requires less bandwidth than watching a high-resolution movie. But if someone turns on a faulty microwave oven in the kitchen, the noise level P_N increases, and the WiFi-router, also located in the kitchen, increases the transmission power P_S to maintain the channel capacity, so it is still possible to enjoy the movie in the living room.

At this point we know how to characterize information—our bank details—and how to transmit it from one point to another, both through discrete and through continuous channels. But we still have to ensure that nobody falsifies the data on the way, which can be done using cryptographic methods, the topic of the next section.

12.5 Cryptography Fundamentals

When sending data across a network, anyone can listen to the communication and, unless special precautions are taken, even change the transmitted information. Some level to guarantee the integrity of data is built into the lower levels of network communication, such as the *internet protocol* (IP) and the *transfer control protocol* (TCP), but it guarantees neither privacy nor security against tampering, which makes it unsuitable to transfer money, unless extra security measures are added. We therefore have to consider methods to authorize and authenticate transfers across communication channels. In classical banking we went to the bank, used an identification card to authenticate, and then sign a paper to authorize the transfer. Note that both id-cards and personal signatures are "secrets" that are unique for each person. Let us therefore look at sending secrets across communication channels.

A generic encrypted communication channel is shown in Fig. 12.5, where two parties, commonly called Alice and Bob, secretly exchange a message m, which is encrypted with a key k to give the encrypted message c, also called the *ciphertext*. The latter is then sent over public communication channel, where it can be picked up by an eavesdropper, commonly called Eve, who does not know the key k. Bob, on the other hand, knows k and can use it to decrypt the message. Note that Alice and Bob have to exchange the secret key k over a second channel, shown as the dashed line in Fig. 12.5. The purpose of the encryption is to introduce entropy into the channel, such that the mutual information $I[x; y]$ between Alice and Bob is large, while the mutual information $I[x; z]$ between Alice and Eve is as small as possible. On the right-hand side in Fig. 12.3 we see that the channel capacity for the binary symmetric

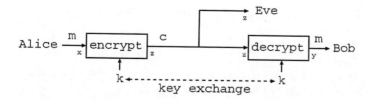

Fig. 12.5 Encrypted communication channel: Alice encrypts the message m with key k and sends the cipertext c to Bob, who decrypts c with the same key. Eve, who does not know the key, cannot recover the message from the ciphertext. Note that the key must be sent across a separate secure channel, indicated by the dahsed line

channel becomes zero, if the probability ε for a flipped bit is $1/2$ in the encryption process. At the same time, we need to ensure that the encryption process is invertible, such that Bob can recover the original message m from the cipertext c.

In the following we will exclusively deal with binary, rather than textual data such that we can use the binary symmetric channel as an example. Let us assume that Alice wants to send the letter "A," which is $m = \text{b'}01000001$ in the ASCII code. One way to encrypt binary numbers is based on the logical *exclusive or*, or $\text{xor}(i, j)$ function, which is defined for one bit $i, j = 0, 1$ to return $\text{xor}(i, j) = 0$ for $i = j$ and $\text{xor}(i, j) = 1$ for $i \neq j$. For multi-bit numbers, such as $m = \text{b'}01000001$, and the secret key k, we define $\text{xor}(m, k)$ to operate on one bit at a time. Using the secret key $k = \text{b'}01101100$, we find the ciphertext $c = \text{xor}(m, k)$ from

$$
\begin{array}{rcl}
m & = & \text{b'}01000001 \\
k & = & \text{b'}01101100 \\
c & = & \text{b'}00101101
\end{array}
$$

whenever m and k have the same bit in a particular position the c contains "0" in that position; if the bits are different, c contains "1." It is easy to convince ourselves that Bob can recover the original message m from the ciphertext c by calculating $m = \text{xor}(c, k)$.

Bob can decode the ciphertext c, but Eve only sees c. Let us therefore calculate the conditional probabilities $p_{y,x}$ and p_{zx} from (12.8), where Alice sends x_i, Bob receives y_i, and Eve observes z_i. In the following MATLAB script, we first define the key k and 2×2 arrays to hold the conditional probabilities. Then we loop over all possible messages m, which implies $\alpha = 1/2$ in Fig. 12.2 and ensures that the same number of "1" and "0" are sent. Inside the loop, we first encode m to obtain the ciphertext c before simulating Bob's action of decoding c with the same key k. Then we compare the eight bits in each byte and update the conditional probabilities. After the loops we normalize the entries in the 2×2 matrices for the conditional probabilities.

```
k=bin2dec('01101100');    % key
pyx=zeros(2,2); pzx=pyx;   % cond. probabilities
for m=0:255                % loop over all messages
  c=bitxor(m,k);           % Alice encodes to ciphertext
  bob=bitxor(c,k);         % Bob decodes with key
  for i=1:8                % compare all eight bits
    bx=bitget(m,i);
    bz=bitget(c,i);
    by=bitget(bob,i);
    pyx(2-by,2-bx)=pyx(2-by,2-bx)+1;   % update probs.
    pzx(2-bz,2-bx)=pzx(2-bz,2-bx)+1;
  end
end
pyx=pyx/1024    % Alice to Bob, pyx=[1,0;0;1]
pzx=pzx/1024    % Alice to Eve, pzx=[0.5,0.5;0.5,0.5]
```

The conditional probabilities p_{yx} for the channel from Alice to Bob, compared with (12.8), are consistent with $\varepsilon = 0$, which implies that the mutual information $I[x; y]$ is unity. Since we have $\alpha = 1/2$ this implies that the channel capacity C, shown on the right-hand side in Fig. 12.3, is also unity. On the other hand, the probabilities p_{zx} for the channel from Alice to Eve, are consistent with $\varepsilon = 1/2$, which indicates that the channel capacity C is zero and Eve cannot extract information from the stream of bits that travels on the public channel. We point out that this result depends to some extent on the chosen key k. We picked one with four bits set and four bits unset. If we pick a key k with three bits set and five bits unset, ε differs from 1/2.

In the example, we only used an eight-bit key, but we can of course make the key k much longer, say 256 bits and encode blocks of the same length, one at a time. Cryptographic systems that work on such fixed-length blocks of data are referred to as *block ciphers*, on which many encryption systems, discussed below, are based.

We note that simple cryptographic algorithms have two significant problems. First, both the sender and the recipient have to know the key k, which makes it mandatory to send the key over a potentially insecure communication channel. We might consider to encrypt the transfer, but this requires an earlier transmission of another key to encode it. We will later address how the key-distribution, which is how this hen-and-egg problem is commonly called, can be solved. The second problem of the simple xor-cipher is that we can easily uncover the presumably secret key by passing a special messages to the encoding function. The trivial example is that $\text{xor}(b'0, k) = k$, where $m = b'0$ is a messages comprising of only zeros.

This second problem was initially solved by the *Data Encryption Standard* (DES), first introduced in the 1970s. It works on blocks with a length of 64 bits and uses a key k_0 with an effective length of 56 bits, comprising of eight 7-bit sections. From this secret key sixteen sub-keys $k_0, \ldots, k_{15}$ are derived. The encoding proceeds in sixteen steps, called *rounds*, labelled by the index i, in which the message m_i at step i is split in a left and a right part $m_i = (L_i, R_i)$, each with a length of 32 bits. The transformation from one round to the next is given by

$$m_{i+1} = (L_{i+1}, R_{i+1}) = (R_i, \text{xor}(L_i, F(R_i, k_i))) , \qquad (12.27)$$

where $F(R_i, k_i)$ is a so-called *Feistel function*. It scrambles the 32 bits in R_i with the sub-key k_i using an algorithm that is specific to DES. Finally R_{i+1} is calculated as the xor with L_i. Note that in each round one half of m_i remains unchanged, while the other half is xor'ed with a key derived from the other half and a sub-key. After sixteen rounds the original message m_0 has mogrified to the cipher text $c = m_{16}$. It turns out that the same algorithm can be used to decipher an encrypted cipher text c. All we have to do is to reverse the order of the sub-keys and pass the cipher text through the sixteen rounds, where the last round uses sub-key k_0, to recover the original message m_0. Since both encoding and decoding use the same algorithm, DES is considered efficient. Over the years, however, it was shown that the DES encryption can be broken within a reasonably short time and its use is no longer recommended. It is superseeded by other encryption algorithms, for example by *Triple-DES,* which uses three 56-bit long keys. It first encrypts using one key, then decrypts with the second key, and finally encrypts again with the third key. In order to decode cipher text, we have to undo this sequence in the reverse order. We first decrypt with the third key, encrypt with the second, and decrypt with the first key.

The presently up-to-date encryption standard is the *Advanced Encryption Standard* (AES), which was announced 2001. It uses symmetric keys and operates on 128-bit long blocks of data. Encryption keys with a length between 128 and 512 bits are supported. The 128 bits of each block are assembled in a 4×4 array of bytes, which is subsequently scrambled in ten to fourteen rounds with sub-keys that are derived from the primary key. In order to reduce the load on the processors, in particular those used in desktop computers, the processors support AES in hardware, which allows to efficiently encrypt data, either locally on the harddisk, or in transit when viewing websites with https.

Sometimes we do not really want to hide the communication from others, but only want to ensure that the integrity of a message is not compromised, either unintentionally due to transmission errors, or intentionally by a third party falsifying the messages. For this purpose we use a special fingerprint of a message, called a *hash* or a *message digest*. One of the simplest hashes of a message $m = (b_1, b_2, b_3)$ consisting of, for example, three bytes b_3, b_2, and b_1 is the checksum c where all bytes are successively xor'ed, such as $c = \text{xor}(b_3, \text{xor}(b_2, b_1))$. This checksum c can be calculated before and after a transmission in order to ensure that the integrity of the message remains intact. Using a one-byte long fingerprint only permits to distinguish between 256 different cases and *collisions*—finding the same c for different messages m_1 and m_2—are likely. Moreover, testing only 256 different messages suffices to find one that produces a given checksum c. In a cryptographic context, where we use the hash value to guarantee the integrity of the message, this is called a *pre-image attack*. And finally, the calculation of the hash must be *deterministic* in the sense that calculating it on different computer systems yields equal hashes. Thus, we need to find deterministic functions that calculate long hashes and ensure that even rather similar messages produce vastly different hash values c, such that moving a decimal point in a financial transaction changes c drastically.

The *MD5* message digest function, first introduced in 1992, calculates a 128-bit long hash value from an input message that can have any length. The message can be a text string, a binary quantity, or the contents of an arbitrary file. For a discussion of the inner workings of the MD5 hash and other hash functions we refer to the specialized literature [12]. During the past decade, however, it was shown that collisions in the sense discussed above can be constructed with moderate computing power, which makes MD5 unsuitable for cryptographic applications. On the other hand, MD5 is still used to verify that downloaded files were not corrupted in transit.

In 2001, the SHA-2 family of hashing algorithms was published. It supports several lengths of the hashes, including the 256-bit version *SHA-256*, which is used by the Bitcoin cryptocurrency. The algorithm is based on breaking the message into 512-bit chunks and then subdiving each chunk further into sixteen 32-bit blocks that are thoroughly scrambled by bit-shifting and xor'ing to obtain a total of 64 blocks, each 32 bits long. These 64 blocks undergo a second round of scrambling, which includes xor'ing with pre-determined pseudo-random numbers and are finally concatenated to produce a 256-bit long hash value. The prominent use of of SHA-256 in several cryptocurrencies has stimulated the development of special hardware, based on field-programmable gate arrays (FPGA) and application specific integrated circuits (ASIC). These devices calculate hashes at a rate of billions per second, and are often used in parallel, which poses a threat against pre-image attacks that seek to find an input to the hash function, which evaluates to a desired hash value and would thus allow to tamper with the fingerprint.

In order to overcome this weakness against brute-force pre-image attacks, modern cryptocurrencies, such as *Ethereum*, use hashing algorithms that require large amounts of memory for the calculation and therefore call for expensive hardware, which precludes the high degree of parallelization that poses a threat to SHA-256. The scrambling in *ethash* [15] hashes is based on first building a very large data structure of pseudo-random numbers and then picking numbers from this structure in a quasi-random fashion, which makes it impossible to keep only a small part of the structure in memory. The calculation of the *scrypt* hash works similarly. The high cost, both in hardware and computationally, accounts for the popularity of these hashes in modern cryptocurrencies.

Occasionally, we need fingerprints of a large number of digital items, say $n = 100$ files. We can then either send n hashes to verify each file individually, or we can assemble the hashes into a *Merkle tree*. It is based on first calculating the individual hashes and then subsequently hash the bit-concatenation of two hashes to obtain $n/2$ hashes. We then repeat the process until only one hash remains, which is called the *Merkle root*. Note that unpaired hashes are hashed with themselves. In this way we obtain a single hash value, the Merkle root, that is used to verify the integrity of the entire group of n files, or any other digital item. We will encounter Merkle trees again, because they are used to fingerprint large numbers of financial transactions that are entered into a crytocurrency database in order to ensure their integrity.

By now, we have collected a number of cryptographic tools that allow us to encrypt and fingerprint digital assets, but we still face the key-distribution problem, which had to wait for a solution until the mid-1970s.

12.6   Early Public-Key Systems

In DES, AES, and many other encryption methods the key used to encrypt and to decrypt is the same, which is thus called a *symmetric key*. Having only symmetric key systems available implies that the keys must be exchanged by some other means. Prior to the mid-seventies, courier services shuttled attaché cases with sheets of paper containing the keys between the communicating partners, for example, the state department and the embassies in foreign countries. This was expensive and only possible when highly secret information needed to be exchanged, especially during the cold war. In 1976, however, Diffie and Hellman published their landmark report [13], in which they proposed to use *asymmetric keys*.   Alice and Bob, the parties of a two-way communication, each have their private secrets; Alice knows her secret a, but does not know Bob's secret b. Beforehand, they publicly agree on using two large numbers g and p, where p is a prime number and g is a *primitive root* of p, which means that $g^j \pmod p$ for $j = 1, \ldots, p$ cycles through all integer values between 1 to $p - 1$. Here $a \pmod p$ denotes the remainder of an integer division of a by p. Once these preliminaries are established, Alice sends the number $A = g^a \pmod p$ to Bob and Bob sends $B = g^b \pmod p$ to Alice, who calculates the secret key via $k = B^a \pmod p = g^{ab} \pmod p$. Bob arrives at the same key by calculating $k = A^b \pmod p = g^{ba} \pmod p$. All further communication between them can then use k as the key for encrypting and decrypting using one of the algorithms mentioned before. Note that they arrive at the same secret key k without divulging their private secrets to anyone. Moreover, an eavesdropper on the communication channel—Eve—can only pick up g, p, A, and B, but faces the problem to determine Alice's secret a from knowing $A = g^a \pmod p$, which entails testing a large number of trials for a in order to find one that produces A. If the numbers g and p are very large, this is computationally not feasible. Despite the big advantage to agree on a key k, the communication only works between two partners, here Alice and Bob, who have to schedule a common session, when they work out k, before being able to use it to encrypt their further communication. This complexity precludes spontaneously sending of, for example, emails.

This deficiency was overcome in 1977 by the *RSA public-key cryptographic system*, developed by Rivest, Shamir, and Adelman. The algorithm is based on finding positive integer numbers n, e, and d, such that for any message, witten as a positive integer m with $0 \leq m < n$, we have $m^{ed} \pmod n = m$. Here n and e constitute the *public key* and d is the *private key*. Let us consider a standard scenario, where Alice publishes her public key (n, e) on her web site and keeps the private key d to herself. If Bob wants to send a message m to Alice, he looks up her public key (n, e) and calculates the encrypted message $c = m^e \pmod n$, which he sends to Alice over an open communication channel. For different messages m the cipher text c jumps around between 0 and n in a quasi-random fashion, which guarantees that it is practically impossible to guess m from c, unless one has access to the private key d. Such functions, which are easy to calculate one way, but extremely difficult to invert, unless one has access to the private key, are called *trapdoor functions*. Since Alice knows her private key d, all she has to do is to calculate $c^d \pmod n = m^{ed} \pmod n = m$ in order to recover the plaintext message m. Eve, who desperately tries to know what Bob is sending to Alice, has access to c, n and e but she has to try out all possible messages $\bar{m}$ that will actually produce the ciphertext $c = \bar{m}^e \pmod n$. If n and e are large, this is not feasible in a reasonable amount of time. We point out that this public-key cryptographic system is the enabling technology that makes trading over the internet possible. There would be no amazon.com or ebay.com without public-key cryptography. Whenever you see a https prefix of an internet address, public-key cryptography is involved to exchange a symmetric key that allows faster communication and is henceforth used to encrypt further communication, for example, to transmit credit card details.

Secretly sending a message from Bob to Alice is only one possible application. Imagine a scenario, where Eve is known to fake Alice's signature in emails, so Alice has to ensure that she is the genuine sender of an email with text $\hat{m}$. She first calculates a hash value $\hat{c}$ of $\hat{m}$ using one of the hashes discussed before. Then she "signs" this hash $\hat{c}$ with her private key d by calculating $s = \hat{c}^d \pmod{n}$ and transmits s along with the plaintext message $\hat{m}$ to Bob. We assume that Bob receives this message as $\hat{m}'$. It was not encrypted and he does not know whether someone tampered with the message. To find out whether the received message $\hat{m}'$ is really the one that Alice had sent, he uses Alice's public key (n, e) to unpack the signed hash s to find $s^e \pmod{n} = \hat{c}^{ed} \pmod{n} = \hat{c}$, the hash of the message $\hat{m}$. He then verifies that the received message $\hat{m}'$ is genuine by calculating its hash $\hat{c}'$. If that equals $\hat{c}$, the messages are the same. Note that here Alice uses her secret private key d to sign the hash of the message and Bob uses the public key (n, e) to verify that the signature can only come from Alice.

But how do we find the keys, or the numbers n, e, and d? Here we only sketch the construction, which is based on *Euler's totient theorem*. It states that if two numbers m and n have no common factor except 1, then $m^{\varphi(n)} = 1 \pmod{n}$, where $\varphi(n)$ is the *totient function* of n. It is equal to the number of integers that have no common factors with n, which is called they are *relatively prime* or *coprime* with respect to n. If we now choose two large prime numbers p and q and define $n = pq$ then the totient of n can be calculated as $\varphi(n) = (p-1)(q-1)$. First taking the kth power and then taking the modulus on both sides of Euler's theorem leads to

$$m^{k\varphi(n)+1} \pmod{n} = m \quad \text{for an integer } k, \tag{12.28}$$

which gives us the desired form m^{ed}, provided we choose a number e, which must be coprime with $\varphi(n)$ and then find a value of k such that $d = (k\varphi(n) + 1)/e$ is integer. Either, we search for d that fulfills $ed \pmod{\varphi(n)} = 1$ or we rewrite this equation as

$$ed - k\varphi(n) = 1 = \gcd(e, \varphi(n)) \tag{12.29}$$

where the greatest common denominator of e and $\varphi(n)$ is unity, or $\gcd(e, \varphi(n)) = 1$, because e and $\varphi(n)$ are coprime. This expression has the form of *Bézout's equation*, which can be easily solved using the gcd() function in MATLAB, which implements Euclid's extended algorithm.

The following short MATLAB script visualizes this process.

```
p=5; q=11; e=3;
n=p*q; phin=(p-1)*(q-1);
[gcdval,d,kk]=gcd(e,phin);
if gcdval ~= 1 disp('Error: e and phin not coprime'); end
d=powermod(d,1,phin)
message=6
ciphertext=powermod(message,e,n)
decoded=powermod(ciphertext,d,n)
```

First we select the two primes p, q and the encryption key e and then calculate n and $\varphi(n)$ before solving Bézout's equation to find the decrypton key d, while the next line ensures that d is positive. After the call to gcd() we display an error message if e is not coprime to $\varphi(n)$. In the last three lines we first define a message, encode it, and decode it again. Here we use the powermod() function which works well for moderately large values of e and d. We emphasize that the system is secure as long as Eve cannot factor n into its prime factors.

These built-in functions work well for moderately-sized integers, but in practice and in order to prevent factorization of n into the underlying primes p and q, the numbers typically have more than a hundred digits, which requires special functions to handle the arithmetic. These large-integer routines are often not very fast, which is a limiting factor on mobile devices, such as smartphones. Therefore, other algorithms are desired, which use smaller integers, yet provide equal or improved security compared to the original RSA algorithm.

12.7   Elliptic Curve Cryptography

Although already proposed in the 1980s for efficient cryptographic applications, elliptic curves only became popular in the past two decades. This method is based on mapping two points A and B that lie on an elliptic curve, which is given by

$$y^2 = x^3 + ax + b \tag{12.30}$$

onto a third point $C = A \oplus B$. The left-hand side in Fig. 12.6 illustrates the construction. The elliptic curve *secp256k1*, which is used in the Bitcoin cryptocurrency and defined by $y^2 = x^3 + 7$, is shown as the black line. The two points A and B lie on the curve and the straight line that connects them is extended until it intersects the curve again, which defines the point C'. Reflecting C' on the horizontal axis intersects the curve, which is symmetric, at point $C = \ominus C'$ where we define C to be the "negative" of C' in an abstract sense and likewise $C = A \oplus B$ in an equally abstract sense. The operations, denoted by $\ominus$ and $\oplus$, describe how to find a new point, here C, from some previously defined points. We emphasize that finding the third point on the curve is almost always possible. A special case is given by the limiting case $B \to A$ in which case two points that define the slope of the straight become the tangent to the curve. Moreover, we need to artificially add a point $\hat{0}$ at infinity to the curve in order to make the equation $C \ominus C = \hat{0}$ meaningful. The point at infinity thus takes the role as a "zero."

The mapping of two points A and B to the third one is easily found by calculating the intersection of the elliptic curve from (12.30) and the straight line that passes through points A and B. It is given by $y = s(x - x_A) + y_A$ and has slope $s = (y_B - y_A)/(x_B - x_A)$. Equating the equations leads to

$$0 = x^3 - s^2 x^2 + \left[a + 2s^2 x_A - 2sy_1 \right] x + \left[b - s^2 x_A^2 - y_A^2 + 2sx_A y_A \right]$$
$$= (x - x_A)(x - x_B)(x - x_C) \tag{12.31}$$
$$= x^3 - [x_A + x_B + x_C] x^2 + [x_A x_B + x_A x_C + x_B x_C] x - x_A x_B x_C$$

The second equality is valid, because the third-order equation in the first equality has three roots—the horizontal coordinates of the points, where the straight line and the elliptic curve intersect. We already know two of the three roots and can therefore determine the coefficients from the quadratic term alone. We find $x_C = s^2 - x_A - x_B$. Inserting into the equation for the straight line yields the vertical coordinate of C'. Inverting the sign leads to $y_C = -s(x_C - x_A) - y_A$. If the two points A and B are equal with $x_A = x_B$ and $y_A = y_B$ we have to replace the slope s by the derivative $s = dy/dx = (3x_A^2 + a)/2y_A$ at point $A = B$. To summarize, we first have to determine the slope from

$$s = \frac{y_B - y_A}{x_B - x_A} \quad \text{for } A \neq B \quad \text{or} \quad s = \frac{3x_A^2 + a}{2y_A} \quad \text{for } A = B \tag{12.32}$$

and then calculate the coordinates of point C from

$$x_C = s^2 - x_A - x_B \quad \text{and} \quad y_C = -s(x_C - x_A) - y_A . \tag{12.33}$$

In order to visualize this operation, we repeatedly add the same point A, shown as a red asterisk on the right-hand side in Fig. 12.6, and calculate $A_2 = A \oplus A$, $A_3 = A_2 \oplus A, \ldots$ for hundred iterations. The black dots show that points are scattered all over the elliptic curve, but not enough to provide the randomness needed for efficient encryption.

Instead of operating with real numbers when calculating the coordinates of point C we will use modular arithmetic over a *finite field* with base p in (12.32) and (12.33). This will increase the apparent randomness of the "jumping around" dramatically. We choose p to be a prime number, because this guarantees that multiplying the

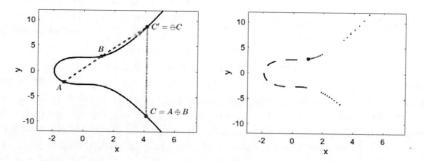

Fig. 12.6 Left: Illustrating the method of adding two points "A" and "B" on the elliptic curve to obtain $C = A \oplus B$. Right: if the underlying field are the real numbers, repeatedly adding the same point, indicated by the red asterisk, jumps to many points on the curve, indicated by black dots

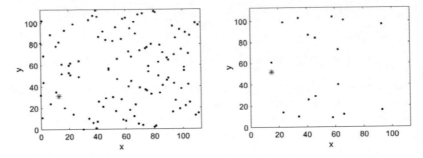

Fig. 12.7 Addition on elliptic curves over a finite field with $p = 113$ and generator point $G = (13, 31)$ on the left and $G = (15, 52)$ on the right

set of number $\mathcal{P} = \{1, \ldots, p - 1\}$ by some arbitrary integer k, thus $\{k, \ldots, k(p - 1)\} \pmod{p}$, results in the same set $\mathcal{P}$ albeit with the order of the numbers scrambled. Most operations, such as addition and multiplication are well-defined when using modular arithmetic; only divisions, used in the calculation of the slope in (12.32), require special attention. We can, however, calculate the inverse of a number in a finite field by exploiting *Fermat's little theorem*, which states that for any number a the equation $a^{p-1} = 1 \pmod{p}$ is valid. Rewriting it as $a^{-1} = a^{-1}1 = a^{-1}a^{p-1} = a^{p-2}$ where all equations are understood modulo p, we find

$$a^{-1} = a^{p-2} \pmod{p} , \qquad (12.34)$$

which provides us with a method to calculate the inverse of a number a in a finite field and allows us to calculate the divisions in (12.32).

We use the MATLAB scripts from Appendix B.11 to iterate (12.32) and (12.33) over a finite field with $p = 113$ which leads to the plots shown in Fig. 12.7. On the left-hand side, the starting position, indicated as a red asterisk, is $G = (13, 31)$; on the right-hand side it is $G = (15, 52)$. Here we denote the starting point by the letter G to signify that it is the *generator* of the group of dots, which show the traversed points. They are scattered all over the admissible range between 0 and $p - 1 = 112$. We observe that there are many more points on the on the left-hand plot than on the right-hand plot. As a matter of fact, the points repeat after a number of additions; on the left-hand side there are $n = 114$ distinct points and on the right-hand side there are only $n = 19$ distinct points, including one point at infinity. It turns out that the "addition" of points on the curve imposes the structure of a *group* $\mathcal{G}_G$ on the points, which are the elements of $\mathcal{G}_G$. The point at infinity serves as a "zero" of the group. Note that the group depends on the starting point G. Moreover, the number of distinct elements n is called the *order of the group*. In order to be useful for cryptograpy n *must be a prime number*. It is the number of distinct ciphertexts that can be encoded in this scheme. The elements in $\mathcal{G}_G$ are indeed a group because it is then possible to reach an element B of the group from any other element A by adding G a finite number j of times, which we denote by $B = A \oplus (j \odot G)$.

Now Alice and Bob can use elliptic curves to determine a shared secret S that is practically impossible to find out by Eve. This method is called *elliptic curve Diffie-Hellman* (ECDH). Alice and Bob agree on an elliptic curve with parameters a and b, a finite field of prime order p, and a generator point G. Alice chooses a private key d_A and Bob chooses d_B and both publish their respective public keys $P_A = d_A \odot G$ and $P_B = d_B \odot G$, which are two points on the curve. Alice then calculates $S_A = d_A \odot P_B = (d_A d_B) \odot G$ and Bob calculates $S_B = d_B \odot P_A = (d_B d_A) \odot G$, where all calculations are performed modulo n. We see that they arrive at the same secret $S = S_A = S_B$ which they can use to encrypt subsequent communication with, for example, AES.

But elliptic curves can also be used to sign documents, for example, the hash value h of message m? This method is called the *elliptic curve digital signature algorithm* (ECDSA). Given the curve parameters a and b, the starting point G, and the order of the group n, Alice chooses her secret, the private key d. Then she calculates a point $P = d \odot G$ on the curve. This is her public key, which she makes publicly known. Knowing a point P on the curve, it is practically impossible to determine how many iterations d are needed to get there from the generator G, which ensures the security of the system. In order to sign the hash h, she picks a random number k in the range from 1 to $n - 1$, which determines another point $R = k \odot G$ on the curve. We stress that k must be truly random, and, in particular, must not be reused in a second transaction, which would make it possible to determine the private key d. The horizontal coordinate $r \pmod{n}$ of the point $R = (r, \cdot)$ is one part of the signature. The second part of the signature s is constructed from the hash h, the private key d, the random value k, and coordinate r by calculating $s = (h + rd)/k \pmod{n}$. Alice then transmits the message m and the signature (r, s) to Bob.

Bob verifies the signature by calculating the hash h of m using the same hashing function as Alice. Using h and the signature (r, s), he then calculates $Q = (h/s) \odot G + (r/s) \odot P \pmod{n}$. Let us determine what he should find, if the signature is valid

$$Q = \frac{h}{s}G + \frac{r}{s}P = \frac{h}{s}G + \frac{rd}{s}G = \frac{h + rd}{(h + rd)/k}G = kG = R, \qquad (12.35)$$

where all calculations are performed modulo n and we omitted the $\odot$. In the second equality we used $P = d \odot G$ and in the third we inserted the definition of the signature s. We find that the signature is valid, if the horizontal coordinate of Q is the same as the horizontal coordinate of R, which is r. Note that the purpose of constructing the signature (r, s) in such a convoluted way is to make reverse-engineering the private key practically impossible, while making the validation through the calculation of Q rather easy.

In this way Alice can send messages m to Bob and he can verify that they actually came from her, and nobody else. Also, once a message is signed, any changes to m will invalidate the old signature, and we have to sign again. This is one of the key feature that prevents falsifying the records in the database of transactions in the bitcoin infrastructure.

12.8   Bitcoins and Blockchains

The cryptocurrency "Bitcoin" maintains an openly accessible database—the *block-chain*—that records all transaction of the underlying valuta—the *bitcoins*—from one owner to the next. The structure of the blockchain database only permits adding new transactions to the blockchain and reading previously written ones. In contast to conventional databases is updating or deleting entries in the database not possible. Moreover, multiple copies of the blockchain are maintained by many *full nodes* of the bitcoin network, shown as the shaded red circles in Fig. 12.8. An elaborate consensus algorithm, implemented by the *miners*, shown as the green shaded ovals in Fig. 12.8, ensures consistency among the multiple copies in a deterministic way. Maintaining a unique state among many identical copies of the blockchain guarantees a high degree of resilience against removed nodes, either due to malfunction or due to censorship. Every owner of bitcoins can therefore always validate his ownership of bitcoins by reading from the blockchain on any node. Trust in the bitcoin cryptocurrency is thus established by the immutability of the blockchain and the consensus algorithm to ensure that all copies are identical. Once a transaction, and therefore ownership of a bitcoin, is recorded in the blockchain, it will stay there forever—it is immutable. The blockchain database therefore qualifies for the abstract definition of "money" from the beginning of this chapter: it maintains a record of purchasing power. Since all transactions since the inception of the system in 2009 are recorded, the database is substantial. At the time of writing in June 2020, it has a size of about 280 GB.

The operation of the blockchain is goverend through the consensus algorithm and the immutability of the records. It is not supervised by an external authority, which appealed to *Satoshi Nakamoto,* the mysterious author of the whitepaper [14] that outlines the bitcoin infrastructure. Satoshi, whose identity remains unknown, was disenchanted with the handling of the US housing crash of 2008 by the federal authorities, who bailed out the banks at the expense of taxpayers. In 2008 he published the description of the bitcoin cryptocurrency on a mailing list [14] and made the

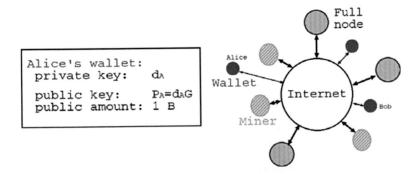

Fig. 12.8 The bitcoin network network consists of full nodes that host the entire bitcoin data base, miners that ensure the integrity of the database, and clients—Alice and Bob—with a wallet that use the network to transfer Bitcoins

source code of an implementation available on a public repository. The system went online in January 2009. The first few bitcoins had a very low value, they are, after all, just bits in a computer's memory, but soon they could be used to buy real items, such as a pizza costing 20 000 bitcoins, which was the first trade that connected bitcoins to the real-world economy. Over the years these connections grew as bitcoins could be used to pay for goods and services and, importantly, be exchanged for conventional currencies. In many countries the gains from trading bitcoins are even taxable. The exchange rate of bitcoins (BTC on foreign exchanges) is very volatile. From an exchange rate of about 400 US$ per Bitcoin early 2016 it rose to almost 20000 US$ in late 2017. It trades, at the time of writing, at around 9000 US$.

In order to understand how the bitcoin system works let us follow the money and consider Alice who, after her graduation, has received a *wallet* with one bitcoin, worth 10^8 *satoshis*, from a rich aunt. A wallet is a computer program that maintains a private and a public key pair, as is illustrated on the left-hand side in Fig. 12.8. Since bitcoin uses *secp256k1* elliptic-curve cryptography, the private key d_A is a 256-bit long integer and her public key is the point $P_A = d_A \odot G$. The wallet can run on a variety of platforms, including her smartphone, which she has to guard carefully such that only she has access to d_A. Now, she wants to pay Bob for her share of a pizza—about 9 $ or 10^5 satoshis. Therefore she initiates her wallet program and asks Bob for his public key P_B before initiating the transfer, which is just a message m_A stating that 10^5 satoshis move from P_A to P_B. The wallet then calculates the SHA-256 hash of m_A, resulting in h_A, which Alice signs with her private key d_A yielding the signed hash $\hat{h}_A$. Finally the wallet sends the package of m_A and $\hat{h}_A$ to the bitcoin system, where it ends up in a pool, waiting to be processed.

This pool is emptied of the messages by *miners*, let's call one of them Marge, who, together with many other miners, works competitively to verify transactions and then enters them into the blockchain database. Marge works through a long list of transaction and when she comes to P_A's transaction m_A, she verifies that wallet P_A actually has sufficient bitcoins, that $\hat{h}_A$ is signed by P_A and that the hash of m_A gives h_A. Finally Marge makes sure that P_B is a valid destination address. If the transaction m_A passes these tests, she puts m_A and its hash h_A in the list of validated transaction, shown on the bottom in Fig. 12.9. After all transactions are validated Marge starts building the next block, here numbered n of the *blockchain*. She successively combines two hashes to form a Merkle tree and stores the root of the tree in the block, next to a timestamp and the hash of the completed previous block, which ensures the continuity and the unique linking of the blocks into the blockchain. Finally, Marge has to validate her work by varying a dummy variable, called a *nonce*, to ensure that the hash of block n begins with a specified number of zeros. This step is required to ensure that the competitor among the miners with the fastest hardware first completes this step. Once block n is complete, work on the next block $n + 1$ can begin. This *proof of work* scheme ensures that the transaction m_A is part of a block that is quickly buried under a large number of successive blocks. Reversing the transaction would entail to recalculate all subsequent transactions and blocks and would require even more powerful hardware. This scheme therefore provides an essential contribution to the integrity of the bitcoin network and prevents *double*

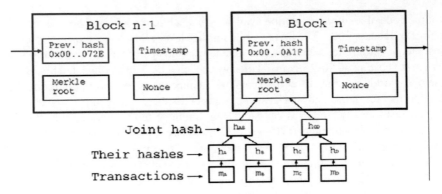

Fig. 12.9 Two consecutive blocks on the blockchain are linked by incorporating the hash of the previous block into the next (red). Each block contains merkle root, constructed from the hashes of a number of transactions, a time stamp, and a nonce. The value of the nonce is adjusted to obtain a hash of the entire block that is smaller than a given limit, which demonstrates the miner's proof-of-work

spending of bitcoins. The system regularly adapts the difficulty of calculating the final hash for a block such a new block is produced every 10 min, which also adjusts the system to increasingly more powerful hardware.

Provided that Marge is the first to complete block n, she posts it to the bitcoin network, such that all other nodes attach it to their copy of the blockchain and subsequent blocks link to it. If the block chain accidentally splits into two branches the consensus is that the longest chain represents the most powerful mining system and will be used henceforth, while the other, *stale* branches die out. The transactions on the stale branch revert to the pool. In this way all copies of the blockchain will stay synchronized and represent a unique state of the bitcoin network. Marge is rewarded for her service to maintain the integrity of the bitcoin network by receiving a number of bitcoins, presently 6.25 bitcoins, for a completed block.

Bob can be sure that the 10^5 satoshis, which Alice has sent, are actually in his pocket, after the block that contains the transaction m_A is buried under six blocks with a later timestamp. At this point the ownership of the 10^5 satoshis is deeply buried in the database—the blockchain. Since the message m_A, which is publicly accessible, only contains the public keys P_A and P_B the system provides some degree of anonymity, unless it is possible to link the public key P_A to Alice, the person.

The award mechanism for the miners is the only way to create new bitcoins up to a maximum of 21×10^6 bitcoins. To reach this limit asymptotically, mining becomes progressively more difficult, because more leading zeros have to be present in the hash of a block, while the number of bitcoins rewarded is progressively reduced. This ensures that bitcoins remain scarce and will not suffer from inflation due to oversupply. However, at some point the received bitcoins will not suffice to finance the running of mining operations. Then the miners will request the initiator of a transaction to pay a transaction fee. This is voluntarily, but miners will likely choose to only process transactions that provide a fee. Furthermore, the large amount of computing and electric power needed to mine bitcoins is subject to criticism, in particular, when considering the many unsuccesfully competiting miners. The resulting carbon footprint of the bitcoin network is rather large, mainly due to the proof-of-work scheme used in the competitive mining process.

Modern blockchain-based networks address some of the criticism and additionally provide more features than the bitcoin network. *Ethereum* is one of these modern networks.

12.9 Ethereum and Smart Contracts

The Ethereum cryptocurrency [15, 16] maintains its own blockchain as a database. The infrastructure with nodes to maintain and store the blockchain, miners to attach new blocks to the blockchain, and clients that initiate transactions, is similar to that of Bitcoin. However, apart from only recording transactions in the underlying currency, called *ether*, the Ethereum blockchain stores program code and data. The data can be changed under the control of the code or through regular transactions, but only after recording the changes on the blockchain through the mining process. The code stored on the blockchain can never be changed retroactively, which makes it particularly trustworthy. One can, for example, use it to automatically transfer funds, provided that certain conditions, which are laid out in the program code, are fulfilled. Such a coded set of rules is commonly referred to as a *smart contract*. It is publicly accessible, anyone can see the rules, yet nobody can change them.

Like Bitcoin, Ethereum uses the *secp256k1* elliptic curve to sign the transactions, but uses the modern *Keccak256* hashing function to calculate hashes. We point out that Keccak256 has some minor differences compared to the SHA-3 hash, which contains modifications by US government organizations [16]. Moreover, the proof-of-work algorithm is based on *ethash* [15], which requires substantial memory and bandwidth. The distinguishing feature of Ethereum is the *Ethereum virtual machine* (EVM). It provides a runtime environment for the smart contracts, which are stored in an intermediate byte-code format. In this way the code runs on the EVM, independent of the hardware of a computer. This is analogous to the way that the JAVA® runtime environment allows the execution of compiled JAVA programs The instructions supported by the EVM are *Turing complete;* any algorithm that can be encoded in computer instructions can also be coded for the EVM. This includes, for example, infinite loops. They are, however, prevented by requiring the caller of the smart contract, either a human-controlled wallet or another contract, to pay for the execution with *gas,* which is directly convertible to ether. Every instruction, such as adding two numbers and every access to memory costs some gas. Thus, an infinite loop would come to a halt, because it runs out of gas. An out-of-gas event reverts all actions of the contract, while the miner keeps all initially allocated gas. Under normal circumstances, once a contract is triggered, a miner runs the contract without problems, updates the blockchain, and receives the used gas as reward.

In order to illustrate the use of smart contracts let us consider a pub that accepts ether as payment for drinks. The publican simply displays a *QR-code* to the customer who has an Ethereum wallet on his smartphone. The QR-code shows the price of the drink and the address of a smart contract to which the payment is directed. This publican is very law abiding and maintains a smart contract that splits the incoming payments into a tax account and a business account. The source code of this smart contract, written in the *solidity* programming language, is the following.

```solidity
pragma solidity ^0.5.17;
contract TaxCollector {
  uint16 drinks;
  constructor() public {
    drinks=0x0;
  }
  function drinks_total() view public returns (uint16) {
    return drinks;
  }
  function () external payable {
    address payable  business=0x3a..;
    address payable tax=0xF2..;
    uint256 t=2*msg.value/10;
    tax.transfer(t);
    business.transfer(msg.value-t);
    drinks += 1;
  }
}
```

The `pragma` directive specifies which version of the compiler to use and inside `contract` definition we first define a global variable `drinks`, that will count the total number of drinks ever sold. As a global variable it is stored on the blockchain and updates every time the contract is executed. The `constructor` is executed when the contract is stored on the blockchain and initiates the variable `drinks` to zero. The function `drinks_total` allows us to query how many drinks have been sold so far. The last function has no name; it is called the fallback function. Moreover, it has the attribute `payable` which allows it to receive funds from the customer who bought the drink and initiated the transaction. The received amount is accessible in the variable `msg.value`. Inside the function the addresses of the tax and business accounts are hard-coded. Then the tax is calculated and stored in the variable `t`, before it is sent to the tax account and the remainder sent to the business account. Note that this contract is only illustrative, but can serve as a starting point for further explorations using the test environment from Appendix B.12. On the other hand, it should not be deployed on the main Ethereum blockchain.

Once the publican has written the smart contract, she compiles it to bytecode using a solidity compiler, signs the bytecode with her private key, and the sends it to a reserved address from where it is scheduled for inclusion on the blockchain. Henceforth, her smart contract will be accessible under an address that is given to the contract during the submission process. She displays this address and the amount to pay on the QR-code shown to the customer, who then transfers ether to the contract. This transaction is included in the next block, at which point the contract runs, updates the `drinks` variable, and distributes the received amount into the tax and the business account. Note that including the contract in the blockchain will cost gas, which the publican pays, and later running the contract costs gas again, which the customer, as the initiator of the transaction, pays. Typically the gas required for a transaction is equivalent to cents of Euro or US$.

The structure of smart contracts with a constructor that runs at submission time and a number of functions that have access to information about the initiator of the calling transaction and the global state of the blockchain, is very flexible. It can be used, for example, to introduce *tokens,* to serve as a unit of trade. One can think of the gambling chips purchased at the entrance of a casino and subsequently used for all gamling in-house. A second example are stocks of a company, which give the holders a stake in the company an entitles them to a share of the annual dividends. Even secondary cryptocurrencies on top of the Ethereum blockchain have been introduced. Libraries facilitate the introduction of new tokens: ERC20 [16] for exchangeable, fungible, tokens and ERC721 [17] for non-fungible tokens, such as virtual collectibles that are unique, for example, cryptokitties [18].

At this point it should be obvious that the initiator of a transaction has the role of a "client," who interacts with a "server," whose role is taken by the blockchain together with the miners. The communication is based on standard network protocols and well-documented interfaces of the smart contracts. This infrastructure facilitates the writing of *decentralized applications* (DApps), where the client software, one can think of enhanced wallets, sends messages and ether back and forth to the contracts. Add-ons to browsers, command-line interfaces, or standalone applications are feasible.

Building smart contracts on top of Ethereum is not the only way to utilize blockchains. Frameworks, such as *Hyper-ledger* [19], can be used to deploy blockchains, either private or public, to store information and company-specific rule sets in smart contracts. Keeping a permanent record of quality-control data in a database, where they can never be changed but trigger mitigation actions if an item is faulty, is one application. Going one step further, smart contracts can formalize the operational rules of a joint venture of mutually non-trusting partners. Crowd-sourcing funds for a novel project, may serve as an example for a so-called a *decentralized autonomous organization* (DAO). Presently such ventures operate outside the jurisdiction of most countries, but that is about to change in the future [20].

The integrity of these organizations relies on the underlying cryptographic methods, which motivates to look into modern developments in physics an computing and how they can compromise the cryptography. Most cryptographic methods are based on trap-door functions that are easy to calculate, but extremely difficult to invert. The classic example is factoring a very large number into its prime factors. The difficulty depends on the available computing power. Since quantum computers use completely different algorithms compared to today's computers, we briefly discuss their inner workings and how much of a threat they pose to cryptography.

12.10   Quantum Computing

The security of many cryptography algorithms is based on the fact that it is practically impossible to find the factors p and q of a very large number $N = pq$. This problem, however, is equivalent to finding the period r, such that $x^r = 1 \pmod{N}$ for a randomly chosen $x < N$, coprime with N. Finding the period r is as difficult as finding the prime factors and might practically take forever, if N is large. Quantum computers, on the other hand, promise to do so efficiently, as we will discuss below. For the time being, let us assume that some technical conditions are fulfilled, that the period r is known and turns out to be an even number. This allows us to rewrite $x^r = 1 \pmod{N}$ as

$$(x^{r/2} - 1)(x^{r/2} + 1) = kN \qquad \text{for some integer } k \qquad (12.36)$$

from which we conclude that $p = \gcd(x^{r/2} - 1, N)$ and $q = \gcd(x^{r/2} + 1, N)$ are factors of N. Trying to factor $N = 15$ provides an example with the smallest non-trivial integers. Let's chose $x = 7$, which leads to the sequence $x^k = \{7, 4, 13, 1, 7, 4, \ldots\}$ for $k = 1, 2, \ldots$ and we find $x^4 = 1$, which implies $r = 4$. For the factors we then obtain $p = \gcd(x^{r/2} - 1, N) = \gcd(48, 15) = 3$ and $q = \gcd(x^{r/2} + 1, N) = 5$, both of which divide $N = 15$. Note that most steps, such as finding the greatest common denominator, take little time. Only finding the period r is extremely time-consuming with classical computers. In his ground-breaking report [21] Shor showed how using a quantum computer can reduce this time dramatically, which, on the long run, might compromise many cryptographic algorithms.

Before discussing quantum computers, let us briefly recall the operation of classical computers, which are based on logic gates that implement operations of Boolean algebra. Since boolean "states" are either true or false, they are represented by "1" and "0," respectively. The top of the left-hand side in Fig. 12.10 shows a gate that implements a logical *not* operation, which inverts the input A and makes it available

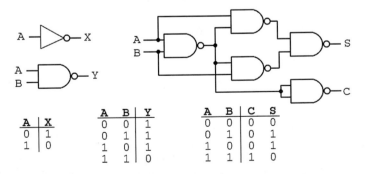

Fig. 12.10 Left: NOT (top) and NAND (below) gate and the table describing the relation of the outputs X and Y to the inputs A and B. Right: Half-adder, composed of NAND gates, which calculates the sum S and a carry-bit C of the inputs A and B

at its output X. The circuit below shows a logical *AND* gate with inverted output, which results in a *not-and*, or NAND-gate. The tables below the two circuits shows the outputs X and Y that correspond to the input values A and B. More complex logical operations, such as adding two binary digits in the *half-adder*, shown on the right-hand side in Fig. 12.10, can be constructed with suitably connected NAND gates. The table below the circuit illustrates how the inputs A and B are combined to produce the sum S and a carry-bit C. All digital devices, including the processors that drive modern computers, are composed of many such elementary building blocks. We emphasize that a state must be in one of the logic states "0" and "1," which represents one bit.

Quantum computers [22], on the other hand, operate on *qubits*, which are quantum-mechanical states that may be visualized as spins pointing up or down, commonly represented as $|0\rangle$ and $|1\rangle$. In contrast to classical bits can qubits be in superposition states $\alpha|0\rangle + \beta|1\rangle$ with $\alpha^2 + \beta^2 = 1$; Schrödinger's cat is in such a state. Instead of using Dirac's notation, we can equally well write the qubits as column vectors, such as

$$|0\rangle = \begin{pmatrix} 1 \\ 0 \end{pmatrix}, \quad |1\rangle = \begin{pmatrix} 0 \\ 1 \end{pmatrix}, \quad \text{and} \quad \alpha|0\rangle + \beta|1\rangle = \begin{pmatrix} \alpha \\ \beta \end{pmatrix}. \tag{12.37}$$

Quantum-mechanical operators take the role of the logic gates and the Pauli-matrix σ_x [23], shown on the left-hand side in (12.38), acts similar to an inverter; it exchanges the "0" and "1" component of a qubit.

$$\sigma_x = \begin{pmatrix} 0 & 1 \\ 1 & 0 \end{pmatrix}, \quad H = \frac{1}{\sqrt{2}} \begin{pmatrix} 1 & 1 \\ 1 & -1 \end{pmatrix}, \quad R_k = \begin{pmatrix} 1 & 0 \\ 0 & e^{2\pi i/2^k} \end{pmatrix} \tag{12.38}$$

The *Hadamard* operator H, shown in the second equation creates a superposition of the "0" and "1" components with $\alpha = 1/\sqrt{2}$ and $\beta = \pm 1/\sqrt{2}$. R_k, shown on the right-hand side in (12.38), adds a phase factor to one of the components. Note that these operators transform one qubit at their input to one qubit at their output. Moreover, all operators are represented as unitary matrices, because they describe the transformation of one spin state to a different spin state, but they do not change the magnitude of the spin.

In order to describe multi-input gates we introduce states that describe multiple qubits q_1 and q_2 as the tensor product of the individual qubits: $|q_1 q_2\rangle = |q_1\rangle \otimes |q_2\rangle$. Using this notation we introduce the *controlled-not*, or CNOT gate, which operates on two qubits $|q_1 q_2\rangle$ and flips q_2, but only if $q_1 = 1$. We can therefore describe it through

$$\text{CNOT} = |00\rangle\langle00| + |01\rangle\langle01| + |10\rangle\langle11| + |11\rangle\langle10|, \tag{12.39}$$

where $\langle q|$ denotes the transpose of $|q\rangle$. It is straightforward to understand the action of CNOT on, for example, $|11\rangle$, because only the third term in (12.39) is non-zero and leads to $|10\rangle$ at the output. Since the first qubit is "1," the second qubit is flipped.

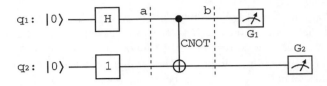

Fig. 12.11 Preparing an entangled state by first preparing two qubits in state "0" and then passing them through a quantum circuit, composed of a Hadamard and a CNOT gate. The small full circle is connected to the control qubit q_1 and the $\oplus$ to the qubit that is changed. The gauges G_1 and G_2 are used to measure the final state

Analogously we introduce the *controlled-R_k* gate: it applies R_k, provided that a control qubit is "1".

The Hadamard and CNOT operators, together with several other operators, take the role of the logic gates in classical circuits and they are used to construct quantum circuits, such as the one shown in Fig. 12.11. We start operation of the circuit with two qubits q_1 and q_2, each prepared in state $|0\rangle$. Nothing happens to the second qubit in the first time step, which is indicated by the unit operator in the lower track. At the same time, a Hadamard operator changes q_1 in the upper track. At this point the system is in state $|q_1 q_2\rangle_a = (|00\rangle + |10\rangle)/\sqrt{2}$. The following CNOT gate will only flip q_2 in the second term, such that the system enters the state $|q_1 q_2\rangle_b = (|00\rangle + |11\rangle)/\sqrt{2}$. This state is called *entangled*, because it is a superposition of qubits that are always equal. Consider the gauge G_1 in the upper track in Fig. 12.11, which will measure either "0" or "1," each with probability $1/2$. Thus, performing the computation with this circuit many times, either result appears with approximately equal frequency. But whatever the result, subsequently measuring q_2 with G_2 always shows the same $G_2 = G_1$. Before the measurement with G_1, the state was undecided and the probabilities are equal, but once we start measuring with G_1, "the system decides" and then G_2 has no other option than to follow G_1. Note how entanglement links the outcome of G_1 and G_2. Here the determinism of classical circuits is replaced by the probabilistic nature of quantum mechanics, where multiple qubits are entangled.

More complex quantum circuits can be constructed from multiple quantum gates that operate on multiple qubits; an important example is the *Quantum Fourier transform* (QFT). It transforms the $N = 2^m$ states that can be expressed by m qubits at the input to qubits at the output, which have additional phase factors $e^{2\pi ijk/N}$. Here j and k with $0 \le j, k \le N - 1$ label the N different states. To illustrate the algorithm we carefully analyze the $m = 3$ qubit version. A crucial trick in the analysis is to express j and k by their binary representations $j = q_1 4 + q_2 2 + q_1$ and $k = \tilde{q}_1 4 + \tilde{q}_2 2 + \tilde{q}_3$, where q denotes the three qubits at the input and $\tilde{q}$ the three qubits at the output of the circuit.

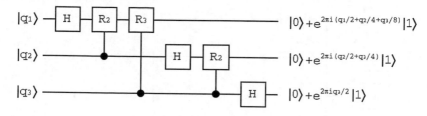

Fig. 12.12 Circuit for the three-qubit quantum Fourier transform. The normalizing factors $1/\sqrt{2}$ are omitted

$$|q_1 q_2 q_3\rangle \rightarrow \frac{1}{\sqrt{8}} \sum_{k=0}^{7} e^{2\pi i jk/8} |\tilde{q}_1 \tilde{q}_2 \tilde{q}_3\rangle$$

$$= \frac{1}{\sqrt{8}} \sum_{\tilde{q}_1=0}^{1} \sum_{\tilde{q}_2=0}^{1} \sum_{\tilde{q}_3=0}^{1} e^{2\pi i j(\tilde{q}_1 4 + \tilde{q}_2 2 + \tilde{q}_3)/8} |\tilde{q}_1 \tilde{q}_2 \tilde{q}_3\rangle$$

$$= \frac{1}{\sqrt{8}} \sum_{\tilde{q}_1=0}^{1} e^{2\pi i j\tilde{q}_1/2} |\tilde{q}_1\rangle \otimes \sum_{\tilde{q}_2=0}^{1} e^{2\pi i j\tilde{q}_2/4} |\tilde{q}_2\rangle \otimes \sum_{\tilde{q}_3=0}^{1} e^{2\pi i j\tilde{q}_3/8} |\tilde{q}_3\rangle$$

$$= \frac{1}{\sqrt{8}} \left(|0\rangle + e^{2\pi i j/2}|1\rangle\right) \otimes \left(|0\rangle + e^{2\pi i j/4}|1\rangle\right) \otimes \left(|0\rangle + e^{2\pi i j/8}|1\rangle\right)$$

$$= \frac{1}{\sqrt{8}} \left(|0\rangle + e^{2\pi i q_3/2}|1\rangle\right) \otimes \left(|0\rangle + e^{2\pi i (q_2/2 + q_3/4)}|1\rangle\right)$$

$$\otimes \left(|0\rangle + e^{2\pi i (q_1/2 + q_2/4 + q_3/8)}|1\rangle\right) \tag{12.40}$$

When evaluating the first bracket we use that $e^{2\pi i(2q_1+q_2)} = 1$ for all values of q_1 and q_2. Moreover, the factor $1/\sqrt{8}$ ensures that the transformation is unitary. We point out that it is straightforward to generalize the method to more qubits. The last equality from (12.40) can be easily translated into the assembly of quantum gates shown in Fig. 12.12. The Hadamard gates H provide the reversed signs for $|1\rangle$ states, consistent with (12.38). The R_k-gates cause a phase shift of the "1" state, provided that the conditioning qubit, indicated by the black dot, is "1." We point out that all quantum gates and consequently also composite circuits are unitary, they are invertible, which allows us to use the QFT circuit backwards to obtain the inverse QFT. Moreover, note how these gates entangle three qubits to perform the Fourier transform on $N = 8$ states with only six gates. This appears to be very efficient considering that a classical Fourier transform of N data points requires $N^2 = 64$ multiplications. Even a fast Fourier-transfrom requires on the order of $N \log_2 N = 24$ multiplications. On the other hand, extracting the desired spectral information from the qubits at the output is difficult, because only phase information is present.

Even though the QFT does not directly provide spectral information, it is very useful in other ways, such as determining of the period of a cyclic process, called *order finding*, and *estimating the phase* ϕ of an unknown unitary operator U in eigen-

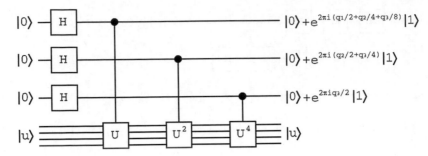

Fig. 12.13 Circuit to determine the binary representation $\phi = q_1/2 + q_2/4 + q_3/8$ of the phase ϕ of the eigenstate $|u\rangle$ of the operator U

state $|u\rangle$, given by $U|u\rangle = e^{2\pi i \phi}|u\rangle$. The operator U is controlled by a qubit, which entangles phase and qubit. Moreover, we assume that the phase is approximated by $\phi = q_1/2 + q_2/4 + q_3/8$, where q_1, q_2, and q_3 provide a binary representation of ϕ. Figure 12.13 illustrates the phase-estimation circuit, which consists of three qubits, initialized to $|0\rangle$, and subsequently put into a superpositioned state with the Hadamard operators H. The second register describes the eigenstate $|u\rangle$, which experiences a phase shift of $e^{2\pi i k\phi}$ when passing the operator U^k. Entangling with the controlling qubit causes the phase of the $|1\rangle$ component to acquire the additional phase factor $e^{2\pi i k\phi}$. If we now express the phase ϕ through its binary representation, we recover the description on the right-hand side in Fig. 12.13. Note that these additional phase factors look exactly like the result of the QFT in Fig. 12.12. Since the QFT is invertible, we simply patch the circuit from Fig. 12.12, but in reverse order, onto the right-hand side of Fig. 12.13. Adding measuring gauges to the output then allows us to recover the bit-pattern that describes the phase ϕ.

Now we have all the tools available to determine the order r from the beginning of this section and defined through $x^r = 1 \pmod{N}$. All we need to do is to construct a unitary operator U which allows us to translate the search for r to a search for a phase ϕ. Let us consider the states $|x^k \pmod{N}\rangle$ and the operator U defined by $U|x^k \pmod{N}\rangle = |xx^k \pmod{N}\rangle = |x^{k+1} \pmod{N}\rangle$. Guided by the idea that the sum of all states within one period r repeats itself under the operation of U we realize that $\sum_{k=0}^{r-1} |x^k \pmod{N}\rangle$ is an eigenvector of U with eigenvalue 1. Likewise, it is easy to show that

$$|u_s\rangle = \frac{1}{\sqrt{r}} \sum_{k=0}^{r-1} e^{-2\pi i k s/r} |x^k \pmod{N}\rangle \tag{12.41}$$

is an eigenvector of U with eigenvalue $e^{2\pi i s/r}$ for all $0 \le s \le r - 1$. Note that the phase we are looking for is here given as $\phi = s/r$. The problem is that we do not know r and therefore cannot determine the eigenvector to directly apply the phase estimation method from the previous paragraph. It turns out, however, that the sum of the r different $|u_s\rangle$ adds up to $|1\rangle$, or $(1/\sqrt{r})\sum_{s=0}^{r-1}|u_s\rangle = |1\rangle$, such that we simply initialize the vector $|u\rangle$ for the phase estimation as $|1\rangle$. Running the phase

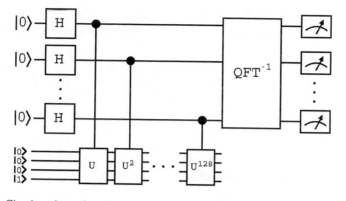

Fig. 12.14 Circuit to determine the order r from (12.36). The operator U is constructed to incre-
ment the power k of a state $|x^k (\text{mod } N)\rangle$. The left-hand part of the circuit determines a binary
representation of $\phi = s/r$ and the inverse quantum Fourier-transfrom makes the individual qubits
visible on the gauges. See the text for details and for definitions of the symbols

estimation multiple times then produces phases with different ratios s/r, but the
values corresponding to the correct period r show up more often.

Figure 12.14 illustrates the algorithm to find r for $N = 15$, which requires $m = 4$ qubits to describe the eigenstate $|u\rangle$ that is initialized to $|1\rangle$, as shown on the
bottom left of the figure. The eigenstate $|u\rangle$ passes through eight controlled unitary
transformations U^{2^j} for $j = 0, \ldots, 7$ and the corresponding phase increments $2^j \phi$
are entangled with the qubits of the phase-measurement circuit, shown in the upper
part of the figure. In order to obtain an accurate measurement of the phase $\phi = s/r$,
which is the ratio of two four-bit numbers, we choose $2m = 8$ qubits for the phase-
estimation circuit, which also explains the number of unitary transformations. The
rest of the phase-estimation circuit is, as before, composed of Hadamard gates, the
inverse QFT, and the measuring gauges to determine the bit-pattern that describes
the phase $\phi = s/r$. Note that ϕ is given in terms of a fractional binary expansion. In
a final step we therefore have to find a fraction s/r that is close to the measured ϕ
and has a denominator r' that is smaller than N, which can be done using *continued
fraction expansion*. Such a value of r' is then a candidate for the period r that
determines the factors of N and thus solves the factoring problem as discussed in the
first paragraph of this section. We emphasize that the number of quantum gates that
are required can be shown to scale with the number of bits m required to describe N
as the m^3 [22]. In contrast, testing all prime numbers smaller than $\sqrt{N}$ scales with
$2^{m/2}$ and is not feasible for very large numbers.

Note that we need $3m = 12$ qubits to factor the four-bit number $N = 15$. Finding
the factors with a large number of bit, say $N = 300$ would require a circuit with 900
qubits, which well beyond what is currently feasible. Moreover, the construction
of the controlled-U gates is very difficult and presently poses a limitation to scale
the algorithm to larger m. Presently the largest number factored is 21. So, for the
time being, algorithms such as RSA, appear to be safe. And should more powerful

quantum computers become available in the future, Alice and Bob can always use entangled states to agree on a secret key that even quantum computers cannot figure out.

Exercises

1. Show that for $p_i = 1/n$ for $i = 1, \ldots, n$, (12.2) reverts to (12.1).
2. Determine the probabilities $p_x(x_i)$ and $p_y(y_j)$, the joint probabilities $p_{xy}(x_i, y_j)$, as well as the conditional probabilities $p_{xy}(x_i|y_j)$ for the binary channel depicted in Fig. 12.2.
3. Determine $H[x]$, $H[y]$, $H[x, y]$, and $H[y|x]$ for the binary channel from Fig. 12.2.
4. If x_i and y_j are statistically independent, the joint probability distribution function factorizes and can be written as $p_{xy}(x_i, y_j) = p_x(x_i)p_y(y_j)$. Show that this implies $H[x, y] = H[x] + H[y]$.
5. Show that $H[y|x] = H[y]$ if x_i and y_j are statistically independent.
6. Alice and Bob agree to use the Diffie-Hellman key exchange mechanism with $p = 9967$ and $g = 33$. Alice sends her public number $A = 7025$ to Bob, who has private key $b = 557$. He now wants to send a single number, the hour $1 < h < 24$ when to meet Alice, in encrypted form to her. (a) Determine the Diffie-Hellman encryption key k that Bob then uses to produce the ciphertext $c = \text{xor}(k, h)$. (b) Alice receives the ciphertext $c = 4148$. When will they meet? Hint: In MATLAB you can use the `powermod()` and `bitxor()` functions.
7. Alice and Bob agree to use the RSA algorithm with the public key $(n, e) = (5561, 5)$. Alice wants to meet Bob and sends the hour, encoded with her private key, which turns out to be $c = 5343$. Eve desperately wants to know the time they meet and intercepts the message such that she also knows c. She realizes that the number n in the public key is rather small, which allows her to determine the decryption key d. Help Eve to find out when Alice and Bob meet? Hint: MATLAB has a built-in command `factor()`.
8. Alice and Bob agree to use ECDSA with a hashing function where they xor the numerical values of the ASCII codes of the characters in a message. For example, the code of the letter "A" is 65. Moreover, they use the elliptic curve $y^2 = x^3 + 7$, an integer field based on $p = 113$, and the generator point $G = (15, 52)$, the same one used to prepare the curve on the right-hand side in Fig. 12.7. Alice, also known by her public point $P = (93, 16)$, sends the message "YES" to an invitation and signs the message with $(r, s) = (66, 11)$. Please help Bob to verify whether her message is authentic. Hint: check out the MATLAB code in Appendix B.11.
9. Show that $U|u_s\rangle = e^{2\pi i s/r}|u_s\rangle$, where $|u_s\rangle$ is defined in (12.41) and U is defined just before.

References

1. R. Hartley, Transmission of information. Bell Syst. Tech. J. **7**, 535 (1928)
2. C. Shannon, The mathematical theory of communication. Bell Syst. Tech. J. **27**, 623 (1948). Reprinted in book form by C. Shannon and W. Weaver with the same title, University of Illinois Press, 1998
3. C. Mackezie, *Coded Character Sets* (Addison-Wesley Publishing Company, Reading, History and Development, 1980)
4. King James bible: available at https://www.gutenberg.org/ebooks/10
5. Project Gutenberg web site: https://www.gutenberg.org
6. W. Shakespeare, *Hamlet*: available at https://www.gutenberg.org/ebooks/1787
7. D. Huffman, A method for the construction of minimum-redundancy codes. Proc.IRE (Institute of Radio Engineers, now IEEE) **40**, 1098 (1952)
8. C. Kittel, H. Kroemer, *Thermal Physics*, 2nd edn. (W.H. Freeman, San Francisco, 1980)
9. S. Moser, P. Chen, *A student's Guide to Coding and Information Theory* (Cambridge University Press, Cambridge, 2012)
10. T. Cover, J. Thomas, *Elements of Information Theory*, 2nd edn. (Wiley, Hoboken, 2006)
11. H. Nyquist, Certain topics in telegraph transmission theory. Trans. AIEE. **47**(2), 617 (1928)
12. B. Schneier, *Applied Cryptography* (J. Wiley, Indianapolis, 2015)
13. W. Diffie, M. Hellman, New directions in cryptography. IEEE Trans. Inf. Theory IT-22, 644 (1976)
14. S. Nakamoto, *Bitcoin: A Peer-to-Peer Electronic Cash System* (2008). Available from https://bitcoin.org/en/bitcoin-paper
15. G. Wood, Ethereum: A secure decentralised generalised transaction ledger. Petersburg version 4ea7b96, (2020)
16. A. Antonopoulos, G. Wood, *Mastering Ethereum* (O'Reilly Media, Sebastopol, 2018)
17. ERC-721 non-fungible tokens on the Ethereum blockchain, http://erc721.org
18. Cryptokitties website, https://www.cryptokitties.co
19. Open-source blockchain Hyperledger, https://www.hyperledger.org
20. Federal Ministery for Economic Affairs and Energy, *Blockchain Strategy of the Federal Government* (2019). https://www.bmwi.de/Redaktion/EN/Publikationen/Digitale-Welt/blockchain-strategy.html. Accessed 9 July 2020
21. P. Shor, Algorithms for quantum computation: discrete logarithms and factoring, in *Proceedings 35th Annual Symposium on Foundations of Computer Science* (IEEE Comput. Soc. Press, 1994), p. 124
22. M. Nielsen, I. Chuang, *Quantum Computation and Quantum Information* (Cambridge University Press, Cambridge, 2010)
23. C. Cohen-Tannoudji, B. Diu, F. Laloe, *Quantum Mechanics*, vol. 2 (Wiley, New York, 1977)

Chapter 13
Solutions for Selected Exercises

Abstract This chapter presents the solutions to many of the end-of-chapter exercises. Numerical solutions with sample code in MATLAB, where applicable, can be found in the Electronic Supplementary Material.

In this chapter we present solutions to most end-of-chapter exercises. For many solutions, the accompanying MATLAB code is available in the external supplementary material (ESM) from this book's web page at https://doi.org/10.1007/978-3-030-63643-2_13.

The solutions below are referenced by the chapter and exercise number, separated by a dot.

Exercise 2.1

1992, G. Soros shorted the pound by selling 10^{10} £ with the promise to buy them back later. The British bank had to buy all those pounds in order to stay within certain margins of the exchange rate with respect to the German Mark. It was required to do so after having previously joined the European exchange rate mechanism. The British central bank could not counter this onslaught and, in the end, had to devalue the British pound by almost 10%, which brought Soros a 10% profit on this 10 billion £ gamble.

Exercise 2.5

DAX: Deutsche Börse Aktiengesellschaft (AG), is a publicly owned joint stock company. NYSE: the Intercontinental Exchange (ICE) is a company that owns stock exchanges, the NYSE among them. They are listed at the NYSE. Stockholm: owned by NASDAQ, Inc, which is a financial company that operates stock exchanges, the NASDAQ in New York among them, where NASDAQ Inc. is listed.

Exercise 2.6

A person who owns a fully paid item, say a stock, has the *long* position. Conversely, a person who owes the stock to someone else or has sold it, is *short* of the stock. Thus

Electronic supplementary material The online version of this chapter (https://doi.org/10.1007/978-3-030-63643-2_13) contains supplementary material, which is available to authorized users.

"not having" corresponds to "short". One might hypothesize that the origin of this nomenclature comes from medieval times, when debt was recorded with *tally sticks*. These are specially marked pieces of wood that were broken in two. The shorter piece was given to the borrower and the longer piece to the lender as a record of the transaction.

Exercise 3.1

The Lagrange function $L(x, y, \dot{x}, \dot{y})$, including the constraint, is given by

$$L(x, y, \dot{x}, \dot{y}) = \frac{m}{2}\left(\dot{x}^2 + \dot{y}^2\right) - \frac{k}{2}\left(x^2 + xy + y^2\right) + \lambda(x + y - 1), \quad (13.1)$$

where λ is a Lagrange multiplier. The Euler-Lagrange equations then lead to the following equations of motion

$$
\begin{aligned}
0 &= \frac{d}{dt}\frac{\partial L}{\partial \dot{x}} - \frac{\partial L}{\partial x} = m\ddot{x} + k\left(x + \frac{y}{2}\right) - \lambda \\
0 &= \frac{d}{dt}\frac{\partial L}{\partial \dot{y}} - \frac{\partial L}{\partial y} = m\ddot{y} + k\left(y + \frac{x}{2}\right) - \lambda \\
0 &= \frac{\partial L}{\partial \lambda} = x + y - 1 .
\end{aligned}
\quad (13.2)
$$

The difference of the first two equations leads to

$$0 = m(\ddot{x} - \ddot{y}) + \frac{k}{2}(x - y) \quad (13.3)$$

and using the third equation to eliminate y gives us

$$0 = 2m\ddot{x} + k\left(x - \frac{1}{2}\right). \quad (13.4)$$

Dividing by $2m$ and substituting $z = x - 1/2$ allows us to read off the eigenfrequency $\omega = \sqrt{k/2m}$ and the equilibrium point at $z = 0$ or $x = 1/2$.

Exercise 3.2 and 3.3

After reading the data file and assigning the stock values to variables S1 and S2, we calculate the daily returns

```
r1=(S1(2:end)-S1(1:end-1))./S1(1:end-1);
r2=(S2(2:end)-S2(1:end-1))./S2(1:end-1);
```

The rest of the analysis closely follows the MATLAB script, discussed in Appendix B.1, adapted to two stocks. See the the files ex3_2.m and ex3_3.m in the ESM for the complete solutions, respectively.

Exercise 3.4

Reading and preparing the daily returns is done with the code already used in Exercise 3.3. In the following code snippet, we illustrate the calculation of the covariance matrix $\hat{C}$, which gives us the volatilities σ_1 and σ_M, and the beta β of the stocks. These quantities are used to evaluate the left and right-hand side of (3.38).

```
C=[rp1'*rp1, rp1'*rp2; rp1'*rp2, rp2'*rp2]/N
sig1=sqrt(C(1,1))   % volatility Apple
sigM=sqrt(C(2,2))   % volatility SP500
beta=C(1,2)/sigM^2
rf=0.05/N;  % per trading day
lhs=rm1      % of (3.38)     % 6.9e-4
rhs=rf+beta*(rm2-rf)         % 3.9e-4
```

The left-hand side of (3.38) is larger than the right-hand side, which implies that the Apple stock is undervalued. The full solution is available as ex3_4.m in the ESM.

Exercise 3.5

After initializing all variables with the values stated in the text of the exercise, we first calculate the $WACC$ and then use it in the sum to determine the discounted cash flow D.

```
k=1:6       % years
WACC=(V-B)*re/V+B*rd*(1-t)/V
D=sum(C./(1+WACC).^k)   % 7.79E6
```

D turns out to be 7.8 M€, which is less than the asking price of 8.1 M€, which indicates that it will be difficult to recover the initial investment. See ex3_5.m, available in the ESM, for the full solution.

Exercise 4.1

Figure 13.1 shows the two-layer binomial tree with increments f_1 and f_2 given by (4.1) and the probability p from (4.3) that the stock actually increases in value.

Fig. 13.1 Exercise 4.1: Two-layer binomial tree

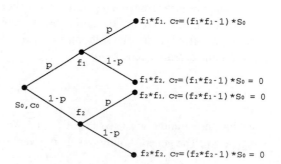

We have to keep in mind that the year is divided into two segments and the annual growth rate ρ has to be divided by two and the volatility σ by $\sqrt{2}$. Using these values, we calculate the stock values after two segments, as shown in the Fig. 13.1. We then compare the values with the strike price K and find that only the upper-most branch contributes, because $f_1 = 1/f_2$ and the payoff function for a call option is $\max(S_T - K, 0)$. Thus we find that, after applying the discount factor $e^{-\rho T}$, the value of the call option at the start of the contract is $c_0 = e^{-\rho T} p^2 (f_1^2 - 1) S_0 = 0.129 S_0$, where T is one year, such that $\rho T = 0.05$. See ex4_1.m from the ESM for the numerical details.

Exercise 4.2

Since $G(r)$ has radial symmetry, we only need the radial part of the Laplace operator in three dimensions Δ_r. Using spherical coordinates in three dimensions, it is given by

$$\Delta_r G(r) = \frac{1}{r^2} \frac{\partial}{\partial r} \left(r^2 \frac{\partial G(r)}{\partial r} \right) \tag{13.5}$$

For non-zero r, inserting $G(r) = 1/4\pi r$ immediately shows that bracket becomes a constant, such that the "outer" derivative results in zero. In order to justify the 4π in the denominator, we employ Gauss' theorem and write the Laplace operator as the scalar product of the divergence ∇ and the gradient ∇ applied to the function G, which is related to the field $E = -\nabla G$. Integrating the field over a small sphere with radius ε around the origin yields

$$\int_{V(\varepsilon)} (\nabla \cdot \nabla) G dV = - \int_{\partial V(\varepsilon)} E dS = \int_{\partial V(\varepsilon)} \frac{1}{4\pi r^2} dS \tag{13.6}$$

where $V(\varepsilon)$ is the volume of the small sphere and $\partial V(\varepsilon)$ its surface. In the second equality we insert the radial gradient of $E = -\nabla G = 1/4\pi r^2$. Since we have $r = \varepsilon$ on the surface of the sphere and $dS = 4\pi \varepsilon^2$ we see that the integral over ΔG is unity. Summarily, the value is zero, except at the origin, where the integral amounts to unity.

Exercise 4.3

The response of the system to an initial velocity perturbation v_0 can be calculated by matching the initial values of the homogeneous solution of the differential equation to the initial velocity. The trial solution $x = Ae^{i\omega t}$ leads to the following equation to determine the eigenfrequencies ω

$$\omega^2 - 2i\alpha\omega - \omega_0^2 = 0 \tag{13.7}$$

which has the solutions $\omega_{\pm} = i\alpha \pm \sqrt{\omega_0^2 - \alpha^2}$. We assume that the damping is weak with $\alpha < \omega_0$. The general solution is then given by

$$x = Ae^{i\omega_+ t} + Be^{i\omega_- t} \qquad \text{and} \qquad \dot{x} = i\omega_+ Ae^{i\omega_+ t} + i\omega_- Be^{i\omega_- t} . \tag{13.8}$$

Matching the initial values $x(0) = 0$ and $\dot{x}(0) = v_0$, we find

$$A = -\frac{v_0}{2i\sqrt{\omega_0^2 - \alpha^2}} \qquad \text{and} \qquad B = -A . \tag{13.9}$$

The response of the system to a velocity impulse at $t = 0$ then turns out to be

$$x(t) = \frac{v_0 e^{-\alpha t}}{\omega_0^2 - \alpha^2} \sin\left(\sqrt{\omega_0^2 - \alpha^2}\, t\right) \tag{13.10}$$

Exercise 4.4

The temperature-diffusion equation is given by the indicated substitutions and has the solution

$$T(x, t; d) = \frac{T_0}{\sqrt{4\pi Dt}} \exp\left[-\frac{(x - d)^2}{4Dt}\right] , \tag{13.11}$$

where d denotes the position where the local temperature rise occurs. Since the insulated end dictates that $\partial T/\partial x = 0$ at $x = 0$, we place an additional image source at $x = -d$, which then satisfies the boundary conditions at $x = 0$. The temperature distribution is therefore described by $T_b(x, t) = T(x, t; d) + T(x, t; -d)$, such that the temperature profile at the insulated end is given by $T_b(0, t)$. Inserting suitable numbers and plotting the temperature profile shows an initial temperature rise up to a maximum value and a subsequent asymptotic decay. See Fig. 13.2 for an illustration that was generated by ex4_4.m from the ESM.

Exercise 4.5

(a) After substituting $y = (z - \hat{\rho}t)/\sqrt{2\sigma^2 t}$ the integral over $\psi(z, t)dz$ becomes

$$\int_{-\infty}^{\infty} \psi(z, t)dz = \frac{1}{\sqrt{\pi}} \int_{0}^{\infty} e^{-y^2} dy = 1 , \tag{13.12}$$

which shows that the integral is normalized.

(b) In order to simplify the notation we introduce $g(z, t) = (z - \hat{\rho}t)^2/2\sigma^2 t$ and $A = 1/\sqrt{2\pi\sigma^2}$ and write $\psi(z, t) = At^{-1/2}e^{-g(z,t)}$. Showing that $\psi(z, t)$ solves (4.30) we need to calculate its derivatives with respect to t and to z. For the derivatives of $g(z, t)$ we find

$$\frac{\partial g}{\partial t} = \frac{-z^2 + \hat{\rho}^2 t^2}{2\sigma^2 t^2} , \quad \frac{\partial g}{\partial z} = \frac{z - \hat{\rho}t}{\sigma^2 t} , \quad \text{and} \quad \frac{\partial^2 g}{\partial z^2} = \frac{1}{\sigma^2 t} . \tag{13.13}$$

For the derivatives of $\psi(z, t)$ with respect to z we then obtain

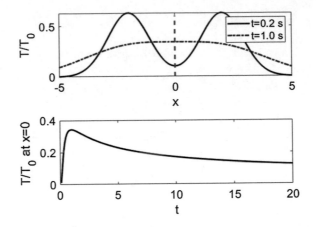

Fig. 13.2 Exercise 4.4: Top: the temperature profile at $t = 0.2$ s and $t = 1$ s along the slab, which extends from its insulated end at $x = 0$ towards the right. Note that at $t = 0.2$ s the temperature rise is large near $x = d = 2$ before spreading out. Bottom: the temperature at $x = 0$ as a function of time. We see that it takes a short time for the temperature to rise, followed by a long-time decay

$$\frac{\partial \psi}{\partial z} = -\psi g' \quad \text{and} \quad \frac{\partial^2 \psi}{\partial z^2} = \left(g'^2 - g''\right) \psi \tag{13.14}$$

where we use the notation $g' = \partial g / \partial z$. For the left-hand side of (4.30) we find

$$\frac{\partial \psi}{\partial t} = -\frac{At^{-3/2}}{2} e^{-g} - At^{-1/2} e^{-g} \frac{\partial g}{\partial t} = \left[-\frac{1}{2t} + \frac{z^2 - \hat{\rho}^2 t^2}{2\sigma^2 t} \right] \psi . \tag{13.15}$$

For the right-hand side of (4.30) we calculate

$$-\hat{\rho}\psi' + \frac{\sigma^2}{2}\psi'' = \hat{\rho}g'\psi + \frac{\sigma^2}{2}\left(g'^2 - g''\right)\psi = \left[\frac{\hat{\rho}(z - \hat{\rho}t)}{\sigma^2 t} + \frac{(z - \hat{\rho}t)^2}{2\sigma^2 t^2} - \frac{1}{2t} \right] \psi$$

$$= \left[\frac{z^2 - \hat{\rho}^2 t^2}{2\sigma^2 t} - \frac{1}{2t} \right] \psi , \tag{13.16}$$

which equals the expression we found for $\partial \psi / \partial t$ in (13.15).

Exercise 4.6 and 4.7

This sought probability $P_{>2}$ is given by

$$P_{>2} = \int_{2S_0}^{\infty} \Psi(S, t) dS , \tag{13.17}$$

Fig. 13.3 Exercise 4.6: The red area gives the probability that the stock value has more than doubled its value after two years. The yellow area gives the probability that the stock halves in two years

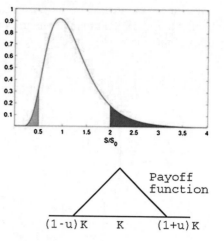

Fig. 13.4 Exercise 4.8: The payoff function

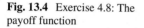

Payoff function

$(1-u)K$ $\quad K \quad$ $(1+u)K$

where $\Psi(S,t)$ is given by (4.32) with $t = 2$ years and $\hat{\rho} = \rho - \sigma^2/2 \approx 0.055$. Substituting $z = \ln(S/S_0)$ and $dz = dS/S$ leads to the integral

$$P_{>2} = \frac{1}{\sqrt{2\pi\sigma^2 t}} \int_{\ln 2}^{\infty} \exp\left[-\frac{(z - \hat{\rho}t)^2}{2\sigma^2 t}\right] dz. \tag{13.18}$$

The substitution $y = (z - \hat{\rho}t)/\sqrt{2\sigma^2 t}$ with $dy/dz = 1/\sqrt{2\sigma^2 t}$ then gives us

$$P_{>2} = \frac{1}{\sqrt{\pi}} \int_{\frac{\ln(2)-\hat{\rho}t}{\sqrt{2\sigma^2 t}}}^{\infty} e^{-y^2} dy = \frac{1}{2} \operatorname{erfc}\left[\frac{\ln(2) - \hat{\rho}t}{\sqrt{2\sigma^2 t}}\right] \tag{13.19}$$

For the numbers given in the exercise, we find $P_{>2} \approx 0.085$, which is shown as the red area in Fig. 13.3.

The probability $P_{<1/2}$, asked for in Exercise 4.7, is given by

$$P_{<1/2} = 1 - P_{>1/2} = 1 - \frac{1}{2}\operatorname{erfc}\left[\frac{\ln(1/2) - \hat{\rho}t}{\sqrt{2\sigma^2 t}}\right]. \tag{13.20}$$

Inserting numbers yields $P_{<1/2} \approx 0.029$, which is shown as the yellow area in Fig. 13.3.

Exercise 4.8

The payoff function f is sketched in Fig. 13.4 and its functional form is given by

$$f(S) = \begin{cases} 0 & \text{for } S < 1 - u)K, \\ S - (1-u)K & \text{for } (1-u)K < S < K, \\ uK - (S - K) & \text{for } K < S < (1+u)K, \\ 0 & \text{for } (1+u)K < S. \end{cases} \tag{13.21}$$

Averaging with $\Psi(S, t)$ from (4.32) and discounting with $e^{-\rho t}$ yields an equation similar to (4.33). Splitting the integral into the separate domains, we obtain for the price of the option O

$$
O = e^{-\rho t} \int_{(1-u)K}^{K} [S - (1-u)K]\Psi(S, t)dS + e^{-\rho t} \int_{K}^{(1+u)K} [(1+u)K - S]\Psi(t)dS
$$

$$
= \frac{e^{-\rho t} S_0}{\sqrt{2\pi\sigma^2 t}} \left\{ \int_{\ln(1-u)K/S_0}^{\ln K/S_0} e^{-(z-\hat{\rho}t)^2/2\sigma^2 t} \left[e^z - \frac{(1-u)K}{S_0} \right] dz \right. \tag{13.22}
$$

$$
\left. + \int_{\ln K/S_0}^{\ln(1+u)K/S_0} e^{-(z-\hat{\rho}t)^2/2\sigma^2 t} \left[\frac{(1+u)K}{S_0} - e^z \right] dz \right\}
$$

where we use the substitution $z = \ln(S/S_0)$. The exponent of the term with e^z can be simplified to

$$
-\frac{(z - \hat{\rho}t)^2}{2\sigma^2 t} + z = -\frac{(z - \hat{\rho}t - \sigma^2 t)^2}{2\sigma^2 t} + \hat{\rho}t + \frac{\sigma^2}{2}t , \tag{13.23}
$$

which allows us to complete the square in the exponent and calculate the integrals. For the first one we get

$$
\frac{1}{\sqrt{2\pi\sigma^2 t}} \int_{a}^{b} e^{z-(z-\hat{\rho}t)^2/2\sigma^2 t} = e^{\hat{\rho}t + \sigma^2 t/2} \left[N\left(\frac{b - \hat{\rho}t - \sigma^2 t}{\sqrt{\sigma^2 t}} \right) \right.
$$

$$
\left. - N\left(\frac{a - \hat{\rho}t - \sigma^2 t}{\sqrt{\sigma^2 t}} \right) \right] \tag{13.24}
$$

and for the second integral

$$
\frac{1}{\sqrt{2\pi\sigma^2 t}} \int_{c}^{d} e^{-(z-\hat{\rho}t)^2/2\sigma^2 t} = N\left(\frac{d - \hat{\rho}t}{\sqrt{\sigma^2 t}} \right) - N\left(\frac{c - \hat{\rho}t}{\sqrt{\sigma^2 t}} \right) \tag{13.25}
$$

where $N(z)$ is the cumulative normal distribution, defined in (4.39). Inserting these expressions into (13.22), we finally arrive at

$$
O = S_0 \left\{ N\left(\frac{\ln(K/S_0) - \hat{\rho}t - \sigma^2 t}{\sqrt{\sigma^2 t}} \right) - N\left(\frac{\ln((1-u)K/S_0) - \hat{\rho}t - \sigma^2 t}{\sqrt{\sigma^2 t}} \right) \right.
$$

$$
\left. - N\left(\frac{\ln((1+u)K/S_0) - \hat{\rho}t - \sigma^2 t}{\sqrt{\sigma^2 t}} \right) + N\left(\frac{\ln(K/S_0) - \hat{\rho}t - \sigma^2 t}{\sqrt{\sigma^2 t}} \right) \right\}
$$

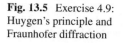

Fig. 13.5 Exercise 4.9:
Huygen's principle and
Fraunhofer diffraction

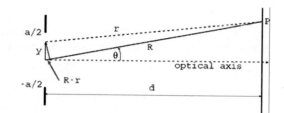

$$-Ke^{-\rho t}\left\{(1-u)N\left(\frac{\ln(K/S_0)-\hat{\rho}t}{\sqrt{\sigma^2 t}}\right)\right.$$

$$-(1-u)N\left(\frac{\ln((1-u)K/S_0)-\hat{\rho}t}{\sqrt{\sigma^2 t}}\right) \qquad (13.26)$$

$$\left.-(1+u)\left[N\left(\frac{\ln((1+u)K/S_0)-\hat{\rho}t}{\sqrt{\sigma^2 t}}\right)-N\left(\frac{\ln(K/S_0)-\hat{\rho}t}{\sqrt{\sigma^2 t}}\right)\right]\right\}$$

We refrain from simplifying this further.

Exercise 4.9

The geometry is shown in Fig. 13.5. The mid-point of the aperture is placed on the
optical axis and we consider the contributions of points on the aperture to an image
point P, which subtends an angle θ with respect to the optical axis. The distance r
from points on the aperture varies when compared to the distance R of the ray that
comes from the mid-point. For small angles θ the difference between R and r is
approximately given by $R - r \approx y \sin \theta$. All source points on the aperture therefore
have a small path-length difference to the image point P. The superposition of all
rays from the aperture then gives the field strength E at P

$$E = \int_{-a/2}^{a/2} \frac{e^{2\pi i r(y)/\lambda}}{r(y)} dy \approx \frac{e^{2\pi i R/\lambda}}{d} \int_{-a/2}^{a/2} e^{-2\pi i y \sin\theta/\lambda} dy. \qquad (13.27)$$

Evaluating the integral leads to

$$E \approx \frac{e^{2\pi i R/\lambda}}{d} \frac{a \sin((ka/2)\sin\theta)}{(ka/2)\sin\theta} \qquad (13.28)$$

with the abbreviation $k = 2\pi/\lambda$. The intensity $I(\theta)$ at point P is proportional to the
squared absolute value of the field strength. We thus obtain

$$I(\theta) \approx \left(\frac{a}{d}\right)^2 \left(\frac{\sin((ka/2)\sin\theta)}{(ka/2)\sin\theta}\right)^2, \qquad (13.29)$$

the well-known Fraunhofer diffraction pattern of a single slit.

Exercise 5.1

Replacing the price for a call option c with that of a put option p in the steps from (5.3) to (5.6) leads to the Black-Scholes equation for p.

Exercise 5.2

From Fig. 5.4 we read that the payoff is $1.2S_0$ if the stock price at maturity exceeds the strike price K. A straightforward way to calculate the option price O is based on using $\Psi(S,t)$ from (4.32) with ρ replaced by the risk-free rate r_f to calculate the expectation value of the pay off and then applying the discount factor $e^{-r_f t}$ to back-propagate the future payoff to today. This leads to

$$O = 1.2S_0 e^{-r_f t} \int_K^\infty \Psi(S,t)dS = \frac{1.2S_0 e^{-r_f t}}{\sqrt{2\pi\sigma^2 t}} \int_{\ln(K/S_0)}^\infty e^{-(z-\hat{r})^2/2\sigma^2 t}dz$$

$$= \frac{1.2S_0 e^{-r_f t}}{2}\, \mathrm{erfc}\left[\frac{\ln(K/S_0) - \hat{r}t}{\sqrt{2\sigma^2 t}}\right], \tag{13.30}$$

where we use the same substitutions as in Exercise 4.6 and $\hat{r} = r_f - \sigma^2/2$.

Exercise 5.3

The expectation value of the payoff is given by

$$O = e^{-r_f t}\int_{K_1}^{K_2} 2S_0\Psi(S,t)dS = \frac{2S_0 e^{-r_f t}}{\sqrt{2\pi\sigma^2 t}} \int_{\ln(K_1/S_0)}^{\ln(K_2/S_0)} e^{-(z-\hat{r}t)^2/2\sigma^2 t}dz, \tag{13.31}$$

where we used the substitution $z = \ln(S/S_0)$ in the second equality and $\hat{r} = r_f - \sigma^2/2$. A second substitution $y = (z - \hat{r}t)/\sigma^2 t$ leads to

$$O = \frac{2S_0 e^{-r_f t}}{\sqrt{2\pi}} \int_{\frac{\ln(K_1/S_0)-\hat{r}t}{\sigma^2 t}}^{\frac{\ln(K_2/S_0)-\hat{r}t}{\sigma^2 t}} e^{-y^2/2}dy \tag{13.32}$$

$$= 2S_0 e^{-r_f t}\left[N\left(\frac{\ln(K_2/S_0) - \hat{r}t}{\sigma^2 t}\right) - N\left(\frac{\ln(K_1/S_0) - \hat{r}t}{\sigma^2 t}\right)\right]$$

Exercise 5.4

The equation, equivalent to (5.13), that describes the price for this option is given by

$$O(x,\tau) = K\int_0^\infty (e^{nx'} - 1)\frac{1}{\sqrt{2\pi\sigma^2\tau}}e^{-r_f\tau-(x'-x)^2/2\sigma^2\tau}dx'$$

$$= \frac{Ke^{-r_f\tau}}{\sqrt{2\pi\sigma^2\tau}} \int_0^\infty \left[e^{nx'-(x'-x)^2/2\sigma^2\tau} - e^{-(x'-x)^2/2\sigma^2\tau} \right] dx' . \quad (13.33)$$

where we substituted $x' = \ln(S/K)$ and note that the option from the exercise corresponds to $n = 2$. The exponent in the first exponential can be rewritten as

$$-\frac{(x'-x)^2 - 2n\sigma^2\tau x'}{2\sigma^2\tau} = -\frac{(x'-x-n\sigma^2\tau)^2}{2\sigma^2\tau} + nx + \frac{n^2}{2}\sigma^2\tau \quad (13.34)$$

which allows us to express the option price as

$$O(x,\tau) = \frac{Ke^{-r_f\tau}}{\sqrt{2\pi\sigma^2\tau}} \left[e^{nx+n^2\sigma^2\tau/2} \int_0^\infty e^{-(x'-x-n\sigma^2\tau)^2/2\sigma^2\tau} dx' \right.$$

$$\left. - \int_0^\infty e^{-(x'-x)^2/2\sigma^2\tau} dx' \right] \quad (13.35)$$

$$= \frac{Ke^{-r_f\tau}}{\sqrt{2\pi\sigma^2\tau}} \sqrt{\sigma^2\tau} \left[e^{nx+n^2\sigma^2\tau/2} \int_{-\frac{x+n\sigma^2\tau}{\sqrt{\sigma^2\tau}}}^\infty e^{-y^2/2} dy - \int_{-\frac{x}{\sqrt{\sigma^2\tau}}}^\infty e^{-y^2/2} dy \right]$$

$$= Ke^{-r_f\tau} \left[e^{nx+n^2\sigma^2\tau/2} N\left(\frac{x+n\sigma^2\tau}{\sqrt{\sigma^2\tau}} \right) - N\left(\frac{x}{\sqrt{\sigma^2\tau}} \right) \right] .$$

Substituting back $x = z + \hat{r}\tau$ with $z = \ln(S/K)$ and $\hat{r} = r_f - \sigma^2/2$, we finally obtain

$$O(S,\tau) = \frac{S^n}{K^{n-1}} e^{-(n-1)r_f\tau+n(n-1)\sigma^2\tau/2} N\left(\frac{\ln(S/K) + [r_f + (n-1/2)\sigma^2]\tau}{\sqrt{\sigma^2\tau}} \right)$$

$$- Ke^{-r_f\tau} N\left(\frac{\ln(S/K) + [r_f - \sigma^2/2]\tau}{\sqrt{\sigma^2\tau}} \right), \quad (13.36)$$

which gives the requested result for $n = 2$.

Exercise 5.7
From the definition of $N(z)$ in (4.39), we see that $N'(z) = e^{-z^2/2}/\sqrt{2\pi}$ is a Gaussian, which is a symmetric function in z. Conversely, $N(z)$ is the integral over $N'(z)$. Since $N(\infty) = 1$ we can write

$$1 = \int_{-\infty}^{-z} N'(y)dy + \int_{-z}^\infty N'(y)dy = N(-z) - \int_z^{-\infty} N'(-x)dx$$

$$= N(-z) + \int_{-\infty}^{z} N'(x)dx = N(-z) + N(z) \tag{13.37}$$

where we split the integral at $-z$ and substituted $y = -x$ in the second integral. Intuitively, we find the same result, because the area under the tails of $N'(z)$ on the left and right side are equal. Therefore the area under the left tail $N(-z)$ is just what is missing to complete $N(z)$ to unity.

Exercise 5.8

The Black-Scholes equation for the forward contract f from (5.23) is given by

$$0 = \frac{\partial f}{\partial t} + \frac{1}{2}\sigma^2 S^2 \frac{\partial^2 f}{\partial S^2} + r_f S \frac{\partial f}{\partial S} - r_f f \tag{13.38}$$

where we bring all terms to one side of the equation. Inserting the derivatives of (5.23)

$$\frac{\partial f}{\partial t} = -r_f K e^{-r_f t}, \qquad \frac{\partial f}{\partial S} = 1, \qquad \text{and} \qquad \frac{\partial^2 f}{\partial S^2} = 0 \tag{13.39}$$

in (13.38), we see that the right-hand side

$$- r_f K e^{-r_f t} + r_f S - r_f f = -r_f \left(S - K e^{-r_f t}\right) - r_f f = 0 \tag{13.40}$$

is indeed zero.

Exercise 6.1

The arguments of the cumulative normal distribution $N(d_i)$ are given by (4.43) with ρ replaced by r_f, such that the call option from (5.17) can be written as $c = SN(d_1) - K e^{-r_f \tau} N(d_2)$, where we use $\tau = T - t$ and $N(y)$ is defined in (4.39). The requested Δ_c is then given by

$$\Delta_c = \frac{\partial c}{\partial S} = N(d_1) + SN'(d_1)\frac{\partial d_1}{\partial S} - K e^{-r_f \tau} N'(d_2)\frac{\partial d_2}{\partial S} \tag{13.41}$$

with $N'(y) = e^{-y^2/2}/\sqrt{2\pi}$ and

$$\frac{\partial d_1}{\partial S} = \frac{\partial d_2}{\partial S} = \frac{1}{\sigma\sqrt{\tau}S}. \tag{13.42}$$

Inserting in the expression for Δ_c and using $d_1 = d_2 + \sigma\sqrt{\tau}$ we obtain

$$\Delta_c = N(d_1) + \frac{1}{\sqrt{2\pi\sigma^2\tau}S}\left[S e^{-(d_2+\sigma\sqrt{\tau})^2/2} - K e^{-r_f \tau} e^{-d_2^2/2}\right]$$

$$= N(d_1) + \frac{1}{\sqrt{2\pi\sigma^2\tau}S}e^{-d_2^2/2}\left[S e^{-\ln(S/K)-r_f\tau} - K e^{-r_f \tau}\right]. \tag{13.43}$$

Since the square bracket is zero, we find the result from (5.22), namely $\Delta_c = N(d_1)$. Differentiating Δ_c once again with respect to S gives us $\Gamma_c = \partial \Delta_c/\partial S = \partial N(d_1)/\partial S$ for the call option

$$\Gamma_c = N'(d_1)\frac{\partial d_1}{\partial S} = \frac{e^{-d_1^2/2}}{\sqrt{2\pi\sigma^2\tau}S}. \tag{13.44}$$

Exercise 6.2

From (4.45), as a consequence of the call-put parity, we know that $p = Ke^{-r_f\tau} - S + c$. Differentiating with respect to S, we find $\Delta_p = \partial p/\partial S = -1 + \partial c/\partial S = N(d_1) - 1$, where we used the result from Exercise 6.1. As a second option, we differentiate $p = Ke^{-r_f\tau}N(-d_2) - SN(-d_1)$ and obtain

$$\Delta_p = \frac{\partial p}{\partial S} = -Ke^{-r_f\tau}N'(-d_2)\frac{\partial d_2}{\partial S} - N(-d_1) + SN'(-d_1)\frac{\partial d_1}{\partial S}. \tag{13.45}$$

Using similar reasoning as in Exercise 6.1 and $N(y) + N(-y) = 1$, we find the same result as before: $\Delta_p = N(d_1) - 1$.

Exercise 6.4

The solution is a straightforward application of (6.9). The numerical solution can be found in ex6_4.m in the ESM.

Exercise 6.5

Figure 13.6 shows the profit diagram of the butterfly spread. The script ex6_5.m is available in the ESM.

Exercise 7.1 and 7.2

After setting up the $n \times 2$ matrix, shown in (7.1), and using (7.4) to solve it once with all error bars σ set to unity. The slope $x_1 = a$ and intercept $x_2 = b$ with their respective error bars are $a = 1.5 \pm 0.15$ and $b = -2.8 \pm 0.4$, where we use (7.7) to determine the error bars. In a second fit we take the error bars, given in the table from the exercise into account, and find $\tilde{a} = 1.7 \pm 0.2$ and $\tilde{b} = -3.2 \pm 0.6$. We determine the R^2 from the predicted measurement values $\hat{y}_i = \sum_j A_{ij}x_j$ of the first fit and find $R^2 = 0.89$.

Exercise 7.3

$C(\mathbf{y})$ is given by $C(\mathbf{y}) = (\Lambda^2)^{-1}$ and, since Λ is diagonal, it equals its transpose: $\Lambda = \Lambda^t$. Straightforward evaluation then leads to

$$\begin{aligned}C(\mathbf{x}) &= JC(\mathbf{y})J^t = (A^t\Lambda^2A)^{-1}A^t\Lambda^2(\Lambda^2)^{-1}\Lambda^2A(A^t\Lambda^2A)^{-1}\\&= (A^t\Lambda^2A)^{-1}(A^t\Lambda^2A)(A^t\Lambda^2A)^{-1} = (A^t\Lambda^2A)^{-1},\end{aligned} \tag{13.46}$$

which proves the statement.

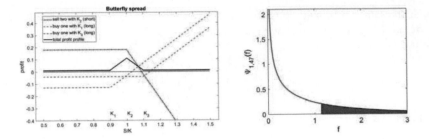

Fig. 13.6 Left: Profit diagram of a butterfly spread from Exercise 6.5. Right: The F-distribution from Exercise 7.6 with the red are indicating the probability to find an even larger F-value

Exercise 7.4

With the Gaussian $g(x) = e^{-x^2/2\sigma^2}/\sqrt{2\pi}\sigma$, we find for part a)

$$f_a(y) = \int_{-\infty}^{\infty} g(x)\delta(y - (x - a))dx = g(y + a) . \qquad (13.47)$$

Likewise, we obtain for the part b)

$$f_b(y) = \int_{-\infty}^{\infty} g(x)\delta(y - bx))dx = \int_{-\infty}^{\infty} g(x)\frac{\delta(x - y/b)}{|b|}dx = \frac{1}{|b|}g(y/b) . \quad (13.48)$$

Here we use the property of the delta function $\delta(f(x)) = \sum_i(x - x_i)/|f'(x_i)|$, where the sum extends over all zeros x_i of $f(x)$. Applying this to part c) yields

$$f_c(y) = \int_{-\infty}^{\infty} g(x)\delta(y - cx^2))dx = \int_{-\infty}^{\infty} g(x)\left[\frac{\delta\left(x - \sqrt{y/c}\right) + \delta\left(x + \sqrt{y/c}\right)}{|2cx|}\right]dx$$

$$= \frac{g\left(\sqrt{y/c}\right) + g\left(-\sqrt{y/c}\right)}{|2\sqrt{cy}|} = \frac{1}{|\sqrt{cy}|}g\left(\sqrt{y/c}\right) \qquad (13.49)$$

Here we have to keep in mind that the square has two solutions, one positive and one negative. The last equality is valid, because the Gaussian $g(x) = g(-x)$ is a symmetric function.

Exercise 7.5

The solution for parts (a) through (d) follows the discussion from Sect. 7.1. The following code snippet illustrates the calculation of the F-value and the probability to find an even larger value—the p-value. The latter is shown on the right-hand plot in Fig. 13.6.

```
n=1; m-N-4;
Fval=(chi2_p-chi2_q)/(chi2_q/m)
xhat=n*Fval/(m+n*Fval)
p_tail=1-betainc(xhat,n/2,m/2)
```

There are $N = 51$ data points, such that $m = 47$. The values are $\chi_q = 51$ by con-struction from part c), $\chi_p = 52.24$, which results in the value $\hat{f} = 1.14$ that is used to prepare the right-hand plot in Fig. 13.6.

Exercise 7.6

The test results are stored in the array x and analyzed in the following code snippet, which calculates the sample average and variance from (7.24) and then uses $A(\hat{t}, v)$ from (7.34) to calculate the 90% confidence interval.

```
X=mean(x);  dx=x-X;
S=sqrt((1/(n-1))*sum(dx.^2));
tt=X/(S/sqrt(n));
A=@(that,nu)1-betainc(nu./(nu+that.^2),nu/2,0.5);
g=@(t)A(t,nu)-0.90;     % 90 % confidence
that=fzero(g,[0,20]);   %                  =2.13
mu1=X-that*S/sqrt(n);   % lower bound =12.7
mu2=X+that*S/sqrt(n);   % upper bound =17.7
```

The returned values are shown in the code.

Exercise 8.1

After reading the data from file and inspecting that the trend is indeed a straight line, we determine the parameters of the line and subtract it from the data. Inspection of the residual oscillation shows that the period is between 45 and 50. Testing several values reveals the 47 is the likely period and differencing provides the residuals. The following code snippet illustrates the procedure.

```
A=[x,ones(n,1)];
p=inv(A'*A)*A'*y
plot(x,y,'*',x,polyval(p,x),'r')
y2=y-polyval(p,x);
figure; plot(x,y2); title('After removing the trend')
figure; y3=y2(48:end)-y2(1:end-47);
subplot(2,1,1); plot(y3); title('Residuals')
subplot(2,1,2); histogram(y3,30)
```

Exercise 8.2

After reading in the data file, we calculate γ_0 and the auto-covariances γ_j from (8.8). In the end we divide by γ_0 to determine the autocorrelations ρ_j, whose values we display in a histogram.

```
y=dlmread('ex8_2.dat'); n=length(y);
gamma0=y'*y/n;
m=30;   % maximum coefficients
acf=zeros(1,m);
for j=1:m
  acf(j)=y(1:end-j)'*y(j+1:end);   (8.8)
end
acf=acf/n; rho=acf/gamma0; bar(rho(1:m))
```

Only ρ_1 and ρ_2 are significantly above the noise level, which indicates that the data come from a MA(2) process.

Exercise 8.3

After loading the data and calculating the autocorrelations ρ_j in the same way as in Exercise 8.2, we calculate the PACF in the following code snippet, which construct the matrix from (8.21) for increasing size k. Then it solves for the coefficients ϕ_j and always records the highest coefficient ϕ_{jj} as the PACF.

```
pacf(1)=rho(1);
for j=2:m      % up to some maximum, say m=4
  a=eye(j);
  for k=1:j   % loop over lines
    a(k,k+1:j)=rho(1:j-k);
    a(k+1:j,k)=rho(1:j-k)';
  end
  b=a\rho(1:j)';   % (8.22)
  pacf(j)=b(j);
end
```

Plotting the PACF shows that the data is generated by an AR(3) process that has only ϕ_3 significantly different from zero.

Exercise 8.4

The analysis follows the Box-Jenkins procedure outlined in Sect. 8.6 and the code closely follows the code already shown in Exercises 8.1 to 8.3. As a result of the analysis, shown on the left-hand side in Fig. 13.7, the residuals can be described by an AR(2) process, despite the PACF having significant components at positions 12 and 13, albeit with opposite signs, which suggests they are an artefact of the noise. In the spirit of building parsimonious models, we dare to ignore them, because the model fits the data reasonably well, as shown on the right-hand side in Fig. 13.7.

Exercise 8.5

We use the following code snippet to symbolically evaluate the coefficients. Here we treat the symbolic variable x as the lag operator, which was denoted $\hat{L}$ in Sect. 8.7.

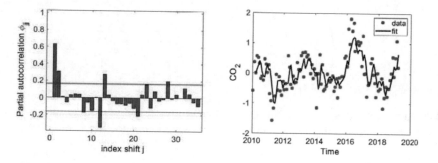

Fig. 13.7 The PACF of the data (left) and the AR(2)-model together with the data points after removal of trend and seasonality

```
syms f1 f2 f3 t1 t2 t3 x
p=1-t1*x-t2*x^2-t3*x^3;
q=1-f1*x-f2*x^2-f3*x^3;
y=simplify(taylor(p/q,'Order',4))
```

After interpreting the coefficients of the polynomial y as $-\pi_j$, we find the following results

$$
\begin{aligned}
\pi_1 &= \theta_1 - \phi_1 \\
\pi_2 &= \theta_2 - \phi_2 + \theta_1\phi_1 - \phi_1^2 \\
\pi_3 &= \theta_3 - \phi_3 - 2\phi_1\phi_2 + \phi_1\theta_2 + \phi_1\theta_2 + \phi_2\theta_1 + \phi_1^2\theta_1 - \phi_1^3 .
\end{aligned}
\tag{13.50}
$$

Exercise 8.6

We first load the n data points from `ex8_6.dat` into the variable x and then use the following code snippet to implement the EWMA filter.

```
m=3     % or 1, 10, 30
u(1)=x(1);
for k=2:n
    u(k)=(m*u(k-1)+x(k))/(m+1);
end
```

Plotting both x and u shows that increasing m reduces the amplitude of the filtered signal and shifts the oscillation phase of the filtered signal with respect to that of the un-filtered data.

Exercise 9.1

The generating function is the Fourier transform $\tilde{R}(k)$ of the distribution function $R(x)$ and is given by

$$\tilde{R}(k) = \frac{1}{a} \int_{-a/2}^{a/2} e^{ikx} = \frac{1}{ika} \left(e^{ika/2} - e^{-ika/2} \right) = \frac{\sin(ka/2)}{ka/2} . \qquad (13.51)$$

Exercise 9.2

The cumulant expansion is defined in (9.16). Introducing $y = ka/2$ and $d/dk = (a/2)d/dy$ and defining $f(y) = \ln(\sin(y)/y)$, we find the coefficients c_m from

$$c_m = \left(\frac{a}{2} \right)^m (-i)^m \left. \frac{d^m f(y)}{dy^m} \right|_{y=0} . \qquad (13.52)$$

We use the following code to symbolically differentiate $f(y)$ and subsequently set y to zero. Here n is the order of the derivative.

```
syms y
n=4;                          % order of derivative
fun=(-i)^n*log(sin(y)/y)
f=diff(fun,y,n)               % symbolic differentiation
q=double(subs(f,eps))         % substitute y=0
[num,den]=rat(q,1e-8)         % write as fraction
```

Repeatedly running the code for different values of n, we find the cumulants to be

$$c_1 = 0, \quad c_2 = \frac{1}{3} \left(\frac{a}{2} \right)^2, \quad c_3 = 0, \quad \text{and} \quad c_4 = -\frac{2}{15} \left(\frac{a}{2} \right)^4 . \qquad (13.53)$$

Exercise 9.3

Convoluting a function with itself in real space is equivalent to transforming the square of its Fourier transform back to real space. We note that the Fourier transform $\tilde{f}(k)$ of the Cauchy distribution from (9.4) is given by (9.26) with $\mu = 1$, such that its square is

$$\tilde{f}(k)^2 = e^{-2a|k|} , \qquad (13.54)$$

which is the Fourier transform of (9.4) with a replaced by $2a$.

Exercise 9.4

he code below fills the array x with 10^4 groups of ten random numbers each. Then it adds the ten numbers to produce an array y that is used to produce a histogram, which shows a distribution close to a Gaussian, as predicted by (9.23). Then we calculate the rms value of the histogram and find about 1.8, which is close to the second cumulant $c_2 = a^2/12$ with $a = 2$ from Exercise 2 multiplied by $\sqrt{N} = \sqrt{10}$. This is a consequence of the central limit theorem, discussed in Sect. 9.7.

Fig. 13.8 Exercise 9.6: The average maximum expected value for $D_\nu(x)$, a Gaussian and an exponential distribution

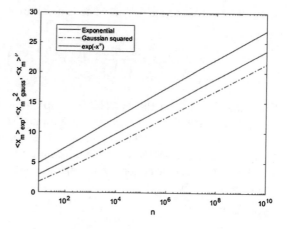

```
x=-1+2*rand(10,10000); y=sum(x);
histogram(y); xlabel('sum of ten numbers')
rms_of_histogram=rms(y)
```

Exercise 9.5

After taking out the quarter from the middle, there are two remaining segments with a length of 3/8, each. When zooming into one of the segments, only half the points remain, thus $N \sim 2^{-m}$, while the scale is reduced by $\varepsilon \sim (3/8)^{-m}$. The fractal dimension is therefore $D = \ln(2^{-m})/\ln\left((3/8)^{-m}\right) = \ln(2)/\ln(3/8) \approx 0.7067$.

Exercise 9.6

Integrating e^{-x^ν} and the substitution $z = x^\nu$ leads to

$$\int_0^\infty e^{-x^\nu} dx = \frac{1}{\nu} \int_0^\infty \frac{e^{-z} dz}{z^{1-1/\nu}} = \frac{1}{\nu} \int_0^\infty e^{-z} z^{(1-\nu)/\nu} dz = \frac{1}{\nu} \Gamma\left(\frac{1}{\nu}\right) \tag{13.55}$$

such that $A(\nu) = \nu/\Gamma(1/\nu)$ and $D_\nu(x) = \nu e^{-x^\nu}/\Gamma(1/\nu)$. The cumulative distribution function $C(y)$ of $D_\nu(x)$, requested in part b, is given by

$$C(y) = \int_0^y D_\nu(x) dx = \frac{\nu}{\Gamma(1/\nu)} \int_0^{y^\nu} e^{-z} \frac{1}{\nu} z^{(1/\nu)-1} dz = P\left(\frac{1}{\nu}, y^\nu\right), \tag{13.56}$$

where we used the substitution $z = x^\nu$ and $dx = \frac{1}{\nu} z^{(1/\nu)-1} dz$. Here $P(a, b)$ is the incomplete gamma function. For $\Xi(x)dx$, shown in (9.32), we find

$$\Xi(x)dx = nP\left(\frac{1}{\nu}, x^\nu\right)^{n-1} \frac{\nu}{\Gamma(1/\nu)} e^{-x^\nu} dx \tag{13.57}$$

such that the average maximum value is given by

$$\langle x_n \rangle = \frac{n\nu}{\Gamma(1/\nu)} \int_0^\infty x P\left(\frac{1}{\nu}, x^\nu\right)^{n-1} e^{-x^\nu} dx . \tag{13.58}$$

Choosing $\nu = 1/2$ in the MATLAB script `ex9_6.m` from the ESM produces Fig. 13.8, which shows the average maximum values, suitably scaled to show a linear behavior, for D_ν, a Gaussian, and an exponential distribution. Note that the arguments in MATLAB's incomplete gamma function `betainc()` are reversed with respect to the convention used in Abramowitz and Stegun's *Handbook of mathematical functions*.

Exercise 9.7

The solution $N(t)$ strives towards an equilibrium defined by $dN/dt = 0$, which leads to the equilibrium level $N_\infty = \delta/\beta$. Consequently, there is no divergence and $\tau_c \to \infty$.

Exercise 10.1

Let's calculate that straight away using the concepts and definitions presented in Sect. 10.1

$$\langle f| \left(x\frac{d}{dx}\right)^t g \rangle = \langle g| \left(x\frac{d}{dx}\right) f \rangle^* = \left[\int dx \langle g|x\rangle \langle x| \left(x\frac{d}{dx}\right) f \rangle\right]^*$$

$$= \left[\int dx g(x)^* x f'(x)\right]^* = \left[\int dx (x g(x)^*) f'(x)\right]^*$$

$$= -\left[\int dx f(x) \frac{d}{dx}(x g(x)^*)\right]^*$$

$$= -\left[\int dx f(x) \left(g(x)^* + x g'(x)^*\right)\right]^* \tag{13.59}$$

$$= -\int dx f(x)^* g(x) - \int dx f(x)^* x g'(x)$$

$$= -\int dx \langle f|x\rangle \langle x|g\rangle - \int dx \langle f|x\rangle \langle x| \left(x\frac{d}{dx}\right) g \rangle$$

$$= -\langle f|g\rangle - \langle f| \left(x\frac{d}{dx}\right) g \rangle .$$

The presence of the first term $-\langle f|g\rangle$ proves that the operator is neither hermitian not anti-hermitian. Note that in the step from the second line to the third we use partial integration and drop the term at the (infinite) boundaries, where we assume the functions f and g to be zero.

Exercise 10.2

Inserting the definition of the momentum operator into the Hamiltonian, we find

$$H = -\frac{\hbar^2}{2m}\frac{\partial^2}{\partial x^2} + \frac{m\omega^2}{2}x^2 . \tag{13.60}$$

We therefore need to calculate the derivatives of the wave function $\psi(x)$ with respect to x and find

$$\psi'(x) = -\beta^2 x\psi(x) \quad \text{and} \quad \psi''(x) = \left(\beta^4 x^2 - \beta^2\right)\psi(x) . \tag{13.61}$$

For the matrix element $\langle\psi|H|\psi\rangle$, requested in (a), we then find

$$\langle\psi|H|\psi\rangle = \int_{-\infty}^{\infty}\left[-\frac{\hbar^2}{2m}\left(\beta^4 x^2 - \beta^2\right) + \frac{m\omega^2}{2}x^2\right]\psi(x)^2 dx$$

$$= -\frac{\hbar^2}{2m}\left(\beta^4\frac{1}{2\beta^2} - \beta^2\right) + \frac{m\omega^2}{2}\frac{1}{2\beta^2} = \frac{\hbar\omega}{4} + \frac{\hbar\omega}{4} , \tag{13.62}$$

where we used $\int x^2\psi^2(x)dx = 1/2\beta^2$ in the second equality. Incidentally, the first term in the last equality equals the expectation value of the kinetic energy, requested in part (c). The second term equals that of the potential energy, requested in (d). The expectation value of the position operator, requested in (b), is given by $\langle\psi|x|\psi\rangle \propto \int x\psi^2 dx = 0$. Likewise, we find the expectation value of the momentum operator from (c) to be zero.

Exercise 10.3

Inserting the substitution $x = z + \hat{r}t$ and $x' = z'$, mentioned below (5.11), causes (5.11) to have the same form as (10.22), only the symbols for x and z are exchanged.

Exercise 10.4

The MATLAB code ex10_4.m from the ESM illustrates how to repeatedly calculate the integral for increasing number N of random numbers. It turns out that we need around N random numbers for the integral I to stabilize at the $\Delta I/I \approx 1/\sqrt{N}$ level. Thus about $N = 10^6$ samples are required to reach the 10^{-3} level.

Exercise 10.5

The following code snippet first defines the Cauchy distribution cauchy() from (9.4) with $a = 1$ as an anonymous function. After initializing the starting value x_0 and the β-parameter for the Metropolis-Hastings algorithm, it passes the function cauchy() to the metropolis routine from Appendix B.5. Finally, the histogram of the generated random numbers is shown with the function cauchy(), suitably scaled, superimposed.

```
cauchy=@(x)(1/pi)./(1+x.*x);
N=1000000; x0=1; beta=1;
y=metropolis3(cauchy,beta,1000,x0);      % burn in
x=metropolis3(cauchy,beta,N,y(1000));
dx=0.1; xx=-30:dx:30;          % for plotting
hist(x,xx); hold on; plot(xx,N*cauchy(xx)*dx,'r')
```

Exercise 11.1

The full MATLAB code is available in the file ex11_1.m from the ESM. Varying the investment fraction σ we find that the utility $V[p]$ has a maximum at about 2.78 for $\sigma = 0.43$.

Exercise 11.2

The following code snippet implements the recursion relation for the random process λ. Plotting the values shows that they slowly approach a constant value.

```
lambda=zeros(1,n);
for t=2:n
   lambda(t)=gamma*lambda(t-1)+(2-gamma)+a*randn;
end
```

The asymptotic value λ_∞ is determined by the condition $\lambda_\infty = \lambda_{t+1} = \lambda_t$ and replacing ε_t by its average value, which is zero. The equation defining the stochastic process λ then becomes $\lambda_\infty = \gamma \lambda_\infty + (2 - \gamma)$, which we solve for $\lambda_\infty = (2 - \gamma)/(1 - \gamma)$. For $\gamma = 0.9$ we find $\lambda_\infty = 11$.

Exercise 11.3

The canonical momenta for the Lagrangian are given by

$$p_x = \frac{\partial L}{\partial \dot{x}} = 3m\dot{x} + mr\dot{\theta} \quad \text{and} \quad p_\theta = \frac{\partial L}{\partial \dot{\theta}} = mr\dot{x} + mr^2\dot{\theta} . \tag{13.63}$$

Solving for the velocities $\dot{x}$ and $\dot{\theta}$ leads to

$$\dot{x} = \frac{rp_x - p_\theta}{2mr} \quad \text{and} \quad \dot{\theta} = \frac{3p_\theta - rp_x}{2mr^2} \tag{13.64}$$

which allows us to write the Hamiltonian as the Legendre transform of the Lagrangian

$$\mathcal{H}(x, \theta, p_x, p_\theta) = p_x\dot{x} + p_\theta\dot{\theta} - L(x, \theta, \dot{x}, \dot{\theta})$$
$$= \frac{3m}{2}\dot{x}^2 + mr\dot{x}\dot{\theta} + \frac{mr^2}{2}\dot{\theta}^2 + kx^2 + \frac{mgr}{2}\theta^2 \tag{13.65}$$
$$= \frac{1}{4m}p_x^2 - \frac{1}{2mr}p_x p_\theta + \frac{3}{4mr^2}p_\theta^2 + kx^2 + \frac{mgr}{2}\theta^2$$

where we used (13.64) to replace $\dot{x}$ and $\dot{\theta}$.

Exercise 11.4

The Riccati equation for this process is given by (11.57), which in our case reads

$$\dot{\kappa} = -q - 2a\kappa + \frac{b^2}{r}\kappa^2 . \tag{13.66}$$

In steady-state, we have $\dot{\kappa} = 0$ and the equation becomes a quadratic equation for κ

$$\kappa^2 - 2\frac{ar}{b^2}\kappa - \frac{qr}{b^2} = 0 \tag{13.67}$$

which has the solutions

$$\kappa = \frac{ar}{b^2}\left[1 \pm \sqrt{1 + \frac{b^2 q}{a^2 r}}\right] . \tag{13.68}$$

Inserting this value of κ in (11.54), we arrive at the result shown in (11.62).

Exercise 11.5

The cost function from (11.41) is $g(x, u) = (qx^2 + ru^2)/2$ such that the Hamiltonian becomes

$$\mathcal{H}(x, u, p) = \frac{1}{2}\left(qx^2 + ru^2\right) + p\left(\alpha x^\beta + u\right) . \tag{13.69}$$

Applying Hamilton's equation yields

$$
\begin{aligned}
\dot{p} &= -\frac{\partial \mathcal{H}}{\partial x} = -\left(qx + \alpha p\beta x^{\beta-1}\right) \\
\dot{x} &= \frac{\partial \mathcal{H}}{\partial p} = \alpha x^\beta + u \\
0 &= \frac{\partial \mathcal{H}}{\partial u} = ru + p
\end{aligned}
\tag{13.70}
$$

Solving the last equation for $u = -p/r$ allows us to eliminate u from the second equation and results in the following set of two coupled non-linear differential equations

$$
\begin{aligned}
\dot{p} &= -qx - \alpha p\beta x^{\beta-1} \\
\dot{x} &= \alpha x^\beta - \frac{1}{r}p .
\end{aligned}
\tag{13.71}
$$

Exercise 12.1

Inserting $p_i = 1/n$ for all i in (12.2), we find

$$H = -\sum_{i=1}^{n} \frac{1}{n} \log_2(1/n) = -\frac{1}{n} \log_2(1/n) \sum_{i=1}^{n} 1 = -\log_2(1/n) \qquad (13.72)$$

because the sum evaluates to n.

Exercise 12.2

The probabilities $p_x(1) = 1 - \alpha$ and $p_x(0) = \alpha$ are shown in Fig. 12.2 and the conditional probabilities $p_{yx}(y_j|x_i)$ in (12.8), which allows us to calculate the joint probabilities

$$p_{xy}(x_i, y_j) = \begin{pmatrix} p_{xy}(1,1) & p_{xy}(0,1) \\ p_{xy}(1,0) & p_{xy}(0,0) \end{pmatrix} = \begin{pmatrix} p_{yx}(1|1)p_x(1) & p_{yx}(1|0)p_x(0) \\ p_{yx}(0|1)p_x(1) & p_{yx}(0|0)p_x(0) \end{pmatrix}$$

$$= \begin{pmatrix} (1-\varepsilon)(1-\alpha) & \varepsilon\alpha \\ \varepsilon(1-\alpha) & (1-\varepsilon)\alpha \end{pmatrix}. \qquad (13.73)$$

Summing the joint probabilities $p_{xy}(x_i, y_j)$ over the different x_i gives us

$$p_y(y_j) = \begin{pmatrix} p_y(1) \\ p_y(0) \end{pmatrix} = \begin{pmatrix} \sum_{x_i} p_{xy}(x_i, 1) \\ \sum_{x_i} p_{xy}(x_i, 0) \end{pmatrix} = \begin{pmatrix} p_{xy}(1,1) + p_{xy}(0,1) \\ p_{xy}(1,0) + p_{xy}(0,0) \end{pmatrix}$$

$$= \begin{pmatrix} 1 - \varepsilon - \alpha + 2\varepsilon\alpha \\ \varepsilon + \alpha - 2\varepsilon\alpha \end{pmatrix} = \begin{pmatrix} 1 - \gamma \\ \gamma \end{pmatrix} \qquad (13.74)$$

with $\gamma = \varepsilon + \alpha - 2\varepsilon\alpha$. The backwards conditional probabilities $p_{xy}(x_i|y_j)$ follow from Bayes' theorem (12.12). Inserting the respective probabilities on the right-hand side, we find

$$p_{xy}(1|1) = \frac{p_{yx}(1|1)p_x(1)}{p_y(1)} = \frac{(1-\varepsilon)(1-\alpha)}{1 - \varepsilon - \alpha + 2\varepsilon\alpha}$$

$$p_{xy}(1|0) = \frac{p_{yx}(0|1)p_x(1)}{p_y(0)} = \frac{\varepsilon(1-\alpha)}{\varepsilon + \alpha - 2\varepsilon\alpha}$$

$$p_{xy}(0|1) = \frac{p_{yx}(1|0)p_x(0)}{p_y(1)} = \frac{\varepsilon\alpha}{1 - \varepsilon - \alpha + 2\varepsilon\alpha} \qquad (13.75)$$

$$p_{xy}(0|0) = \frac{p_{yx}(0|0)p_x(0)}{p_y(0)} = \frac{(1-\varepsilon)\alpha}{\varepsilon + \alpha - 2\varepsilon\alpha}.$$

Exercise 12.3

With the abbreviation $H_b(x) = -x \log_2(x) - (1-x) \log_2(1-x)$ for the entropy of a binary system, and using the probabilities from the previous exercise, we find

$$H[x] = H_b(\alpha) \quad \text{and} \quad H[y] = H_b(\gamma) \quad \text{with} \quad \gamma = \varepsilon + \alpha - 2\varepsilon\alpha . \quad (13.76)$$

The joint entropy $H[x, y]$, defined in (12.14), depends on the joint probabilities $p_{xy}(x_i, y_j)$ that we calculated in the previous exercise and can therefore be written as

$$
\begin{aligned}
H[x, y] &= -\big\{ (1 - \varepsilon)(1 - \alpha) \log_2 ((1 - \varepsilon)(1 - \alpha)) + \varepsilon\alpha \log_2(\varepsilon\alpha) \\
&\quad + \varepsilon(1 - \alpha) \log_2((\varepsilon(1 - \alpha)) + (1 - \varepsilon)\alpha \log_2((1 - \varepsilon)\alpha)\big\} \\
&= -\big\{ (1 - \varepsilon) \log_2(1 - \varepsilon) + (1 - \alpha) \log_2(1 - \alpha) \quad\quad (13.77) \\
&\quad + \varepsilon \log_2(\varepsilon) + \alpha \log_2(\alpha)\big\} \\
&= H_b(\alpha) + H_b(\varepsilon)
\end{aligned}
$$

where the calculations that take us from the first to the second line of the equation are rather lengthy. The conditional entropy $H[y|x]$ follows from (12.15) as $H[y|x] = H[x, y] - H[x] = H_b(\varepsilon)$ and inserting $H[x]$ and $H[x, y]$ from this exercise. We note that it can also be calculated by inserting the joint and conditional probabilities in the definition of $H[y|x]$ given on the right-hand side in (12.15), though the ensuing calculations are lengthy.

Exercise 12.4

We use the definition of the joint entropy from (12.14) and insert the factorized probabilities. This leads to

$$
\begin{aligned}
H[x, y] &= -\sum_i \sum_j p_x(x_i) p_y(y_j) \log_2(p_x(x_i) p_y(y_j)) \\
&= -\sum_i \sum_j p_x(x_i) p_y(y_j) \big[\log_2(p_x(x_i)) + \log_2(p_y(y_j))\big] \\
&= -\bigg[\sum_j p_y(y_j) \sum_i p_x(x_i) \log_2(p_x(x_i)) \quad\quad\quad (13.78) \\
&\quad + \sum_i p_x(x_j) \sum_j p_y(y_j) \log_2(p_y(y_j)) \bigg] \\
&= H[x] \sum_j p_y(y_j) + H[y] \sum_i p_x(x_j) .
\end{aligned}
$$

Since the probabilities $\sum_i p_x(x_i) = 1$ add up to unity and likewise $\sum_j p_y(y_j) = 1$, it follows that $H[x, y] = H[x] + H[y]$.

Exercise 12.5

Equation 12.15 tells us that $H[y|x] = H[x, y] - H[x]$ and from the previous exercise we know that $H[x, y] = H[x] + H[y]$ for statistically independent variables x and y. Combining these two equations leads to $H[y|x] = H[y]$.

Exercise 12.6

The following code snippet illustrates the communication between Alice and Bob. After defining p and g, Bob uses Alice's public number A and his secret number b to calculate the encryption key $k = A^b (\mathrm{mod}\ p)$ that he subsequently uses to decode the ciphertext c. He finds that the plan is to meet at 17, or 5 pm.

```
p=9967; g=33;
A=7025;                 % Alice's public number
b=557;                  % Bob's secret
k=powermod(A,b,p)       % secret key, =4133
c=4148;                 % encrypted hour
h=bitxor(k,c)           % 17
```

Exercise 12.7

Eve knows the public key (n, e) and tries to find the prime factors of n with MAT-LAB's built-in function factor() to find the prime factors, here $p = 67$ and $q = 83$. Note that this would practically take forever for very large numbers n. It works her, because n is rather small. Knowing p and q, Eve then calculates the totient ϕ_n and uses it to solve Bezout's equation to determine the private key d, which she subsequently uses to decode the message and finds that the meeting is at 11.

```
c=5343;    % encoded hour
n=5561; e=5;
a=factor(n); p=a(1); q=a(2);   % crack
phin=(p-1)*(q-1)                  % totient
[gcdval,d,kk]=gcd(e,phin) % coprime -> gcdval must be 1!
if gcdval ~= 1 disp('Error: not coprime'); return; end
d=powermod(d,1,phin)            % decryption key
decoded=powermod(c,d,n)        % decoded message, =11
```

Exercise 12.8

Bob has to collect all variables needed to calculate Q from (12.35) and then compare its x-coordinate with r, provided as part of the signature. The following code snippet shows how Bob first converts the message m to numerical form and then determines the hash h. He then uses (12.34) to calculate the inverse of the other part of the signature s and finds h/s and r/s before he uses the function ECCadd_p() from Appendix B.11 to calculate $(h/s) \odot G$ and $(r/s) \odot P$. Adding both contributions yields Q. The first component of Q turns out to be equal to $r = 66$ and therefore the message is authentic.

```
p=113; a=0; b=7; G=[15,52]; n=19;   % elliptic curve
P=[93,16];                          % Alice's public point
r=66; s=11;                         % signature
m='YES';                            % message
nm=double(m);                       % numeric form
h=bitxor(nm(3),bitxor(nm(2),nm(1)));   % hash
sinv=powermod(s,n-2,n);             % 1/s
hs=powermod(h*sinv,1,n);            % h/s
rs=powermod(r*sinv,1,n);            % r/s
Q1=G; for j=2:hs, Q1=ECadd_p(Q1,G,a,b,p); end   % (h/s)*G
Q2=P; for j=2:rs, Q2=ECadd_p(Q2,P,a,b,p); end   % (r/s)*P
Q=ECadd_p(Q1,Q2,a,b,p)             % (h/s)*G+(r/s)*P
```

Alice's script ex12_8a.m to sign the message and Bob's script ex12_8b.m to validate it are available in the ESM.

Exercise 12.9

Just before (12.41) the operation of the operator U on the state $|x^k (\mod N)\rangle$ is defined as $U|x^k (\mod N)\rangle = |x^{k+1} (\mod N)\rangle$, such that we find

$$
\begin{aligned}
U|u_s\rangle &= \frac{1}{\sqrt{r}} \sum_{k=0}^{r-1} e^{-2\pi i k s/r} |x^{k+1} (\mod N)\rangle \\
&= e^{2\pi i s/r} \frac{1}{\sqrt{r}} \sum_{k=0}^{r-1} e^{-2\pi i (k+1)s/r} |x^{k+1} (\mod N)\rangle \qquad (13.79) \\
&= e^{2\pi i s/r} \frac{1}{\sqrt{r}} \sum_{k'=1}^{r} e^{-2\pi i k' s/r} |x^{k'} (\mod N)\rangle \\
&= e^{2\pi i s/r} |u_s\rangle .
\end{aligned}
$$

In the last step we realize that the sum over k' still extends over all the r element of the set that successive x^k cycle through. Only the starting point is a different element. As a matter of fact, the element with $k' = r$ is the same as that with $k = 0$. Therefore the sum still describes $|u_s\rangle$.

Appendix A
On the Independence of Certain Random Variables

In the derivation of student's t-distribution in Chap. 7 we used the fact that for n normal-distributed random variables $x_1, \ldots, x_n$ with mean zero, the average $\bar{x} = (x_1 + \cdots + x_n)/n$ and the variance $S^2 = \left((x_1 - \bar{x})^2 + \cdots + (x_n - \bar{x})^2\right)/(n-1)$ are statistically independent. The latter is equivalent to the ability to factor the joint distribution function in two factors, one for the average and one for the rest. We start by showing that the variance S^2 can be rewritten such that the first random variable x_1 is replaced by the average $\bar{x}$. In the following argument, we follow Casella and Berger's *Statistical Inference* and write

$$
\begin{aligned}
S^2 &= \frac{1}{n-1} \sum_{i=1}^{n} (x_i - \bar{x})^2 = \frac{1}{n-1}\left[(x_1 - \bar{x})^2 + \sum_{i=2}^{n} (x_i - \bar{x})^2 \right] \\
&= \frac{1}{n-1}\left[\left(\sum_{i=2}^{n}(x_i - \bar{x})\right)^2 + \sum_{i=2}^{n} (x_i - \bar{x})^2 \right]
\end{aligned}
\tag{A.1}
$$

where we used that $\sum_{i=1}^{n}(x_i - \bar{x}) = 0$ which leads to

$$
x_1 - \bar{x} = - \sum_{i=2}^{n}(x_i - \bar{x})
\tag{A.2}
$$

that we use to replace the first term in the sum above. Thus we can write the variance S^2 in terms of the average $\bar{x}$ and the x_i with $i \geq 2$.

In order to prove the statistical independence, we need to show that the joint probability function of the independent random variables x_i can be written as a factor that depends only on the average $\bar{x}$ and the x_i with $i \geq 2$. The joint distribution function of the x_i is

V. Ziemann, *Physics and Finance*, Undergraduate Lecture Notes in Physics,
https://doi.org/10.1007/978-3-030-63643-2

$$\psi(x_1, \ldots, x_n) = \frac{1}{(2\pi)^{n/2}} \exp\left[-\frac{1}{2}\sum_{i=1}^{n} x_i^2\right] \tag{A.3}$$

In order to rewrite the exponent in an explicitly factored form we introduce new variables

$$y_1 = \bar{x} = \frac{1}{n}\sum_{i=1}^{n} x_i$$

$$y_2 = x_2 - \bar{x} = x_2 - y_1 \tag{A.4}$$

$$\vdots$$

$$y_n = x_n - \bar{x} = x_n - y_1 \, .$$

The Jacobian J of this coordinate transformation has the following form

$$J = \frac{\partial(y_1, \ldots, y_n)}{\partial(x_1, \ldots, x_n)} = \begin{pmatrix} 1 & -1 & -1 & -1 & \ldots & -1 \\ 1 & 1 & 0 & 0 & \ldots & 0 \\ 1 & 0 & 1 & 0 & \ldots & 0 \\ \vdots & \vdots & \vdots & \ddots & \ldots & \vdots \\ 1 & 0 & 0 & 0 & \ldots & 1 \end{pmatrix} \tag{A.5}$$

and its determinant is $\det J = n$. Now we have to replace $x_1, \ldots, x_n$ by $y_1, \ldots, y_n$ in the joint distribution function. To this end we split the exponent up into one parts with x_1 only and the remaining terms. From the definition of $y_1 = \bar{x} = (1/n)\sum_{i=1}^{n} x_i$ we find

$$y_1 = \frac{1}{n}x_1 + \frac{1}{n}\sum_{i=2}^{n} x_i = \frac{1}{n}x_1 + \frac{1}{n}\sum_{i=2}^{n}(y_i + y_1) \tag{A.6}$$

and

$$x_1 = ny_1 - \sum_{i=2}^{n}(y_i + y_1) = ny_1 - \sum_{i=2}^{n} y_i - (n-1)y_1 = y_1 - \sum_{i=2}^{n} y_i \, . \tag{A.7}$$

For the exponent we then obtain

$$x_1^2 + \sum_{i=2}^{n} x_i^2 = \left[y_1 - \sum_{i=2}^{n} y_i \right]^2 + \sum_{i=2}^{n} (y_i + y_1)^2$$

$$= y_1^2 - 2y_1 \sum_{i=2}^{n} y_i + \left(\sum_{i=2}^{n} y_i \right)^2 + \sum_{i=2}^{n} (y_i^2 + 2y_1 y_i + y_1^2)$$

$$= ny_1^2 + \left(\sum_{i=2}^{n} y_i \right)^2 + \sum_{i=2}^{n} y_i^2 , \qquad (A.8)$$

which shows that the exponent is a sum containing one term depending on y_1 only and the remaining terms are assembled in the form that characterizes the variance S^2 in (A.1). The distribution function therefore factors and that shows that $y_1 = \bar{x}$ and the rest of the variables are statistically independent.

Following the definition of χ_p^2 in (7.36), we introduced the test statistics f in (7.37) and used that $\chi_p^2 - \chi_q^2$ in the numerator and χ_q^2 in the denominator are statistically independent. Here χ_q^2 is given by (7.36) with the $N \times p$-matrix A replaced by the $N \times q$-matrix B, which is given by A with $q - p$ additional columns corresponding to the additional fit parameters. The statistical independence made it possible to write the distribution function $\Psi_{n,m}(f)$ as the product of two χ^2-distributions in (7.39). In order to motivate why the two distributions are independent, we introduce the random variables $\mathbf{s} = \mathbf{y} - A\mathbf{x}$ and $\mathbf{t} = \mathbf{y} - B\mathbf{w}$, whose squares add up to the respective χ^2-distributions. Here we assume all σ_i to be unity to simplify the notation. Note that the dimension of $\mathbf{s}$, $\mathbf{t}$ and $\mathbf{y}$ is N.

Let us explicitly calculate the random variables $\mathbf{s}$ and investigate their relation to the "measurements" $\mathbf{y}$. We find

$$\mathbf{s} = \mathbf{y} - A\mathbf{x} = \left(1 - A \left(A^t A \right)^{-1} A^t \right) \mathbf{y} = P_A \mathbf{y} , \qquad (A.9)$$

where we determine the fit parameters $\mathbf{x} = \left(A^t A \right)^{-1} A^t \mathbf{y}$ with the help of the pseudo inverse from (7.4). It is easy to show that the matrix P_A is a projection operator with the properties $P_A^2 = P_A$ and $P_A^t = P_A$. Being a projector implies that its eigenvalues λ_A are either zero or one, where the number of ones determines the number of degrees of freedom of the χ^2-distribution $\chi_p^2 = \sum_{j=1}^{N} s_j^2$, because fitting with A introduces p constraints among the N normally distributed s_j^2, such that only $N - p$ of them are independent.

Before addressing this point further, let us consider the matrix B with the $q - p$ additional columns. The projector derived from B is $P_B = \left(1 - B \left(B^t B \right)^{-1} B^t \right)$. Without proof, we state that P_A and P_B commute $P_A P_B - P_B P_A = 0$, which implies that both projectors can be diagonalized simultaneously by an orthogonal matrix O, which transforms $\mathbf{s}$ to $\mathbf{r} = O\mathbf{s}$ and $\mathbf{t}$ to $\mathbf{r}' = O\mathbf{t}$. Since O is orthogonal, normally distributed independent random variables are mapped onto normally distributed independent random variables. Before expressing χ_p^2 and χ_q^2 in the new variables, we write the eigenvalues λ_A and λ_B next to each other. For illustrative purposes we chose $p = 2$ when fitting with A and $q = 4$ when fitting with B. Next to the eigenvalues we show the corresponding χ_p^2 and χ_q^2, expressed in the new variables r_k and r_k', the components of $\mathbf{r}$ and $\mathbf{r}'$, respectively

$$\lambda_A = (0, 0, 1, 1, 1, 1, 1, \ldots) \quad \text{such that} \quad \chi_p^2 = \sum_{k=p+1}^{N} r_k^2 ,$$

$$\lambda_B = (0, 0, 0, 0, 1, 1, 1, \ldots) \quad \text{such that} \quad \chi_q^2 = \sum_{k=q+1}^{N} r_k'^2 . \qquad \text{(A.10)}$$

Note that $r_k = r_k'$ where the eigenvalues of P_A and P_B are unity, such that the difference between χ_p^2 and χ_q^2 is given by $\chi_p^2 - \chi_q^2 = \sum_{k=p+1}^{q} r_k^2 = r_3^2 + r_4^2$, whereas $\chi_q^2 = \sum_{k=q+1}^{N} r_k^2 = r_5^2 + \cdots + r_N^2$. This shows that the random variables that make up the numerator and denominator of (7.37) are independent.

Appendix B
Software

In this appendix we briefly discuss some of the MATLAB scripts mentioned through-out the book. They are intended to invite the reader to explore the algorithms by vary-ing parameters and observing how the output changes. The operation of all scripts is briefly discussed. The source code is available from this book's web page at https://www.springer.com/9783030636425.

B.1   Markowitz Simulation

The following script `markowitz.m` encodes the portfolio optimization from Chap. 3. After clearing the workspace, we generate test data for the returns r_1, r_2, and r_3 over a period of 1000 trading days before initializing the vector $\mathbf{e}$ that appears in (3.11), the average returns $\langle r \rangle$, and the covariance matrix C from (3.8). In the next step the matrix, here called A that appears in (3.16) is prepared and inverted. In the following loop over the desired return ρ we first calculate the Lagrange multiplier λ_i from (3.17) and then the portfolio return vector $\mathbf{w}$ from (3.18) before storing ρ and the resulting volatility σ. After the loop, we plot ρ versus σ—the efficient frontier—in Fig. 3.2 and annotate the axes. Finally we plot the return and volatility of the test data, determine the minimum parameters from (3.20) and (3.21) and add the points to the plot.

```
% markowitz.m
clear all; close('all')
N=1000;   % number of trading days
r1=0.01*ones(N,1)+0.01*randn(N,1);          %..make test data
r2=0.02*ones(N,1)+0.02*randn(N,1)-0.3*r1;
r3=0.03*ones(N,1)+0.03*randn(N,1)+0.3*r1-0.5*r2;
ee=ones(3,1);                               % three stocks
ra=[mean(r1); mean(r2) ; mean(r3)];         % average return
```

© The Editor(s) (if applicable) and The Author(s), under exclusive license to Springer Nature Switzerland AG 2021
V. Ziemann, *Physics and Finance*, Undergraduate Lecture Notes in Physics, https://doi.org/10.1007/978-3-030-63643-2

```
rp=[r1-mean(r1),r2-mean(r2),r3-mean(r3)]; % for cov. matrix
C=rp'*rp/N; % covariance matrix, (3.8)
CC=inv(C);   % and its inverse
A=[ra'*CC*ra , ee'*CC*ra ;    % (3.16)
    ra'*CC*ee , ee'*CC*ee]
AA=inv(A);
K=0;
for rho=0:0.0002:0.04;  %...loop over the desired returns
    lambda=AA*[rho;1];                      % (3.17)
    w=lambda(1)*CC*ra+lambda(2)*CC*ee;      % (3.18)
    K=K+1;
    rrho(K)=rho;
    sig(K)=sqrt(w'*C*w);          % definition of sigma
end
plot(sig,rrho)
xlabel('Volatility \sigma')
ylabel('Portfolio return \rho')
hold on
%......plot data points for the three underlying assets
plot(sqrt(C(1,1)),ra(1),'k*',sqrt(C(2,2)),ra(2),'k*', ...
     sqrt(C(3,3)),ra(3),'k*')
rhomin=-AA(1,2)/AA(1,1);      % (3.20)
sigmin=sqrt(det(AA)/AA(1,1))  % (3.21)
plot(sigmin,rhomin,'*')
text(sigmin+0.001,rhomin,'\leftarrow Minimum risk')
%...add a zero risk asset
r0=0.008;  % rate of return of the zero risk asset
deltar=ra-r0*ee; % (3.25)
K=0;
for rho=r0:0.0001:5*r0
  ww=(rho-r0)/(deltar'*CC*deltar)*CC*deltar;  % (3.28)
  K=K+1;
  rrho2(K)=rho;
  sig2(K)=sqrt(ww'*C*ww);                      % (3.29)
end
plot(sig2,rrho2,'r--')
rhot=r0+(deltar'*CC*deltar)/(ee'*CC*deltar)    % (3.30)
sigmat=sqrt((deltar'*CC*deltar)/(ee'*CC*deltar)^2)
plot(sigmat,rhot,'r*')
text(sigmat+0.001,rhot-0.001, ...
      '\leftarrow Tangent/Market portfolio')
text(0.031,0.027,'\leftarrow Efficient frontier');
text(0.015,0.033,'Capital market line \rightarrow')
```

The second part of the script illustrates the analysis of the portfolio with the risk-free asset. First the excess return above the risk-free rate $\Delta\mathbf{r}$ is calculated and used in the loop over ρ to find the portfolio vector $\mathbf{w}$ from (3.28) and the corresponding value of the volatility σ. Plotting ρ versus σ now reveals the capital market line, shown as a red dashed line in Fig. 3.2. Finally the parameters of the market portfolio are determined from (3.30) and added to the figure.

B.2   Functions for Call and Put Options

The functions black_scholes_call() and the corresponding equation for the put option are used in the hedging simulation shown in Fig. 5.3 from Sect. 5.4. The functions receive the current stock price S_0, the strike price K, the time until maturity dt, the risk-free rate r_f, and the volatility σ as input parameters and they return the values of the corresponding option, as well as Δ, Γ, and Θ defined in Sect. 6.1. Both functions first define $N(z)$ from (4.39) and then calculate d_1 and d_2 from (4.43).

The function for the call option uses (5.17) to determine the option value c and (5.22) to calculate Δ_c. Γ and Θ are calculated by differentiating c and Δ_c, respectively.

```
function [c,delta,gamma,theta] ...
        =black_scholes_call(S0,K,dt,rf,sig)
N=@(z)0.5*erfc(-z/sqrt(2));
d1=(log(S0./K)+(rf+0.5.*sig.^2).*dt)./(sig.*sqrt(dt));
d2=(log(S0./K)+(rf-0.5.*sig.^2).*dt)./(sig.*sqrt(dt));
c=S0.*N(d1)-K.*exp(-rf.*dt).*N(d2);
delta=N(d1);
gamma=exp(-0.5*d1.^2)./(S0.*sqrt(2*pi*sig.^2.*dt));
theta=-0.5*(sig.^2).*(S0.^2).*gamma ...
    -rf.*K.*exp(-rf.*dt).*N(d2);
```

The function for the put option works in much the same way, only (4.42) with ρ replaced by r_f is used. Again, Δ_p is taken from (5.22).

```
function [p,delta]=black_scholes_put(S0,K,dt,rf,sig)
N=@(z)0.5*erfc(-z/sqrt(2));
d1=(log(S0./K)+(rf+0.5.*sig.^2).*dt)./(sig.*sqrt(dt));
d2=(log(S0./K)+(rf-0.5.*sig.^2).*dt)./(sig.*sqrt(dt));
p=K.*exp(-rf.*dt).*N(-d2)-S0.*N(-d1);
delta=N(d1)-1;
```

Calculating Θ and Γ is left as an exercise.

B.3   Dynamic Hedging Simulation

The following script `dynamic_hedging.m` is used to prepare the plots in Fig. 5.3.
After initializing the parameters and allocating arrays to hold all intermediary values
in the simulation, the function `black_scholes_call()` is used to calculate the
initial price c_0 and hedging parameter Δ_0. They are used to determine how many
shares must be purchased for the hedge. The money needed for the hedge is borrowed
from the bank. Once all values are initialized, we loop over the remaining days until
maturity of the option. Each day, we calculate how many days are left until maturity,
the interest we need to pay for the borrowed money, the increase in value of the
money I received for the option and the the changed price of the share. Since the
time until maturity and the share price S have changed, we calculate the new Δ in
a call to `black_scholes_call()` and use that information to update the hedge
and the money borrowed from the bank.

```
% dynamic_hedging.m
clear all;  close all
rho=0.09;
T0=1;      % one year
S0=1;      % initial share value
K=1.0;     % strike price, K=1.2 in lower plot
rf=0.05;   % risk free rate
sig=0.3;   % annual volatility
N=252;     % trading days
S=zeros(N,1); hedge=zeros(N,1);
borrow=zeros(N,1); value=zeros(N,1); KK=K*ones(N,1);
[c0,delta0]=black_scholes_call(S0,K,T0,rf,sig);
S(1)=S0;
hedge(1)=delta0*S0;    % initial hedge
borrow(1)=hedge(1);    % borrow money for hedge
value(1)=c0;           % pocket the money for the option
for i=2:N
  dt=T0*(N-i)/N;                    % trading days left
  interest=borrow(i-1)*rf/N;        % interest on borrowed
  value(i)=value(i-1)*exp(rf/N);  % money call option
  S(i)=S(i-1)*exp(rho/N);%*(1+sig*randn/sqrt(N));
  [c,delta]=black_scholes_call(S(i),K,dt,rf,sig);
  hedge(i)=delta*S(i);
  borrow(i)=borrow(i-1)+hedge(i)-hedge(i-1)+interest;
end
if S(i) > K                % in the money
  cost=borrow(i)-K;        % repay borrowed, get strike price
else                       % out of the money
  cost=borrow(i);
```

```
end
nn=1:N;
plot(nn,S,'b-',nn,KK,'b:',nn,hedge,'r',nn,borrow,'g');
xlim([0,255]); xlabel('trading days');
ylabel('S,K,\Delta S,Borrowed')
legend('Stock S','Strike K','Hedge \Delta S','Borrowed')
```

Once the loop completes, we test whether we have to deliver the share to the option buyer and calculate our cost accordingly, before plotting the data and annotating the axes.

B.4   Cantor, Koch, and Mandelbrot

The following MATLAB code produces the plot for the Cantor set in Fig. 9.5. First, we define the number of generations, plot the first generation as a line from zero to one, and define the axes. After initializing the start point for the segment and the number of segments, we loop over the generations. In each iteration, we increase the number of segments by three and reduce their length correspondingly, before plotting all line segments.

```
% cantor.m
generations=6;
plot([0,1],[0,0],'Linewidth',10)
axis([-0.02,1.02, -1-generations 1])
start=[0];
seg=1;
for k=1:generations
   segprev=seg;       % previous number of segments
   seg=seg*3;         % present number of segments
   lll=1/seg;         % length of each segment
   start=[start,start+2*segprev]; % start of each line
   for m=1:length(start)   % plot the line
      line([start(m)*lll,(start(m)+1)*lll],[-k,-k], ...
           'Linewidth',10)
   end
end
ylabel('Generation')
```

The following code is used to generate the plot with the first few generations of the Koch snowflake that is shown in Fig. 9.6. We first define the starting triangle as the polygon PP and start the loop over the four generations, shown in Fig. 9.6. Inside the loop, we store the number of points in the previous generation in the variable NN and allocate space for the next generation of points in PN. In the loop over m, we

add the position of the additional point—the triangular detour—to the list of points
that define the next generation. The position of the additional point is calculated in
the function `nextpt()`, discussed below. Once the loop over m completes and all
points in the next generation are known, they are plotted. Only in the first generation
the initial red triangle is added. Finally the array containing the previous points PP
is updated, such that it can be used when calculating the following generation.

```
% koch.m
PP=[-0.5,1/sqrt(3)-0.5*sqrt(3);
    0,1/sqrt(3);
    0.5,1/sqrt(3)-0.5*sqrt(3);
    -0.5,1/sqrt(3)-0.5*sqrt(3)];   % same as first
for generation=1:4
  subplot(2,2,generation)
  NN=size(PP,1);   % number of points , first=last
  PN=zeros(4*NN-3,2);
  for m=1:NN-1
    mm=4*m-3;
    PN(mm,:)=PP(m,:);
    PN(mm+1:mm+3,:)=nextpt(PP(m:m+1,:));
    PN(mm+4,:)=PP(m+1,:);
  end
  plot(PN(:,1),PN(:,2))
  axis equal
  axis([-0.6 0.6 -0.6 0.6])
  title(['Generation:' num2str(generation) ])
  hold on
  if generation ==1
      plot(PP(:,1),PP(:,2),'r')
  end
  PP=PN;
  pause(0.5)
end
```

The following function calculates the coordinates of the additional points for the
"triangular detour." It receives the coordinates of two points in PP as input and
returns the coordinates of the three points for the detour. Two of the points lie on the
line connecting the start points at 1/3 and 2/3 of the way. The third point is located
perpendicular to the connecting line, which is found with the help of DPP.

```
% nextpt.m
function PPP=nextpt(PP)
PPP=zeros(3,2);
PPP(1,:)=(2*PP(1,:)+PP(2,:))/3;
PPP(3,:)=(PP(1,:)+2*PP(2,:))/3;
```

```
DPP=PP(2,:)-PP(1,:);
PPP(2,:)=0.5*(PP(1,:)+PP(2,:))-[DPP(2),-DPP(1)]/(2*sqrt(3));
```

Running the koch.m script produces Fig. 9.6, but the reader is encouraged to play with the position of the additional point to explore variations of the snowflake.

The synthetic stock charts, shown in Figs. 9.7 and 9.8, rely on the function nextiter4_random() to iteratively calculate intermediary points. As input it receives the start and end point of a line segment in the array p, as well as the control points pa and pb that describe where and how the intermediate points are displaced. The variable scramble controls whether the three line segments should be randomly scrambled. The function returns the coordinates of the intermediate points in the variable q. Inside the function first the additional points are calculated and then the output array q is filled. Note that two points are doubled, which is required, if we intend to scramble the order of the line segments. Scrambling is controlled by the variable scramble and if it is unity, we first determine a random sequence of the numbers 1, 2, and 3 that we use to reshuffle the order of the segments. Note that the horizontal coordinate of the first and last point must agree with those of the initially provided coordinates in p.

```
function q=nextiter4_random(p,pa,pb,scramble)
q=zeros(6,2);
dp=p(2,:)-p(1,:);
q(1,:)=p(1,:);
q(2,:)=p(1,:)+[pa(1)*dp(1),pa(2)*dp(2)];
q(3,:)=q(2,:);
q(4,:)=p(1,:)+[pb(1)*dp(1),pb(2)*dp(2)];
q(5,:)=q(4,:);
q(6,:)=p(2,:);
if scramble
   qq=q;
   a=1+floor(3*rand);
   b=1+floor(3*rand);
   while (a==b)
      b=1+floor(3*rand);
   end
   c=(6/a)/b;
   dax=qq(2*a,1)-qq(2*a-1,1);
   dbx=qq(2*b,1)-qq(2*b-1,1);
   q(1,1)=p(1,1);
   q(1,2)=qq(2*a-1,2);
   q(2,1)=p(1,1)+dax;
   q(2,2)=qq(2*a,2);
   q(3,1)=q(2,1);
   q(3,2)=qq(2*b-1,2);
   q(4,1)=q(2,1)+dbx;
```

```
    q(4,2)=qq(2*b,2);
    q(5,1)=q(4,1);
    q(5,2)=qq(2*c-1,2);
    q(6,1)=p(2,1);
    q(6,2)=qq(2*c,2);
end
```

The driver program to generate Figs. 9.7 and 9.8 is called `monofractal_random.m`. After clearing the workspace, it initializes some variables and calls the function `nextiter4_random()` with `scramble` set to zero, which is used to show the generator in the upper plots in the figures. In the loop over `generation` the points are added and reshuffled ten times, where the array `p` is increased in every iteration to hold the additional points. After the loop completes, the results are displayed.

```
% monofractal_random.m
clear all; close all
hold off
scramble=1    % choose 0 or 1
p=[0,0; 1,1];
dp=p(2,:)-p(1,:);
pa=[4/9,2/3];   % control points for the generator
pb=[5/9,1/3];
q0=nextiter4_random(p,pa,pb,0);   % unscrambled generator
subplot(4,1,1)
plot(q0(:,1),q0(:,2),'b')
for generation=1:10
  N=size(p,1);
  p2=zeros(3*N,2);
  mm=1;
  for m=1:2:N-1    % only every other
    p2(mm:mm+5,:)=nextiter4_random(p(m:m+1,:),pa,pb,scramble);
    mm=mm+6;
  end
  subplot(4,1,2)
  plot(p2(:,1),p2(:,2),'b')
  p=p2;    % copy back for next iteration
  pause(0.1)
end
jump=7;      % steps for the derivative calculation
subplot(4,1,1)    % the generator
plot(q0(:,1),q0(:,2),'b')
subplot(4,1,2)    % chart
plot(p2(:,1),p2(:,2),'b')
subplot(4,1,3)    % derivative of chart, increments
d=p2(jump:end,2)-p2(1:end-jump+1,2);
```

```
plot(p2(jump:end,1),d)
subplot(4,1,4)          % histogram of increments
[height,pos]=hist(d,100);
bar(pos,log(height))
```

The reader is encouraged to explore different control points and observe how the resulting charts change.

B.5   Metropolis-Hastings Algorithm

The function `metropolis3.m` encodes the Metropolis-Hastings algorithm, discussed in Sect. 10.7. As input, the function receives the function f from which the random numbers are sampled, the look-around factor β, the number nmax of random numbers to return and a starting value x0. It returns an array x with random number that are drawn from the supplied function f. The function directly implements the algorithm discussed in Sect. 10.7.

```
% metropolis3.m
function x=metropolis3(f,beta,nmax,x0)
  x=zeros(1,nmax);
  x(1)=x0;
  fx=f(x0);
  for k=1:nmax-1
    y=x(k)+beta*(2*rand-1);   % new candidate
    fy=f(y);
    alpha=fy/fx;         % check if new is better
    if (alpha>1)         % accept, if better
      x(k+1)=y;
      fx=fy;
    else                 % here alpha is smaller than unity
      u=rand;            % get random number
      if (alpha>u)       % compare with random number u
        x(k+1)=y;        % also accept, if larger than u
        fx=f(y);
      else
        x(k+1)=x(k);     % else re-use old value
      end
    end
  end
end
```

The metropolis function is used to calculate some of the path integrals, discussed in the next section.

B.6   Numerical Path Integrals

The script BS_pricing_kernel_from_uniform.m numerically evaluates
the path integral, defined in (10.64) using paths that are based on uniformly distributed random positions along a path. First the parameters of the problem are
defined, followed by the anonymous function pBS(). It is later used to show the
analytic result for comparison. Then we define the number of sample path to prepare
before actually filling the matrix x with the paths; one column for each path. The
variables term1 and term2 are filled with the expression in the square bracket in
(10.62). They are used to calculate eSBS, which is a row vector that contains the
contribution of each path to the path integral. In the following two lines we determine
the positions xx to evaluate the path integral and then loop over all paths to add their
contribution to the integral. After the loop finishes, we normalize the path integral
to unity and add the discount factor $e^{-r_f \tau}$ before plotting both the numerical and the
analytic path integral.

```
% BS_pricing_kernel_from_uniform.m
clear all; close all; tic
rf=0.03;                 % 3 % risk-free rate
sigma=0.3;               % 30 % volatility per year
rfhat=rf-0.5*sigma^2;    % for convenience
t=1;                     % 1 year
N=6;                     % number of time slices
dt=t/N;                  % time per slices
pBS=@(xf)exp(-rf*t-((xf+t*rfhat).^2)/(2*t*sigma^2)) ...
         /sqrt(2*pi*t*sigma^2);
%....define the paths
Npath=20000000;          % sample paths
x=-4*sigma+8*sigma*rand(N,Npath);
%....
term1=sum(((x(2:end,:)-x(1:end-1,:)+dt*rfhat).^2,1);
term2=(x(1,:)+dt*rfhat).^2;         % from start to first
eSBS=exp(-(term1+term2)/(2*dt*sigma^2));
ix=round(x(end,:)/(sigma/5)); ixmin=min(ix); ixmax=max(ix);
xx=(ixmin:ixmax)*sigma/5;
path_integral=zeros(1,ixmax-ixmin+1);
for k=1:Npath
  ipos=ix(k)-ixmin+1;
  path_integral(ipos)=path_integral(ipos)+eSBS(k);
end
i0=(xx(2)-xx(1))*sum(path_integral); % normalize
path_integral=exp(-rf*t)*path_integral/i0;
plot(xx,path_integral,'k*',xx,pBS(xx),'r--') ;
xlim([-1.5,1.5]); toc;
```

```
xlabel('x_f'); ylabel('Pricing kernel p_{BS}')
legend('uniform MC','analytical'); set(gca,'FontSize',16)
```

In order to evaluate p_{BS} with the Metropolis-Hasting algorithm generating the paths, we only have to replace the section where the paths are defined. The following code snippet first defines the number of sample paths to prepare and defines the anonymous function h() from which the random numbers will be drawn. Then it defines the starting point x0 and the β for the metropolis algorithm, runs the algorithm for 1000 burn-in iterations, and finally generates the required number of random numbers to define the paths. Since the random numbers fill a one-dimensional array, we have to reshape() the output in such a way that each column defines one path.

```
%....define the paths
Npath=100000;                          % number of sample paths
h=@(x)exp(-x.^2/(2*(2*sigma)^2));    % for Metropolis-Hastings
x0=0.01; beta=3*sqrt(sigma^2*dt);
y=metropolis3(h,beta,1000,x0);     % 1000 iteration burn-in
x=metropolis3(h,beta,Npath*N,y(1000));
x=reshape(x,[N,Npath]); % x=permute(x,[2,1]);
%....
```

Likewise, we can replace section, where the paths are generated by the following code snippet, which prepares paths that resemble a random walk with rms step size $2\sigma/\sqrt{N}$ and then uses the built-in cumsum() function to cumulatively add up the individual steps in order to obtain the meandering paths that are later used to evaluate the path integral. As already pointed out in the main text, is the step size artificially increased in order to create paths that cover a wider range and thereby more accurately determine the tails of the pricing kernel. Adding paths that probe far-away regions does not affect the accuracy negatively, because their contribution to the path integral is exponentially suppressed by the much increased action S_{BS} on these paths.

```
%....define the paths
Npath=10000;          % sample paths
x=2*randn(N,Npath)*sigma/sqrt(N);
x=cumsum(x,1);
%....
```

Again, we encourage the reader to explore the scripts and vary parameters.

B.7 Macroeconomic Models

The following script solow.m is used to prepare Fig. 11.1. After defining the parameters of the model and allocating space for the variables, we initialize the variables.

In the loop, we use (11.1)–(11.4) to step the variables forward in time. After the
loop completes, the variables are plotted and the axes annotated.

```
% solow.m
clear all; close all;
k0=1;               % initial capital
delta=0.1;          % depreciation of capital
lambda=1;           % technology level
sigma=0.15;         % fraction of output to re-invest
Theta=0.7;          % Cobb-Douglas exponent of production
N=300;              % number of generations
kk=zeros(N,1); yy=kk; ii=kk;   % initialize storage
kk(1)=k0;           % set values at initial time
yy(1)=lambda*k0^Theta;
ii(1)=sigma*yy(1);
for t=1:N-1         % and loop over later times
   kk(t+1)=(1-delta)*kk(t)+ii(t);   %   (11.4)
   yy(t+1)=lambda*kk(t)^Theta;      %   (11.1)
   ii(t+1)=sigma*yy(t+1);           %   (11.3)
end
tt=1:N;
plot(tt,kk,'k',tt,yy,'k-.',tt,ii,'k--','LineWidth',2);
xlabel('Time step t');
ylabel('k_t, y_t, i_t')
legend('Capital k_t','Output y_t','Investment i_t')
```

B.8   The Donkey's Solution

The script `donkeys_solution.m` calculates the optimum trajectory for the donkey as discussed in Sect. 11.5. First the parameters of the model are defined and used
to set up the matrix A, which encodes (11.49). This system of equations is solved in
the next step, which returns the integration constants c_3 and c_4. The other two integration constants c_1 and c_2 are known from the main part of the text. Subsequently,
we insert the integration constants in (11.47) and (11.46) to prepare inline functions
`x1` and `x2` for the position of the donkey and its speed. Plotting both variables and
annotating the axes completes this script.

```
% donkeys_solution.m
clear all; close all
alpha=0.1;    % friction constant
L=100;        % distance to cover in [m]
T=20;         % time for the travel in [s]
t=0:0.1:T;
```

```
aT=alpha*T;
y=[L;0];
A=[(exp(aT)-exp(-aT)-2*aT)/alpha , 1+aT-exp(aT);
   (exp(-aT)+exp(aT)-2)/alpha , 1-exp(aT)]
x=A\y; C3=x(1); C4=x(2);
C1=2*alpha^2*C3-alpha^3*C4
C2=alpha^2*C4
% position
x1=@(t)C4-(C3/alpha)*exp(-alpha*t)-(C1/alpha^2)*t ...
   -(1/(2*alpha^2))*(C2-C1/alpha)*exp(alpha*t);
% velocity
x2=@(t)C3*exp(-alpha*t)-C1/alpha^2 ...
   -(1/(2*alpha))*(C2-C1/alpha)*exp(alpha*t);
subplot(2,1,1); plot(t,x1(t))
xlabel('t [s]'); ylabel('x=x_1 [m]');
subplot(2,1,2); plot(t,x2(t))
xlabel('t [s]'); ylabel('v=x_2 [m/s]');
```

The plots in Fig. 11.5 show the output of running the script two times; once with $\alpha = 0.1$ and once with $\alpha = 1$.

B.9 Character Frequency in the King James Bible

The analysis to determine the entropy per character in the King James bible in Sect. 12.1 is based on the following Unix script, written in bash. After defining the input file, here pg10.txt, we translate all lower-case characters to upper case with tr and then delete numbers and several special characters with the tr -d command. Finally, we fold the input to a format that contains only one character per line, which we then sort to have all equal characters following one another, before removing the duplicates with the uniq -c command. The option -c writes the number of occurrences of each character next to it, such that we obtain the table shown in Sect. 12.1.

```
#!/bin/bash
input=pg10.txt
tr '[:lower:]' '[:upper:]' < $input \
  | tr -d '[0-9:\n;()?!-@$*#%/]' > out.txt
fold -w1 < out.txt | sort | uniq -c > out2.txt
awk '{print $1}' < out2.txt > out3.txt
```

We point out that the end-of-line characters on UNIX and Windows systems differ and running dos2unix pg10.txt might be required on a Unix system to fix that problem.

B.10   Entropy of Audio Files

The probability distribution of sound amplitudes of the first movement of Beethoven's fifth symphony that we show in Fig. 12.4 is based on the following script, which first defines a function g to describe a Gaussian before loading the audio file with MAT-LAB's built-in `audioread()` function, which returns the samples in the variable y and the sampling frequency `fs`. Next, we assign left and right channel to the variables `l` and `r` and calculate the rms and the mean of the sum of the channels. The histogram of amplitudes is returned by the call to the `hist()` function, which prepares a histogram of the input samples, from which we calculate the probabilities tt p of the amplitudes. Before calculating the entropy from (12.20), we remove samples with zero probability, which do not contribute to the entropy. For comparison, we prepare a Gaussian with the same rms as Beethoven's symphony and calculate its entropy, before plotting both.

```
% audio_analysis.m
clear all; close all;
g=@(x,m,sig)exp(-((x-m).^2)./(2*sig.^2))./sqrt(2*pi*sig.^2);
[y,fs]=audioread('beethoven_S5_M1.wav');
l=y(:,1); r=y(:,2); sigma=rms(l+r); avg=mean(l+r);
[N,x]=hist(l+r,1000);
dx=x(2)-x(1);
p=N/(sum(N)*dx); pp=p(p>0);
en_beethoven=-sum(pp.*log(pp))*dx    % entropy, (12.20)
yg=g(x,avg,sigma); en_gauss=-sum(yg.*log(yg))*dx; % gauss
en_analytical=0.5*log(2*pi*exp(1)*sigma^2)    % (12.24)
plot(x,p,'k',x,g(x,avg,sigma),'r--','LineWidth',2)
xlabel('x=u/u_0'); ylabel('dN/dx');
legend('Beethoven','Gauss')
```

B.11   Elliptic Curves

The addition of points that lie on an elliptic curve is discussed in the beginning of Sect. 12.7 and defined in (12.32) and (12.33). The function `ECadd_p()` implements these equations. It receives the coordinates of points two points P_a and P_b, the coefficients a and b that define the elliptic curve in (12.30), and the prime p. The function returns the coordinate of the point $P_c = P_a \oplus P_b$. Inside the function, we first catch the exceptional conditions related to the additional point at infinity that serves as "zero" for the addition of points. Then we code (12.33) with the slope s defined by (12.32), where differentiate the case with $P_a = P_b$ and $P_a \neq P_b$. Note that we use (12.34) to calculate the denominator in (12.32). In order to prevent numer-

ical problems we use MATLAB's built-in powermod() function instead of mod()
function.

```
% ECadd_p.m
function Pc=ECadd_p(Pa,Pb,a,b,p)
if Pa(2) == Inf, Pc=Pb; return; end
if Pb(2) == Inf, Pc=Pa; return; end
if Pa(1) == Pb(1) & Pa(2) ~= Pb(2), Pc=[0,Inf]; return; end
if Pa(1) == Pb(1)
  if Pa(2) == Pb(2) & Pa(2) == 0
    Pc=[0,Inf];
    return
  else
    denominv=powermod(2*Pa(2),p-2,p);  % (12.34)
    s=powermod((3*Pa(1)*Pa(1)+a)*denominv,1,p);
  end
else
  denominv=powermod(Pb(1)-Pa(1),p-2,p);
  s=powermod((Pb(2)-Pa(2))*denominv,1,p);
end
Pc(1)=powermod(s*s-Pa(1)-Pb(1),1,p);  % (12:33)
Pc(2)=powermod(-s*(Pc(1)-Pa(1))-Pa(2),1,p);
```

The following script ECCadd_p_test.m prepares the two plots shown in Fig. 12.7.
After defining the parameters $a, b,$ and p as well as the function f2 for the right-
hand side of the elliptic equation, we choose the x-coordinate of the generator G
and search a suitable y-coordinate that lies on the curve. Note that not all values of
x lead to a suitable value of y. We break the routine, if no point is found. Otherwise,
we plot the G as a red asterisk, and start a loop in which G is repeatedly added
to P_c until the point at infinity appears, which indicates the order to the underlying
group $\mathcal{G}$, as discussed in Sect. 12.7. At this point we break the execution of loop and
plot the points and annotate the axes.

```
% ECadd_p_test.m
clear all, close all
p=113; a=0; b=7;
f2=@(x)powermod(x.^3+a*x+b,1,p);
xstart=15;    % order: 13->114, 15->19
y2=f2(xstart)
for k=1:p    % is it on the curve?
  yy=powermod(k,2,p);
  if yy==y2
    q=[k,yy];
    G=[xstart,k]
    break
```

```
      end
  end
  if ~exist('G','var') disp('no y found'); return; end
  ishow=1;
  Pc=G; data(1,:)=G;
  plot(G(1),G(2),'r*','MarkerSize',10); hold on
  for k=2:2*p
    Pc=ECadd_p(Pc,G,a,b,p);
    data(k,:)=Pc;
    if Pc(2) == Inf
       disp(['Order of group = ',num2str(k)]); ishow=0;
    end
    if ishow==0, break; end
  end
  plot(data(2:end,1),data(2:end,2),'k.','MarkerSize',10)
  xlim([0,p]); ylim([0,p]); xlabel('x'); ylabel('y')
```

Again, the reader is encouraged to explore different parameters $xstart$, but also a, b, and p.

B.12   Ethereum Contract Test Environment

A test environment to develop Ethereum contracts comprises of the *ganache* blockchain from https://www.trufflesuite.com/ganache, which comes as a single binary file for most computer systems. Running the binary sets up a blockchain, ready to be used. In order to interact with it, we need to install the *MetaMask* wallet from https://metamask.io as an extension to a browser—Chrome is recommended. After installing it, assigning passwords, and noting the security phrase in a safe place, we choose the ganache blockchain and import one of the test accounts into Meta-Mask. This allows us to send Ethereum to other test accounts. Programming smart contracts is most easily accomplished using the *Remix* online editor and compiler, available from https://remix-project.org. Once it is loaded into the browser, we select the solidity compiler, open a new file and start coding. All activities are accessible from the menu on the far-left side of the browser window. The second item from the top gives access to the source code of the contract, the third compiles the contract, and the fourth is used to deploy it on a blockchain. Since deployment costs gas, we have to connect the editor to our wallet, by changing "JavaScript VM" to "Injected Web3" under "Environment." After granting MetaMask the permission to interact with the editor, we can press the "Deploy" button, which submits the contract to our test blockchain. There is a basic interface to the contract at the bottom of the page that also provides the address under which the contract is known on the blockchain.

With the contract deployed, we can directly send ether currency from our Meta-Mask wallet to the address of the contract, as described in Sect. 12.9.

Index

© The Editor(s) (if applicable) and The Author(s), under exclusive
license to Springer Nature Switzerland AG 2021
V. Ziemann, *Physics and Finance*, Undergraduate Lecture Notes in Physics,
https://doi.org/10.1007/978-3-030-63643-2

Printed in the United States
by Baker & Taylor Publisher Services